Monsoon and Civilization

The divine mountain of Yulong Xueshan in the Yunnan province is affected both by the Indian monsoon and the Asian monsoon. (Photo by Takeshi Takeda)

Monsoon and Civilization

EDITED BY

Yoshinori Yasuda
International Research Center
for Japanese Studies

Vasant Shinde
Deccan College
Post-Graduate and Research Institute

Asian Lakes Drilling Programme (ALDP)
Geo-genom Project

Lustre Press
•
Roli Books

This book is dedicated to

JÖRG F.W. NEGENDANK
PROFESSOR EMERITUS OF
GEOFORSHUNGZENTRUM IN POTSDAM

Professor Negendank is a pioneer in the high-resolution reconstruction of the environmental history based on analysis of annually laminated sediments in European crater lakes. He also showed great interest in the studies of annually laminated sediments in the Asian monsoon region, and became involved in high-resolution reconstruction of the environmental history of Japan, China, and India. He also made significant contributions to the training of researchers in Asia. The high-resolution reconstruction study of environmental history that Professor Negendank has cultivated is now expanding into a comparative study of the East and West regions of the Eurasian continent.
(Photo by Yoshinori Yasuda)

Research Foundation of the Ministry of Education,
Culture, Sports, Science and Technology in Japan
Asian Lakes Drilling Programme (ALDP)
and Geo-genom Project

ISBN: 81-7436-305-X

© **International Research Center for Japanese Studies, 2004**
3-2 Oeyama-cho, Goryo, Nishikyo-ku, Kyoto 610-1192, Japan
Tel: 81- (75) 335-2150, Fax: 81- (75) 335-2090
E-mail: yasuda@nichibun.ac.jp
Website: http://www.nichibun.ac.jp

First published in 2004 by
Roli Books Pvt. Ltd.
M 75 Greater Kailash II Market,
New Delhi 110 048, India
Tel: (011) 29212271, 29212782,
Fax: (011) 29217185, 29213978
E-mail: roli@vsnl.com,
Website: Rolibooks.com

EDITORS:
Yoshinori Yasuda
Vasant Shinde

ASSOCIATE EDITORS:
Junko Kitagawa
Shweta Sinha Deshpande

PHOTOGRAPHS:
Takeshi Takeda
Yoshinori Yasuda

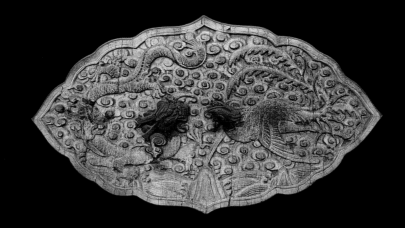

Contents

PART II: INDIAN MONSOON VARIABILITY AND HUMAN ADAPTATIONS

Introduction

Discovery of Riverine Civilizations in Monsoon Asia

YOSHINORI YASUDA

Boundary Zone Between the Monsoon Asia and Dry Asia

The Eurasian continent can be largely divided into Monsoon Asia, Arid Asia, Atlantic Asia, and Boreal Asia (Fig. 1) (Yoshino, 1999). The monsoons (seasonal winds) predominantly affect climate in Monsoon Asia: the prevailing Southwest Monsoons cause the rainy season, while the prevailing northwest monsoons cause the dry season.

Monsoon Asia is characterized by its temperate and wet climate, described by Tetsuro Watsuji as 'a combination that produces an unbearable summer' (Watsuji, 1935). However, this climate was ideal for the growth of forests—the representative feature of Monsoon Asia.

Monsoon Asia encompasses the following regions: India; the southern foot of the Himalayas; Southeast Asia; south China east of Sichuan and the Yunnan Provinces in the eastern margin of the Tibetan Plateau; north China east of the Shianxi and Liaoning Provinces located east of the Dahinganling Mountains; and the Pacific coastal regions south of Sakhalin.

The dominant summer crops in Monsoon Asia such as rice, foxtail millet (*Setaria*), broomcorn millet (*Panicum*), and sorghum, belong to the *Gramineae* family. Rice, especially, is the representative crop of Monsoon Asia.

To the west of the humid Monsoon Asia lies Arid Asia, which consists mainly of deserts and grasslands unsuitable for agriculture. It is therefore a land of pastoralists and nomads. A tongue-shaped region of humid Atlantic Asia penetrates Arid Asia from the west, forming what may be referred to as a 'wet tongue' from the Atlantic. Although Atlantic Asia is not as humid as Monsoon Asia, it is characterized by winter rains in the Mediterranean coastal region. The rainy season in Monsoon Asia falls in summer. Thus rice, a summer crop, is the representative crop of Monsoon Asia. In contrast, in Atlantic Asia where the rainy season falls in the winter, the representative crops are winter crops such as wheat and barley.

The cool Boreal Asia lies north of Atlantic Asia.

When the four great civilizations—Mesopotamian, Egyptian, Indus, and Yellow River Civilizations—are plotted on the climate division map (Fig. 1), an interesting observation can be made. The four ancient civilizations all rose along great rivers that flow at the boundary zone between the wet and dry regions.

The Tigris and Euphrates rivers, which gave rise to the Mesopotamian Civilization, flow in the region where the wet Atlantic Asia comes into contact with Arid Asia. The Indus river, the cradle of the Indus Civilization, flows at the boundary zone between the wet Monsoon Asia and Arid Asia. The Yellow River Civilization rose in the S-shaped region of the Yellow River, which is at the boundary between the Arid Asia and Monsoon Asia.

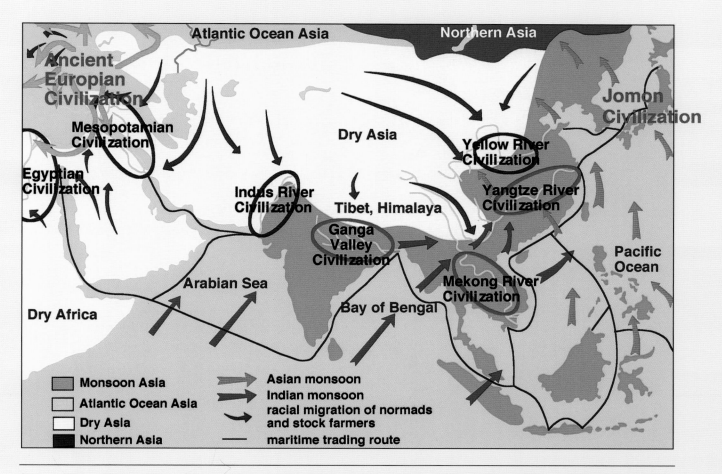

Fig. 1. The relationship between climate divisions of the Eurasian continent and the ancient civilizations and migration of people (adapted from Yasuda, 2000). Blue circle: Riverine civilizations discovered in the 19th-20th century. Red circle: Riverine civilizations that may be discovered in the 21st century.

The Nile river cannot be clearly positioned between such wet and dry regions. However, the banks of the Nile river constitute fertile, humid areas, while dry deserts extend just a short distance away from the Nile river. Besides, the Nile river flowing through the desert can itself be considered a contact zone between wet and dry regions.

There are other rivers in Eurasia that flow in wet-dry boundary zones, such as the Obi and Volga rivers that flow between the humid east Siberian lowlands of Atlantic Asia and Arid Asia.

Ancient civilizations, however, were previously considered to have been absent in these regions due to their cold climate that would have prevented an agricultural revolution. This belief is based on the conventional concept that 'ancient

civilizations rose along great rivers flowing in a wet-dry boundary zone that has experienced an agricultural revolution.' This region stretching from the northern coast of the Caspian Sea to the northern coast of the Black Sea, however, is considered to be one of the homelands of the Indo-European speaking people who played a significant role in the rise and fall of ancient civilizations in Eurasia. They created a powerful, patriarchal and hierarchical society founded on the military strength of horseback warriors bearing lethal weapons. These Indo-European speaking people were the driving force behind the new revolution among agriculturists settled along the great rivers in Eurasia.

According to Okada (1992), the Indo-European speaking people may possibly have originated in the vicinities of the Tarim basin and the Junggar basin, although that remains inconclusive. The wet-dry boundary zone stretching from the northern coast of the Caspian Sea to the northern coast of the Black Sea, one of the homelands of the Indo-European speaking people, served as the base for Scythian activity in the 1st millennium BC. Without doubt, a pastoral civilization must have developed at this boundary zone. However, details of the civilization have yet to be definitively tabulated, and research in the 21st century has only begun to reveal the nature of the pastoral civilization.

In contrast, both hunters and pastoralists inhabited the wet-dry boundary zone in northeast Asia stretching from the Amur river valley to the Liao river valley in China. The former is located between the humid regions of Boreal Asia and Arid Asia, and the latter is located between Monsoon Asia and Arid Asia. Deep mixed forests of *Pinus* and broad-leafed deciduous trees such as deciduous *Quercus*, *Ulmus*, *Acer*, and *Juglans* covered the northern limits of Monsoon Asia until 1000 cal. yrs. BP, comprising the east Inner Mongolia Autonomous Region, Liaoning Province, and Jilin Province. These forests gradually changed to coniferous forest zones (the taiga) of the Boreal Asia north of Amur river.

Okada (1992) states that the Yin Dynasty was founded by hunters from these forests following the invasion of the middle valley of the Yellow River from the north. The subsequent Zhou Dynasty was founded by the pastoral western barbarians who invaded the Yellow River Valley from the wet-dry boundary zone of the west.

Contact and Fusion of Pastoralists and Agriculturalists

In the valleys of the great rivers in humid Monsoon Asia and Atlantic Asia, agriculturalists began to lead a sedentary lifestyle after the agricultural revolution. In contrast, pastoralists inhabited the surrounding Arid Asia. Pastoralists were dependent upon agriculturalists for their staple cereal, which they obtained through trade. Therefore, it can be reasonably concluded that the agricultural revolution preceded the pastoral revolution, which in turn should have preceded the nomadic revolution.

Agriculturists living in the valleys of great rivers neighbouring Arid Asia provided staple cereals and vegetables to pastoralists. Therefore, ever since the agricultural revolution, the great river valleys in the wet-dry boundary zones became the contact zone for the two different cultures of agriculturists and pastoralists. Agriculturists living along the great rivers formed a matriarchal social organization based on consanguine families and were sedentary and conservative. In contrast, pastoralists formed patriarchal societies that necessitated a leader for the herd. They were nomadic traders with access to vast amounts of information. The practice of wearing gold, silver, and treasures during travel is typical of pastoralists.

The village was the unit of activity for the strongly localized agriculturists, while pastoralists moved between several villages and were endowed with the universality to integrate foreign cultures.

The wet-dry boundary zone where agriculturists and pastoralists met corresponds to regions that are extremely vulnerable to the slightest climatic changes. For example, in the peripheral zones of Monsoon Asia, droughts are caused by a small delay in the arrival of the monsoon.

In 1991, an example of severe change in climate, around 5000 ^{14}C yrs. BP (5700 cal. yrs. BP) to a colder and drier climate, was discovered by the author (Yasuda, 1991). The inhabitants of the valleys of great rivers in the wet-dry boundary zones were most seriously affected by this climatic aridification. The environment was drastically transformed by the change in climate and the population expansion that had continued since the agricultural revolution only intensified the devastating situation.

Pastoralists who rapidly fanned out from the Black Sea and Caspian Sea coasts to the rest of the Eurasian continent were also faced with the aridification at 5700 cal. yrs. BP. In the inner regions of the Eurasian continent, significant aridification accompanied climatic cooling, and the demand for water forced pastoralists living in grasslands and semi-desert regions to relocate to the valleys of great rivers. This giant wave of pastoralists invasion from the surrounding dry regions swept into the valleys where agriculturists had originally inhabited.

The rise of the four great civilizations was triggered by this co-mingling of agriculturists and pastoralists, brought about by climate aridification at 5700 cal. yrs. BP, and the resulting population concentration in the great river valleys.

Agriculturists are by nature strongly localized and have a small activity range. In contrast, pastoralists moved across long distances and had also established a trade network. Always faced with the danger of an aggressive encounter with other tribes, pastoralists wore their possession as accessories made of precious metals and were equipped with weapons for self-defence. It was also in the pastoralist society that a king emerged as a leader of the herd. Pastoralists shared common mythologies, and as seen in *The Epic of Gilgamesh*, the rational idea of ruling nature to satisfy the needs of man had its roots in pastoralist society (Yasuda *et al.*, 2000).

With climatic aridification, pastoralists stormed into the valleys of the great rivers in search of water, food, and the riches of agriculturists. This climatic change at 5700 cal. yrs. BP triggered the invasion and plundering of the river valleys by pastoralists, and thereby increased the frequency of contact between pastoralists and agriculturists. Pastoralists eventually settled in the river valleys.

As early as in 1991, the author had suggested that this contact and fusion between the different cultures is what gave rise to urban civilization.

The foremost factor in the birth of urban civilizations is the population expansion accompanying pastoralist migration into the great river valleys. The concentration and expansion of population are the primary factors behind the birth of urban civilization. Pastoralists who had migrated to the valleys had established trade routes and relayed much information about other regions that were previously unknown to agriculturists. Pastoralists played an important role in making a city into a centre of trade and commerce. The universal principle arising from the need for a leader to guide the herd and their migratory lifestyle placing them in contact with different ethnic groups and cultures, provided them the impetus to integrate. This resulted in the emergence of a king, thus transforming the agricultural society centred on consanguine families into a society with a systematic and universal hierarchy. The strong orientation towards the collection of gold and silver treasures also had its roots in the pastoralist way of life.

In this way, urban civilization developed as an outcome of conflict and fusion between pastoralists and agriculturists, that took place in the valleys of the great rivers of wet-dry boundary zones as a result of the change in climate at 5700 cal. yrs. BP. There is no doubt that ancient civilizations had risen where there was frequent contact between people from different environments with contrasting trades and cultures. Valleys of great rivers flowing in the wet-dry boundary zones provided ideal environments for such encounters.

The strong connection between the Mesopotamian and Indus Civilizations has been pointed out earlier. The cylindrical seals recovered from the archaeological sites of the Indus Civilization indicate the possibility of the Indus Civilization having shared the mythology of *The Epic of Gilgamesh* of Mesopotamia. Commonalities have also been established between the mythologies of the Yellow River and Mesopotamian Civilizations. It is not surprising

that the structures of the four great civilizations have much in common, since it was pastoralists who had migrated in large numbers from the coasts of the Black and Caspian Seas that had triggered the rise of these civilizations.

These pastoralists, however, were only able to intrude into the peripheral regions. They were not able to penetrate deeper into the wet regions of Monsoon Asia, such as the middle and lower reaches of the Yangtze river and the Ganges valley. These regions were covered by thick forests, and endemic diseases, such as malaria, foreign to pastoralists originating in dry regions, also prevented their penetration. Therefore, as has been believed earlier, the light of civilization did not reach the wet regions of Monsoon Asia. As a consequence, the wet regions remained in a primitive and barbaric state for a long time. However, civilizations with characteristics completely different from the four great civilizations of the wet-dry boundary zones have been discovered in the forests of the wet regions. These are the Yangtze River, the Jomon, and the Ganges River Civilizations.

Discovery of Riverine Civilizations in Wet Regions

The pastoralist invasion of agricultural societies, and the subsequent fusion triggered by the change in climate at 5700 cal. yrs. BP was the driving force behind the rise of urban civilizations. In contrast, the valleys of the Ganges, Mekong, and Yangtze (Fig. 2) flowing deep in Monsoon Asia, were not influenced by the pastoralists of Arid Asia. Therefore, according to conventional theories, ancient civilizations could not have developed in these regions, and the valleys must

Fig. 2. The most striking feature of the Asian monsoon region is water. The Yangtze river basin, with its rich water resource, has recently come to be recognized as the stage for the Yangtze River Civilization. Yangtze River. (Photo by Takeshi Takeda)

have remained covered by thick forests for a long time. The Japanese archipelago, which is separated from Arid Asia by the Japan Sea, and even more difficult to reach, should also have been left in a primitive and barbaric state in the Jomon period, totally isolated from civilization.

But was there really no rise of ancient civilizations in the valleys of the great rivers flowing deep in Monsoon Asia? Is it possible that we had previously recognized only ancient civilizations rising in wet-dry boundary zones as civilizations? Could riverine civilizations also have existed in the valleys of the Ganges, Yangtze, and Mekong flowing deep in Monsoon Asia?

In contrast to the previously recognized pastoral and wheat-cultivating agrarian civilizations (wheat-cultivating pastoral civilizations), the riverine civilizations in the Ganges, Yangtze, and Mekong, comprised probably hunter, fisher folk, rice-cultivating agrarian civilizations (rice-cultivating-fishing civilizations). Hunters and fisher folks played the roles of pastoralists who were non-existent deep in Monsoon Asia. Compared to the sedentary and consanguine agriculturists, hunters and fisher folks moved more freely and had access to trade and information networks. Wheat-harvesting agriculturists in wet-dry boundary zones relied on the livestock of pastoralists to obtain proteins essential for their diet. In turn, agriculturists in deep Monsoon Asia depended on the fish and wild animals caught by fisher folks and hunters (Fig. 3). In a wheat-cultivating pastoral civilization, trade and information networks were established on

Fig. 3. The livelihoods of the people in the Asian monsoon region were based on rice cultivation and fishery. Fish was the main source of proteins for the people of the Asian monsoon region. A fisherman catching fish in Er-Hai lake in Yunnan Province, China (lower). Even today, there is a rich resource of fish in the Asian monsoon region. (Photo by Takeshi Takeda)

land, while in a rice-cultivating-fishing civilization, the networks are believed to have been connected mainly by rivers and over seas.

Compared, however, to pastoralists who keep livestock, hunters and fisher folks tend to form matriarchal societies and require no central leader. They are also less oriented towards collecting gold and silver treasures. Such differences between pastoralists and hunters/fisher folks should result in characteristics and organizations that distinguish rice-cultivating fishing civilizations from wheat-cultivating pastoral civilizations.

Archaeologists trying to find evidence of past civilizations from sites, artefacts and their findings, so far have supported the above view that rice-cultivating fishing civilizations are clearly different from wheat-cultivating pastoral civilizations.

The archaeological sites and artefacts of wheat-cultivating pastoral civilizations are brilliantly decorated in gold, and the evidence suggests that such civilizations rose and fell dramatically. These civilizations have left a deep scar on their environments, and their giant ruins presently stand in the midst of barren lands. In contrast, the sites and artefacts of rice-cultivating fishing civilizations are sober and lack the golden brilliance. Their ruins lie deep in the ground shrouded by the silence of thick forests.

The author first came by the idea that the orientation for gold may be an indicator for identifying pastoralists, during an excavation at a site in the Yangtze river valley. It seemed that, compared to the Yangtze site, more golden artefacts were unearthed from sites in the Sichuan and Yunnan Provinces, which are nearer Arid Asia and more strongly influenced by Tibetan pastoralists.

The conventional concept of civilization was modelled on the gold-decorated and wheat-

cultivating civilizations that developed at the wet-dry boundary zones. However, a different type of riverine civilization may have existed in the history of mankind—the rice-cultivating fishing civilizations along the great rivers flowing deep in the wet regions of Monsoon Asia. The crops cultivated in the valleys of the Ganges, Yangtze, and Mekong were millet and rice.

Wheat-cultivating civilizations at wet-dry boundary zones were discovered by archaeologists in the 19th–20th century and have been studied extensively. These were civilizations of winter rain with their livelihood centred upon winter crops that grew in an environment of semi-arid grasslands and thin forests.

Archaeology in the 21st century will be an age of discovery of the other ancient civilization that developed in deep Monsoon Asia—the rice-cultivating-fishing civilization. These were civilizations of summer rain with their livelihood centred upon summer crops of rice and millet

such as foxtail millet (*Setaria*) that grew in an environment of wet forestland.

Let us now begin our journey in search of the other civilization that developed along the great rivers in the wet Monsoon Asia.

References

Okada, H. (1992): *The Birth of World History (Sekaishi no Tanjyo)*, Chikumashobo, Tokyo, 263 pp.

Watsuji, T. (1935): *Fudo*, Iwanamishorten, Tokyo, 253 pp.

Yasuda, Y. (1991): 'Climatic change at 5000 years BP and the birth of ancient civilizations'. *Bulletin of the Middle Eastern Culture Center in Japan*, Otto Harrasowitza, IV: 203-218.

Yasuda, Y., (2000): *The Birth of the Riverine Civilizations (Taiga Bunmei no Tanjyo)*, Kagokawashoten, Tokyo, 354 pp.

Yasuda, Y., H. Kitagawa and T. Nakagawa (2000): 'The Earliest Record of Major Anthropogenic Deforestation in Ghab valley, Nort-west Syria: A palynological study', *Quaternary International*, 73/74: 127-136.

Yoshino, M. (1999): 'Environmental Change and Rice-producing Societies in Monsoon Asia: A review of studies and elucidation of problems', *Geographical Review of Japan*, 72: 566-588.

A minority group carrying on the tradition of the rice-cultivating piscatory people of the Asian monsoon region. The Lusheng dance performed by the women of the Miao people in Guizhou province, China. (Photo by Takeshi Takeda)

Part I

Asian Monsoon Variability and Human Adaptations

A fisherman on a boat-house.
(Photo by Takeshi Takeda)

Chapter 1

Late Glacial and Holocene Vegetation Changes Recorded in the Pollen Data from the Hangai Mountains, Central Mongolia

PAVEL E. TARASOV, NADEZHDA I. DOROFEYUK, VALENTINA T. SOKOLOVSKAYA,
TAKESHI NAKAGAWA AND MIROSLAW MAKOHONIENKO

Overview

Radiocarbon-dated pollen records from seven freshwater lakes in the Hangai Mountains provide important information on the Late Quaternary vegetation changes from this poorly known region of Central Asia. Both qualitative and quantitative interpretation of the pollen data suggests that steppe was a dominant vegetation type since 11,500 yrs. BP. However, sedge and shrubby tundra-like associations were more important in the region during the Late Glacial period prior to 9500 yrs. BP. *Larix* became the first tree species to form patchy forests in Hangai after that time. A forestation of the area and spread of forest patches dominated by *Pinus* and *Larix* with some admixture of *Picea* characterized the time period from 7500 to 4000 yrs. BP. After 4000 yrs. BP *Artemisia*-Poaceae steppe associations became more important in the vegetation cover than before. Identifying features are more pronounced in the records from the high-elevated Hangai lakes (2450–2650 m), while the differences between the middle and late Holocene pollen spectra in the records from the low-elevated sites (1450–2061 m) are not so visible. Reconstructed vegetation changes can be interpreted in terms of changes in the regional monsoon circulation, reported by the earlier studies from China and Mongolia.

Introduction

Mongolia occupies a large area in the centre of the Eurasian continent (Fig. 1). Situated far from the oceans at elevations ranging between 560 and 4374 m a.s.l., the country has an extreme continental climate, characterized by low precipitation (500–50 mm/yr) and mean annual temperature anomalies exceeding 40°C (Zhambazhamts and Bat, 1985). Precipitation has a well pronounced minimum during the winter season (5–30 mm), when the area is occupied by the Asian High. The mean January temperatures range from –32° to –16° C, and mean summer

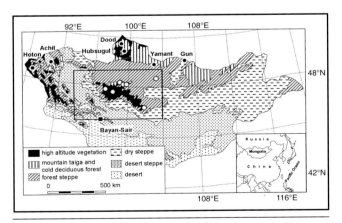

Fig. 1. Map of Mongolian vegetation (after Lavrenko 1979) and location of sites with radiocarbon dated pollen (circle) and plant macrofossil (square) records. Closed circles indicate published records and open circles indicate sites from Hangai presented in this study.

temperatures vary from 12°C to 24°C. Distribution patterns of temperature and precipitation have clear latitudinal and altitudinal gradients. More than half the country lies at an elevation of above 1500 m. Thus, despite the low precipitation the climate in northern and central Mongolia is not arid enough to prevent a development of the forest vegetation (Fig. 1). However, mountain boreal evergreen and deciduous forests and forest-steppe occupy 4% and 23% of the area, while steppe and desert in Mongolia cover 53% and 15% respectively (Hilbig, 1995).

Lakes are widely distributed in Mongolia. Many of them are confined to the transitional forest-steppe zone, which is supposed to be very sensitive to the millennial-scale changes in the regional climate (Whitlock and Bartlein, 1997; Tarasov et al., 2000). The palaeobiological team of the Russian-Mongolian Biological Expedition (Institute of Ecology, Moscow) made great efforts to core over twenty Mongolian lakes (Gunin et al., 1999). Sediments obtained from twenty-five cores have been radiocarbon-dated and results of diatom (Dorofeyuk, 1992; Sevastyanov and Dorofeyuk, 1992; Sevastyanov et al., 1989) and pollen analysis have been partly interpreted and published (Vipper et al., 1976; 1989; Dorofeyuk and Tarasov, 1998; 2000; Gunin et al., 1999; Tarasov et al., 2000).

The Late Glacial and Holocene vegetation history of Mongolia is based largely on undated pollen records (Savina et al., 1981; Golubeva, 1976; 1978; Malaeva, 1989). Only a few radiocarbon-dated pollen diagrams and plant macrofossil records (Fig. 1) have been published. However, they are confined to the western (Vipper et al., 1976; Dinesman et al., 1989; Sevastyanov et al., 1993; Dorofeyuk and Tarasov, 2000; Tarasov et al., 2000) and northern (Dorofeyuk and Tarasov, 1998; Gunin et al., 1999; Krengel, 2000) parts of the country. We applied (Gunin et al., 1999) a quantitative method of 'biomization' (Prentice et al., 1996) adopted for the reconstruction of vegetation in northern Eurasia (Tarasov et al., 1998) to twenty-five radiocarbon-dated pollen records from all over Mongolia. Resulting maps show spatial distribution of Mongolian biomes at twelve key time-slices from 15,000 yrs. BP to today (Gunin et al., 1999). However, this synthesis has demonstrated no change in the vegetation of the Hangai mountains, suggesting that cool steppe has been a dominant vegetation type through the Late Glacial and Holocene time. The conventional biomization method (Prentice et al., 1996) does not allow the reconstruction of transitional vegetation types (e.g., forest-steppe or forest-tundra) and thus is too coarse for the evaluation of small-scale changes in taxa percentages observed in fossil pollen spectra. However, additional information can be obtained by examining changes in taxa percentages and by looking at the relative score values of forest and non-forest biomes (Tarasov et al., 2000).

In this paper we first present the complete results of pollen analysis of radiocarbon-dated sedimentary cores from seven lakes situated in central Mongolia (Fig. 2). Observed changes in the fossil pollen assemblages are then discussed in terms of possible changes in regional vegetation and climate.

Data and Methods

■ The area of study

The Hangai mountain area is located in the central part of Mongolia (Fig. 2). It extends over 700 km from northwest to southeast and over 400 km from northeast to southwest. The maximum elevation (4021 m—Otgon Tenger Uul) is registered in the western part of Hangai. The elevations above 3000 m are common along the main ridge of Hangai and less common in the Tarbagatai ridge passing north of 48° N (Fig. 2). The relief of the area is complex. Numerous valleys cut mountain ridges into sub-longitudinal

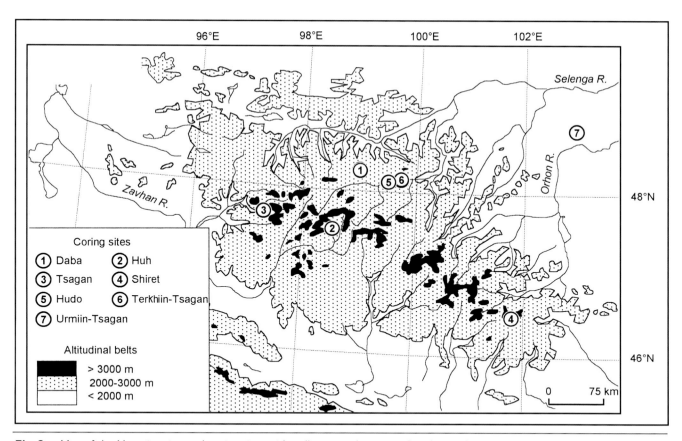

Fig. 2. Map of the Hangai region and coring sites with pollen record presented in this study.

segments. The Ider River Valley and the Selenge River Valley separate Hangai from the mountain area of northern Mongolia. Large dry depressions with many brackish and saline-water lakes separate Hangai from the Mongolian Altai in the west and from the Gobi Altai in the south. Relatively flat land extends eastwards.

Climate stations are rather sparse and confined to the mountain border zone (Zhambazhamts and Bat, 1985). At lower elevations the mean January temperature changes from –24° to –20°C and the mean July temperature is about 16°C. However, extreme temperatures vary from –46.2°C in winter to 33.6°C in summer. Annual precipitation changes from c. 250 mm at low elevation to over 400 mm above 2500 m. It is mainly associated with summer incursions of warm air masses from the Pacific Ocean, interacting with the cold air masses at the Polar front.

Present-day vegetation of Hangai is forest-steppe (Fig. 1) with the dominance of open vegetation communities in the landscape. Dry steppe usually occupies the foothills, while open forests with *Larix* and shrub *Betula* appears at 1750 to 2550 m (Lavrenko, 1979). *Pinus sibirica* trees grow in the upper part of this forest belt. Vegetation is often mosaic. Northern slopes, where the permafrost layer is close to the surface, are most suitable for tree growth, providing an additional source of water during the vegetation season. Tundra communities are characteristic of the vegetation above 2600 m. However, sedge and shrubby associations represented by *Carex* and *Kobresia* species and shrubs of *Betula rotundifolia*, *Betula humilis* and *Salix* may grow at lower elevations, being graded with steppe and forest associations.

Picea does not penetrate in Hangai today. Mountain forests dominated by *Pinus sibirica*

25

with a mixture of *Larix* and *Picea* grow close to Hubsugul lake c. 250–300 km northwards from Terkhiin-Tsagan-Nur and Hudo-Nur lakes (Sites 5, 6 in Fig. 2). The closest habitats with *Pinus sylvestris* are situated c. 260 km to the north and c. 330 km to the northeast from the mentioned lakes at 700–1100 m (Lavrenko, 1983). Although single trees of *Betula pendula* appear today in the river valleys and within larch forests at lower elevation in the Hangai area (Byazrov *et al.*, 1983), the nearest *Betula* patches grow c. 75–100 km eastwards and northeastwards from Terkhiin-Tsagan-Nur (Site 6 in Fig. 2) and c. 30–50 km from Urmiin-Tsagan-Nur (Site 7 in Fig.2).

■ Coring and dating methods

In this section, we present pollen records from seven freshwater lakes situated in the Hangai mountains (Fig. 2). The lakes are different in size and in basin morphology. They are situated at an altitude of 1450 – 2649 m above sea level (Table 1). The Late Glacial-Holocene chronology is based on twenty-seven radiocarbon dates, spanning the interval from 11,180 to 1320 yrs. BP (Table 1). All dates were obtained from bulk sediment, using conventional dating methods (Gunin *et al.*, 1999). Dated material was organic gyttja in all cases, except for the one sample from Daba-Nur core, where dated sediment was peat. While all radiocarbon dates are in the correct sequence, the relatively poor time resolution and the large thickness of dated samples means that only rough age estimation for the recorded pollen events and reconstructed vegetation changes can be assumed.

Coring of the lake sediments has been done manually from the platform established on two rubber boats. The Russian corer (sampler 50 cm in length and 6 cm in diameter) was used to core dense sediment while piston corer (22 cm in length and 3.5 cm in diameter) was applied in the uppermost very soft sediment layer. Changes in the sediment lithology have been visually described in the field.

■ Pollen analysis

Each sample taken for pollen analysis comprised 5 – 10 cm of sediment. This way of sampling was often used in the earlier Soviet studies dealing with reconstruction of general trends rather than short-term events in vegetation development. In the laboratory, each pollen sample was carefully mixed and then treated with the standard procedure (Grichuk and Zaklinskaya, 1948), using heavy liquid separation to extract pollen. All pollen samples usually contain a sufficient number of counted pollen grains to reconstruct changes in vegetation.

We used the Tilia/Tilia-Graph software (Grimm, 1991) to calculate pollen percentages and to draw pollen diagrams. In all diagrams the total sum of arboreal and non-arboreal pollen is taken as 100% for the calculation of the pollen taxa percentages. A definition of the local pollen zone boundaries in the pollen diagrams is supported by CONISS (Grimm, 1987) within Tilia/Tilia-Graph.

■ Interpretation methods

A representative set of 102 modern pollen spectra from Mongolia has been compiled and recently published (Gunin *et al.*, 1999). The main conclusion reported in the latter study was that the spatial patterns in modern pollen data clearly reflect modern vegetation. Thus, the qualitative interpretation of the pollen records from Hangai is based on the known relationships between actual vegetation and modern surface pollen spectra from Mongolia.

Artemisia, Chenopodiaceae and Poaceae species are widely distributed in Hangai today, contributing a lot to the composition of surface pollen spectra. Today *Betula* and *Alnus* pollen in the surface spectra from Hangai is probably

Table 1. Information on studied pollen sites with a list of radiocarbon dates.

Site	Latitude N	Longitude E	Elev. m	^{14}C dates, yr B.P.	Lab. No.	Sample depth, cm	Material
Daba-Nur	48°12'	98°47'40"	2465	3100±120	TA-1355	75–100	gyttja
				5600±80	TA-1356	250–275	gyttja
				7680±100	TA-1357	325–350	gyttja
				9400±100	TA-1358	375–400	gyttja
				10,580±100	TA-1188	460–475	gyttja
				11,180±120	TA-1028	490–505	peat
Huh-Nur	47°32'	98°31'	2649	3430±90	Vib-105	80–85	gyttja
Tsagan-Nur	47°39'	97°15'50"	2236	1050±100	TA-1064	75–90	gyttja
				2660±80	TA-1063	315–340	gyttja
				2850±90	TA-1033	340–370	gyttja
Shiret-Nur	46°32'	101°49'	2500	1320±100	TA-1474	50–75	gyttja
				3150±120	TA-1775	125–150	gyttja
				8360±100	TA-1420	200–220	gyttja
Hudo-Nur	48°08'	99°32'	2061	5450±80	TA-1346	380–400	gyttja
				7740±80	TA-1345	610–630	gyttja
				9230±110	TA-1344	760–780	gyttja
				9800±100	TA-1247	850–875	gyttja
Terkhiin-Tsagan-Nur	48°09'	99°42'	2060	2740±60	TA-1538	175–200	gyttja
				3840±50	TA-1339	275–300	gyttja
				4150±80	TA-1351	300–325	gyttja
				4230±50	TA-1340	325–350	gyttja
				5050±80	TA-1352	400–425	gyttja
				5950±120	TA-1353	450–475	gyttja
				6690±60	TA-1246	525–550	gyttja
				6890±100	TA-1248	550–575	gyttja
Urmiin-Tsagan-Nur	48°50'30"	102°56'	1450	5200±80	TA-1484	240–250	gyttja
				7050±150	TA-1485A	260–280	gyttja

produced by shrubby forms of these taxa, growing in the upper belt of the mountains close to the studied lakes. The conifers are represented only by a few pollen grains. However, the presence of *Larix* pollen in the surface spectra indicates growth of larch close to the sampling site. *Larix* pollen has poor preservation in the sediment and pollen dispersion from the larch tree usually does not exceed several hundred metres (Savina and Burenina,1981). By contrast, pollen of *Pinus*, especially that produced by *Pinus sylvestris*, and *Picea* is more susceptible to long-distance transport. The low content of arboreal pollen indicates dry conditions. However, the absence of boreal evergreen conifers in the vegetation may also suggest low winter temperatures (Prentice *et al.*, 1992), e.g. mean temperature of the coldest month below −35°C.

Our qualitative interpretation has been checked with a quantitative method of pollen-based biome reconstruction (called 'biomization') developed by Prentice *et al.*, (1996). This method is based on the objective assignment of pollen taxa to plant functional types and then to main vegetation types (biomes) on the basis of the known ecology and biogeography of modern plants. Pollen taxa from the former Soviet Union and Mongolia were assigned to plant functional types and to biomes by Tarasov *et al.*, (1998). The test of the method with 102 surface pollen spectra from Mongolia demonstrated that biomes were correctly predicted eighty-five times (Gunin *et al.,* 1999). The equation to calculate the affinity scores for all pollen samples was published by Prentice *et al.,* (1996):

$$A_{ik} = \Sigma_j \delta_{ij} \sqrt{\{\max[0,(p_{jk} - \theta_j)]\}}$$

where A_{ik} is the affinity of pollen sample k for biome i; summation is over all taxa j; δ_{ij} is the entry in the biome *versus* taxon matrix for biome i and taxon j; p_{jk} are the pollen percentages, and θ_j is the universal threshold pollen percentage of 0.5%. Table 2 shows the assignment of pollen taxa from Hangai records to the tundra, steppe and taiga biomes discussed in the present study.

The biome with the highest score or, when several biomes have the same score, the one defined by a smaller number of taxa was then assigned to the given pollen spectrum. Pollen-based Mongolian biome reconstruction at selected time slices (Gunin *et al.,* 1999) suggested that steppe was the dominant biome in Hangai since the Late Glacial. In order to have more complete information on possible changes in the other vegetation types we examined in the present study scores of the main biomes distributed in the

region today (e.g., steppe, tundra, and taiga). Biome reconstruction from the pollen record processed with the PPPBASE software (Guiot and Goeury, 1996).

Pollen Records and Interpretation

▪ Daba-Nur (48°12'N, 98°47'40"E, 2465 m)

Site description: Daba-Nur lake occupies a topographic depression in the Dzagastain-Daba pass through the Tarbagatai ridge (Site 1 in Fig. 2). It has a maximum length of 2.2 km, a maximum width of about 1 km, an average water depth of c. 2 m and a maximum depth of 4.5 m (Dorofeyuk, 1988; Tarasov *et al.,* 1996). There are two small islands in the central part of the lake. The lake is fed by several small streams and by direct precipitation and has no surface outflow. Today flat and swampy shores of the lake are covered with *Carex, Kobresia* and small shrubs of *Betula* and *Salix*. Single *Larix sibirica* and *Pinus sibirica* trees grow on the western and northwestern slopes of the basin at the upper limit of their modern distribution. Mountain tundra communities with *Kobresia, Dryas* and *Empetrum* appear on the eastern slope of the

Table 2. Taiga, steppe and tundra biomes and the pollen taxa from Mongolian records assigned to them.

Biome	Pollen taxa
Taiga	*Abies, Picea, Pinus, Juniperus, Betula, Larix, Alnus, Populus, Salix, Calluna, Cassiope, Empetrum,* Ericales, *Pyrola,* Pyrolaceae
Steppe	*Allium,* Apiaceae, Asteraceae (Asteroideae), Asteraceae (Cichorioideae), Brassicaceae, Campanulaceae, *Cannabis,* Caryophyllaceae, *Centaurea,* Convolvulaceae, Dipsacaceae, *Epilobium,* Euphorbiaceae, Fabaceae, *Filipendula, Galium,* Geraniaceae, *Hippophaë,* Iridaceae, Lamiaceae, *Linaria,* Liliaceae, Onagraceae, Papaveraceae, *Plantago,* Plumbaginaceae, *Potentilla,* Ranunculaceae, Rosaceae, Rubiaceae, Rutaceae, *Scabiosa, Stellera, Taraxacum, Artemisia,* Boraginaceae, Chenopodiaceae, *Kochia,* Poaceae, Scrophulariaceae, Valerianaceae, Polygonaceae
Tundra	*Alnus, Betula, Salix, Dryas, Gentiana, Pedicularis,* Saxifragaceae, Scrophulariaceae, Valerianaceae, Polygonaceae, Poaceae, Cyperaceae, *Calluna, Cassiope, Empetrum,* Ericales, *Pyrola,* Pyrolaceae

basin going up to 3128 m. A 525-cm core was taken in a water depth of 1.75 m, in the central part of the lake, c. 150 m south of the islands.

Pollen record: Daba-Nur pollen diagram (Fig. 3) is divided into six local pollen zones (LPZ). LPZ Db-6 (525–445 cm) is characterized by a very low content of arboreal pollen (less than 5%) represented by *Betula* and by single grains of *Alnus*, *Pinus* and *Picea*. *Artemisia* (31–47%) and Poaceae (23–33%) pollen dominates in the assemblages, while Chenopodiaceae pollen is less abundant (3–9%). The highest values for Cyperaceae (10–20%) and Ranunculaceae (5–7%) pollen occur in this zone. In LPZ Db-5 (445–400 cm) a drastic increase in abundance of *Betula* (10 to 15%) associated with high values of *Artemisia* (50 to 65%) and Brassicaceae (5–7%) is recorded. *Alnus* pollen is systematically registered in this zone. LPZ Db-4 (400–325 cm) shows further increase in *Betula* pollen abundance up to 23%. Small quantities of *Alnus*, *Larix*, *Pinus* and *Picea* pollen are present throughout this zone. In LPZ Db-3 the content of arboreal pollen (*Pinus*, *Betula*, *Larix* and *Picea*) increases to the maximum (34%), while non-arboreal pollen still dominates in pollen assemblages. LPZ Db-2 (225–150 cm) is characterized by a decrease in arboreal pollen to 15%. *Artemisia* is still the most abundant taxon (30–45%). However, Poaceae pollen increases to 30% and several minor herbaceous taxa (e.g., Asteraceae, Fabaceae, Rutaceae and Rubiaceae) appear in this zone. In the uppermost zone (Db-1) a sum of *Betula* and *Pinus* varies from 5 to 15% of the total pollen sum. *Larix* pollen becomes a minor component in the spectra and *Picea* pollen is found only at one level. *Artemisia* reaches a maximum (50–65%) and together with Poaceae and Chenopodiaceae dominates pollen assemblages in this zone.

Reconstructed environments: Steppe associations with *Artemisia* as the most characteristic taxon played the most important role in the surrounding vegetation from 11,500 yrs. BP to the present. However, some changes in the distribution of tundra and forest vegetation around the lake can be detected. The area was probably treeless before 9500 yrs. BP. Pollen of *Betula* and *Alnus* was likely produced by the shrubby forms of these taxa, as it is in the region today. Compared to the present-day vegetation, herbaceous (Cyperaceae-Poaceae) tundra associations occupied a larger area around the lake prior to 10,200 yrs. BP. Co-existence of dry steppe and herbaceous tundra (tundra-steppe) in the vegetation cover would imply dry and cold climate during the Late Glacial period. Peat layers found in the bottom part of the core (Fig. 3) indicate very low lake level and support our interpretation of drier than present Late Glacial climate. Shrubby tundra communities dominated by *Betula* and *Alnus* species were important feature of the local vegetation between 10,200 and 9500 yrs. BP, suggesting conditions becoming wetter than during the Late Glacial. Gyttja sedimentation indicates an increase in the lake level, likely associated with a climate wetter than before. The first appearance of *Larix* close to the lake can be dated to 9500 yrs. BP and *Pinus* (likely *P. sibirica*) appeared slightly later than this date. Between approximately 7500 and 5000 yrs. BP both *Larix* and *Pinus* trees were more abundant around the lake than today, suggesting relatively warm and wet middle Holocene climate. A decay of forest vegetation to the modern level can be reconstructed sometime between 5000 and 4000 yrs. BP. After 4000 yrs. BP position of steppe vegetation close to the lake became strongest since the Late Glacial time.

■ **Huh-Nur (47°32'N, 98°31'E, 2649 m)**

Site description: Huh-Nur (Site 2 in Fig. 2) is an overflowing lake occupying a knee-like depression in the southern macro-slope of the

30

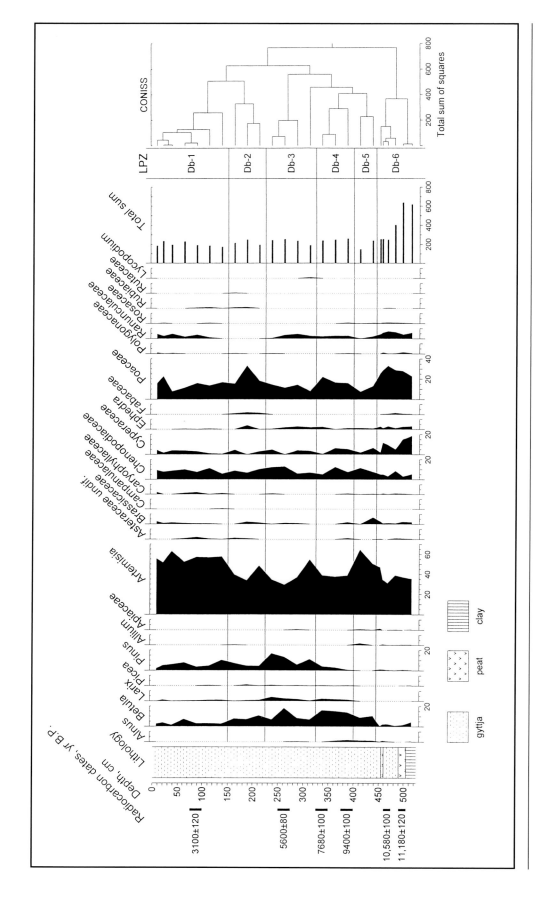

Fig. 3. Lithology column and pollen percentage diagram from Daba-Nur lake. Frequencies are expressed as percentages of the total sum of arboreal and non-arboreal pollen at each level.

main Hangai ridge. The lake has a maximum length of about 15 km, and a maximum width of c. 1.5 km. Huh-Nur is drained by the Shir-Us river, flowing southward to the Zavhan river. Modern vegetation around the lake is represented by tundra and dry steppe associations with *Kobresia* and dwarf-shrub *Salix* species, *Festuca*, *Caragana*, *Potentilla*, etc. A 370-cm core was taken in a water depth of 1.8 m, in the southern lake bay c. 100 m from the western shore.

Pollen record: The Huh-Nur pollen diagram (Fig. 4) has been divided into three LPZ. The lowermost zone (Hh-3: 370–275 cm) is characterized by a highest content of arboreal pollen, including *Pinus* (12–37%), *Betula* (2–5%), *Picea* (2–5%) and *Larix*. However, non-arboreal taxa represented by *Artemisia* (23–38%), *Ephedra* (10–15%) and Chenopodiaceae (8–12%) comprise over 50% of total pollen sum in this zone. A decrease of 10–27% in arboreal pollen is registered in LPZ Hh-2 (275–120 cm). Such drastic change in the pollen composition is mainly due to sharp decrease in abundance of *Pinus* (1–7%) and *Picea* (less than 1%) pollen and may imply a sedimentary hiatus. This suggestion may be partly supported by a change in the colour of the clay from blue to grey at the 260-cm level. *Betula* and *Larix* slightly increase and *Alnus* pollen is systematically registered in LPZ Hh-2. *Abies* pollen is registered in only one spectrum. Among non-arboreal taxa *Artemisia*, Chenopodiaceae, Poaceae and Cyperaceae have slightly higher values compared to Hh-3 zone. In the uppermost LPZ Hh-1 a sum of *Betula* and *Pinus* pollen varies between 5 and 10%. *Picea*, *Larix* and *Alnus* disappear from the pollen assemblages. *Artemisia* pollen reaches its highest content (60 – 77%) in the upper part of this zone.

Reconstructed environments: The earlier part of the record, suggesting a presence of tree/shrub

vegetation around the lake is undated. However, pollen correlation with adequately dated Daba-Nur record provides some ideas about the age of the forest-steppe phase in the Huh-Nur basin. A maximum spread of *Pinus sibirica* around Daba-Nur is dated at 7500–5000 yrs. BP, suggesting that the phase when *Pinus* and *Larix* and probably *Picea* grew close to Huh-Nur lake occurred at that time. Only the presence of *Larix* and probably *Betula* in the Huh-Nur basin is not doubtful between 5000 and 4000 yrs. BP. The latter date is also established by the pollen correlation with the Daba-Nur record. After 4000 yrs. BP. *Artemisia* dominated steppe associations around the lake and gradually gained the importance it has today. However, complete disappearance of *Larix* from the vegetation may have occurred not earlier than 3000 yrs. BP.

■ **Tsagan-Nur (47°39'N, 97°15'50"E, 2236 m)**

Site description: Tsagan-Nur is a small lake situated in the closed basin in the Hangai ridge. Several short streams flow to the lake, but it does not have a surface outflow. Two undated lake terraces are observed at 3 and 10 m above the modern lake level. The southern swampy shore is covered with *Carex* species. Western and eastern steep slopes of the basin are covered with tundra-like *Kobresia* associations. Steppe vegetation with *Artemisia*, *Stipa*, *Poa*, *Aster*, *Veronica*, *Galium*, *Senecio*, *Dianthus* and some other species grows at lower elevations, while open larch forest occupies slopes of the northern exposition c. 100–200 m above the water level. A 440-cm core has been taken in a water depth of 2.6 m, c. 500 m from the southern shore.

Pollen record: Tsagan-Nur pollen record only demonstrates small changes in the pollen assemblages. However, statistical analysis with CONISS provides a possibility to divide the diagram (Fig. 5) into two local pollen zones. LPZ

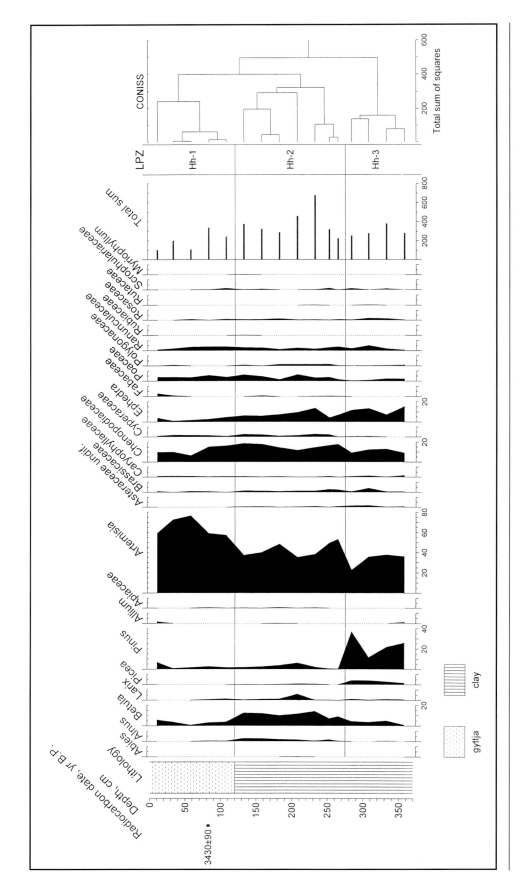

Fig. 4. Lithology column and pollen percentage diagram from Huh-Nur lake. Frequencies are expressed as percentages of the total sum of arboreal and non-arboreal pollen at each level.

Ts-2 (440–170 cm) is characterized by the minor representation of tree/shrub taxa. The total sum of arboreal pollen never exceeds 5% and becomes a zero in the upper part of this zone. Among tree/shrub taxa only *Larix* pollen represents local vegetation, while scars pollen of *Pinus*, *Alnus* and *Betula* is likely wind-transported. Among non-arboreal pollen *Artemisia* is a dominant taxon (50–80%). Pollen of Chenopodiaceae (3–20%) and Poaceae (1–10%) is less abundant. Brassicaceae pollen reaches up to 20% in the bottom part of the record. In Ts-1 (170–0 cm) pollen zone appearance of Lamiaceae and Papaveraceae is registered.

Reconstructed environments: Three available dates from Tsagan-Nur core (Fig. 5) suggest that at least the middle and upper parts of the record represented by the gyttja layer have been deposited during the last 3000 years. During this time period the area around the lake was covered with modern *Artemisia*-Poaceae steppe. The absence of arboreal pollen in the middle part of the record may indicate a deforestation episode which occurred in the region sometime between 1900 and 1500 yrs. BP. However, more data is needed to prove this suggestion at the regional scale.

■ **Shiret-Nur (46°32'N, 101°49'E, 2500 m)**

Site description: Shiret-Nur (Site 4 in Fig. 2) occupies a tectonic depression at the northern slope of one of the southeastern ridges in Hangai. The lake has a maximum length of 4 km, a maximum width of 1.6 km and a maximum water depth of 16 m. Steep slopes are covered with sparse herbaceous vegetation. *Larix* trees grow in the northern part of the lake together with shrubs of *Juniperus sibirica*, *Salix*, *Lonicera altaica* and *Potentilla fruticosa*. Flat surfaces are covered with swampy and steppe associations. The open *Pinus sibirica-Larix* forest with herbaceous communities and shrubs of *Betula*, *Lonicera*, *Rosa*, *Vaccinium* and *Pyrola* occupies the slope exposed to the northeast. A 230-cm core has been taken in the bay in the northern part of the lake c. 100-m from the coast. The water depth there was 2.6 m.

Pollen record: The Shiret-Nur pollen diagram (Fig. 6) is divided into two local pollen zones. In LPZ Sh-2 (230–150 cm) *Betula* reaches 30–45% in the bottom part, *Pinus* becomes more abundant (up to 25%) in the middle part and decrease in both taxa percentages to c. 10% is registered in the upper part of this zone. *Larix* pollen is systematically registered and *Picea* is found at 185-cm level. However, *Artemisia* (25–40%), Chenopodiaceae (15%) and Poaceae (2–5%) pollen dominates in the pollen assemblages. Low contents of *Betula* and *Pinus* pollen and discontinuous curve of *Larix* are characteristics of LPZ Sh-1 (150–0 cm). A minimum in the abundance of arboreal taxa (c. 5%) is registered at c. 75-cm level. However, the peak in the curve of *Pinus* pollen (20%) follows this minimum. Among the non-arboreal taxa, *Artemisia* is a dominant (40–60%) and Chenopodiaceae (15–20%) and Poaceae (5–15%) are co-dominant taxa in this zone.

Reconstructed environments: The earlier (undated) part of the record suggests that tree/shrub vegetation played a more important role around the lake than today. If shrub-like forms of *Betula* produced the major part of its pollen (as it is today), a spread of shrubby associations around the lake can be suggested prior to 8300 yrs. BP. Conifer trees occupied maximal area sometime after 8000 yrs. BP. and before 4000 yrs. BP. However, steppe was a dominant vegetation type even at that time. Since 4000 yrs. BP herbaceous communities dominated by *Artemisia* and Poaceae species strengthened their positions around the lake, suggesting that climate became drier than during the middle Holocene. An extrapolation between two

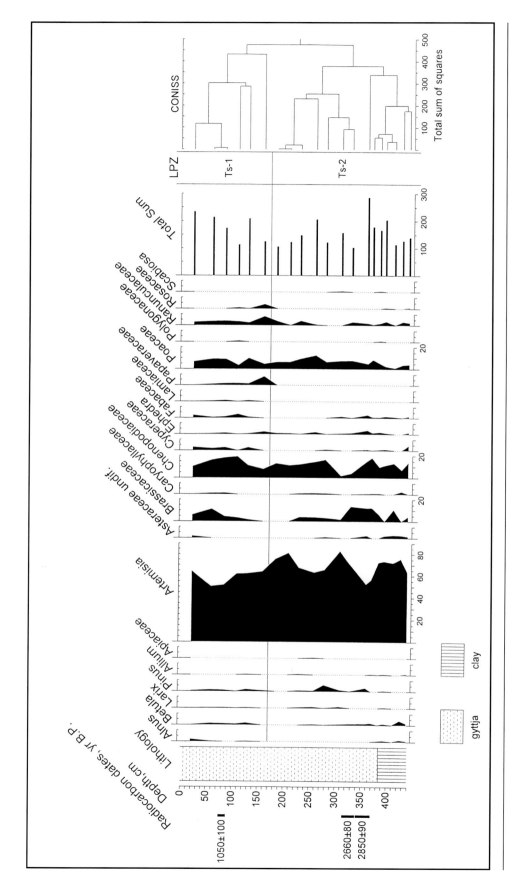

Fig. 5. Litihology column and pollen percentage diagram from Tsagan-Nur lake. Frequencies are expressed as percentages of the total sum of arboreal and non-arboreal pollen at each level.

34

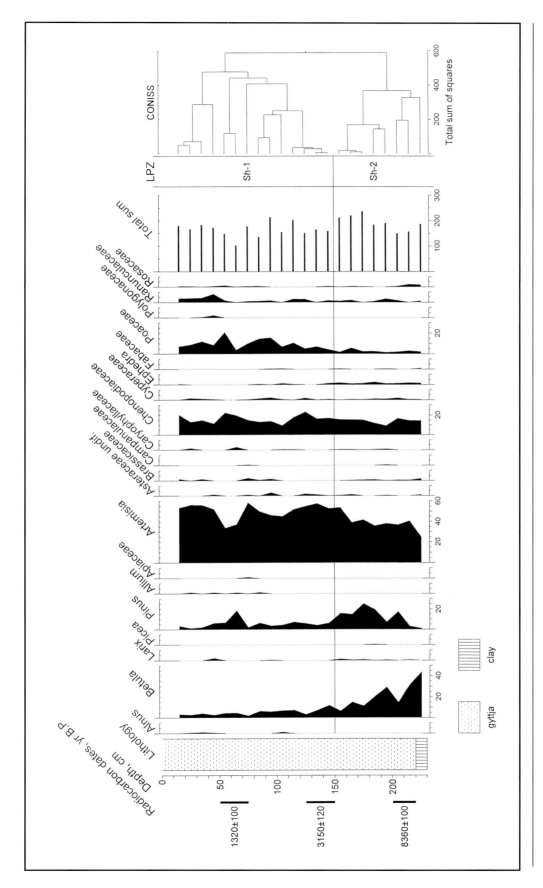

Fig. 6. Lithology column and pollen percentage diagram from Shiret-Nur lake. Frequencies are expressed as percentages of the total sum of arboreal and non-arboreal pollen at each level.

radiocarbon dates would suggest that observed minimum in abundance of arboreal pollen may be attributed to 1600–1700 yrs. BP and a second short-term peak in *Pinus* distribution occurred after that time.

■ Hudo-Nur (48°08'N, 99°32'E, 2061 m)

Site description: Hudo-Nur lake lies in the flat and wide bottom part of the intermountain Tariatskaya depression drained by the Hoit-Terkhiin-Gol river (Site 5 in Fig. 2). The ridges of Hangai and Tarbagatai form a boundary around the Tariatskaya depression. The lake has a length of 5.4 km, a maximum width of 3.5 km and a maximum depth of 3.5 m. A short stream connects Hudo-Nur lake with Terkhiin-Tsagan-Nur lake described later in the present paper. Steppe associations with *Artemisia*, Poaceae and other herbaceous taxa represent modern vegetation around the lake. A southern slope of the depression, going up to above 3000 m is occupied by open *Larix* forest replaced upward by *Larix-Pinus sibirica* forest and then by shrub-sedge-moss alpine vegetation. A 940-cm core has been taken in a water depth of 3.5 m in the bay of the eastern part of the lake. However, pollen was not found in a basal sandy layer.

Pollen record: The Hudo-Nur pollen diagram (Fig. 7) is characterized by very sharp short-term fluctuations in the pollen percentages of the dominant taxa. However, visual comparison shows only small differences between 4 main local pollen zones, selected with CONISS. In LPZ Hd-4 (890–810 cm) *Betula* and *Alnus* dominate among the arboreal pollen taxa (7–27%) and *Larix* pollen is not found. Herbaceous taxa, e.g., *Artemisia* (20–55%), Chenopodiaceae (7–17%), Poaceae (5–17%), Ranunculaceae (3–6%) and Brassicaceae (3–6%) dominate in the pollen assemblages. LPZ Hd-3 (810–560 cm) is characterized by relatively high values of *Pinus* (2–10%). Pollen of *Picea* (up to 6%) and *Larix* is systematically registered and *Abies* pollen only appears in this zone. There is no change in the composition and abundance of herbaceous taxa compared to Hd-4. Furthermore, pollen assemblages of LPZ Hd-2 (560–310 cm) show only a slight difference from Ho-3 (e.g., slightly higher values of Poaceae and Cyperaceae, and slightly lower values of *Ephedra* and Chenopodiaceae). In LPZ Hd-1 (310–0 cm) the abundance of arboreal pollen never exceeds 20%. This corresponds to a visual decrease in *Betula* pollen to less than 5%. In this zone *Artemisia* content reaches 67% and Poaceae becomes generally more abundant than before (up to 28%).

Reconstructed environments: Pollen records show only minor changes in the pollen composition between the main pollen zones, suggesting that vegetation around the site was more or less stable during the Holocene. The earlier part of the record suggests that steppe associations dominated by *Artemisia*, Poaceae and Chenopodiaceae species shared the space with shrubby associations of *Betula* and *Alnus* before 9500 yrs. BP. The non-presence of *Larix* pollen indicates that the area close to the lake was treeless at that time. The first appearance of *Larix* close to the site corresponds to the early Holocene (9500 yrs. BP). The fact that coniferous pollen became more abundant in the middle part of the record would suggest that forest patches in the region occupied a larger area than today between 8000 and 5000 yrs. BP. *Larix* trees were definitely present close to the lake. However, relatively high percentages of *Picea* may indicate that *Larix* trees also grew in Hangai during the middle Holocene when the climate was milder than today. Somewhat drier conditions occurred after 4000 yrs. BP, when percentages of *Betula* became low, *Picea* almost disappeared from the pollen assemblages and *Artemisia* and Poaceae reached highest values.

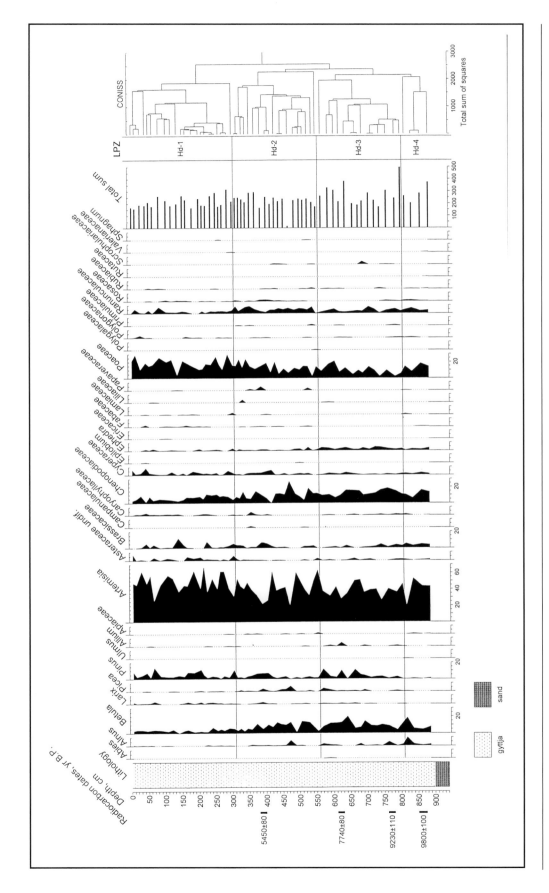

Fig. 7. Lithology column and pollen percentage diagram from Hudo-Nur lake. Frequencies are expressed as percentages of the total sum of arboreal and non-arboreal pollen at each level.

37

■ Terkhiin-Tsagan-Nur (48°09'N, 99°42'E, 2060 m)

Site description: Terkhiin-Tsagan-Nur (Site 6 in Fig. 2) is an overflowing freshwater lake located in the intermountain Tariatskaya depression. The origin of the lake basin is explained by a Horog (or Horgiin-Togo) volcanic eruption, which dammed the valley of the Terkhiin-Gol river by lava flows (Sevastyanov *et al.,* 1989). The volcano is situated 1.5 km northeast of the lake. The modern lake has a length of 16 km, a maximum width of 4.5 km, an average depth of about 6 m and a maximum depth of 20 m. The lake water outflows to the Sumein-Gol river and then via the Orkhon river and the Selenga river reaches Baikal lake. Modern vegetation is steppe, similar to that around Hudo-Nur lake. However, the eastern coast is covered with shrubs of *Betula* and *Salix* and single trees of *Larix* and *Pinus sibirica.* A 610-cm core taken in a water depth of 8.9 m in the northern part of the lake provides a record back to 7000–7300 yrs. BP.

Pollen record: The Terkhiin-Tsagan-Nur pollen diagram (Fig. 8) correlates well with the middle and upper parts of the Hudo-Nur pollen record (Fig. 7). LPZ Te-3 (600–330 cm) is characterized by moderately high values of arboreal pollen (20 to 40%) represented by *Pinus, Betula, Larix, Picea* and *Alnus.* Maximum values of *Artemisia* and Poaceae are 35% and 25% respectively. LPZ Te-2 (330–260 cm) is a transitional zone characterized by a sharp decrease in the content of tree/shrub taxa (mostly *Pinus*) to 10%. Pollen of *Artemisia* and Poaceae becomes more abundant in this zone, reaching 45% and 30% respectively. In the uppermost zone (Te-1) the total sum of arboreal taxa becomes slightly higher (10–25%). However, only pollen of *Pinus, Betula* and *Larix* are systematically counted among the arboreal taxa. Poaceae reaches a maximum (40%) in this upper zone being a dominant taxon together with *Artemisia.*

Reconstructed environments: The Terkhiin-Tsagan-Nur pollen record can be interpreted in the same way as the Hudo-Nur record. *Larix* pollen is continuously present in the pollen assemblages, suggesting that larch grew near the lake during the last 7000 years. However, the density of the open boreal forests and/or number of forest patches was probably higher than today before 4200 yrs. BP. The *Picea* pollen curve is continuous up to 3500 yrs. BP. However *Picea* maximum appears about 6700 yrs. BP. After 4000 yrs. BP tree vegetation became less important in the region than during the middle Holocene interval and the area was generally covered with *Artemisia*-Poaceae steppe similar to today.

■ Urmiin-Tsagan-Nur (48°50'30"N, 102°56'E, 1450 m)

Site description: Urmiin-Tsagan-Nur lake is an overflowing lake situated in the northeastern part of the Hangai region (Site 7 in Fig. 2). It has a length of 1.4 km, a maximum width of about 0.7 km and a maximum depth of 2.2 m. Open *Larix* woodland (forest-steppe) covers the upper parts of the surrounding mountain slopes, and steppe occupies the foothills and the Urmiin-Gol river valley close to the lake. A 290-cm core is taken in a water depth of 1.5 m in the southern part of the lake c. 200 m from the coast.

Pollen record: The most representative pollen taxa do not show great variation in abundance throughout the record (Fig. 9). In LPZ Ur-3 *Pinus* content varies between 7 and 15% and *Betula* varies between 15 and 20% of the total pollen sum. Small quantities of *Larix* and *Picea* are also present. *Artemisia* (35–55%) dominates the pollen specrtra. The lower part of LPZ Ur-2 (220–50 cm) is characterized by higher values of *Pinus* (up to 30%) and *Larix* (up to 5–7%) and by disappearance of *Picea* pollen. Only a slight increase in Poaceae is meaningful when discussing changes in herbaceous taxa. The

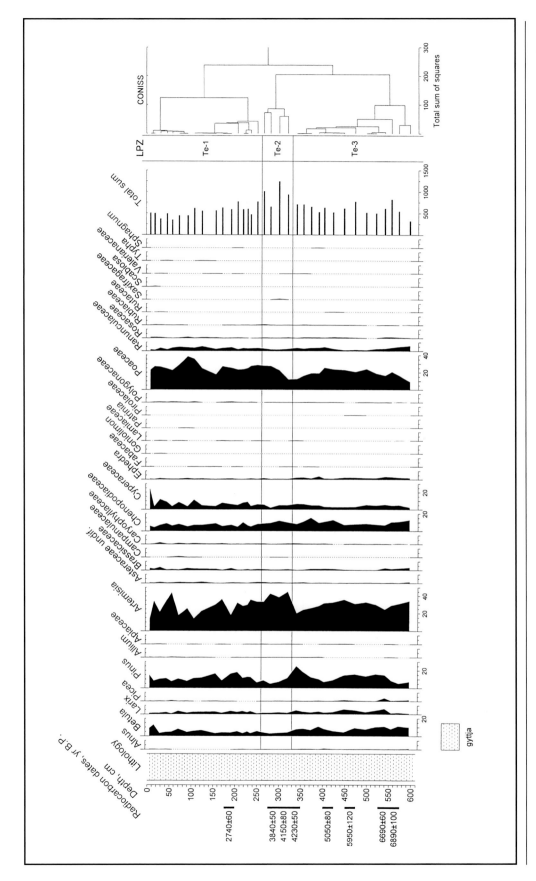

Fig. 8. Lithology column and pollen percentage diagram from Terkhiin-Tsagan-Nur lake. Frequencies are expressed as percentages of the total sum of arboreal and non-arboreal pollen at each level.

39

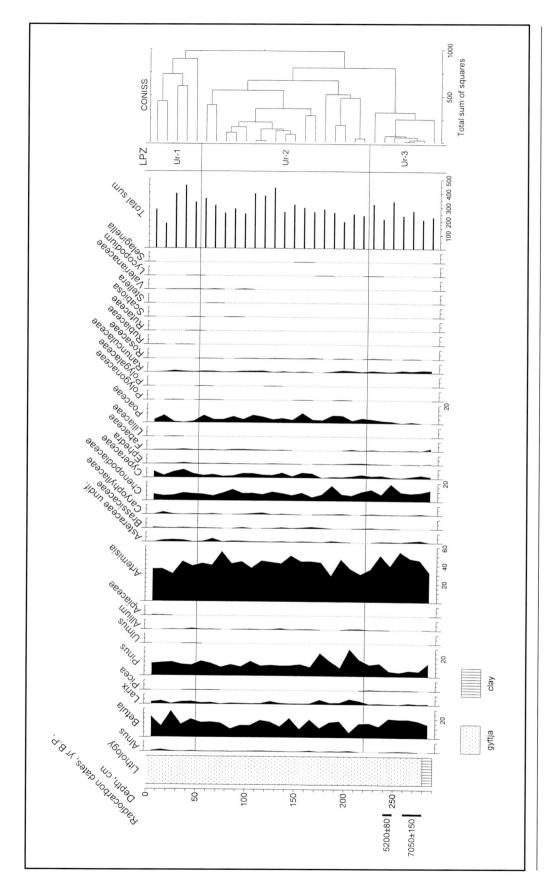

Fig. 9. Lithology column and pollen percentage diagram from Urmiin-Tsagan-Nur lake. Frequencies are expressed as percentages of the total sum of arboreal and non-arboreal pollen at each level.

40

uppermost zone Ur-1 (50–0 cm) does not have any distinct changes in the pollen taxa percentages. However, several additional minor herbaceous taxa have been identified in this zone.

Reconstructed environments: Steppe was a dominant vegetation type around the lake since about 7300 yrs. BP. *Larix* patches have been described close to the site during the coring campaign, but also grew there during the middle and late Holocene. The main part of the record is undated. However, the break in the *Picea* pollen curve occurred soon after 5200 yrs. BP., suggesting that the climate became drier than during the middle Holocene and vegetation became similar to the present day. The same feature in the other records from Hangai is dated to about 4000 yrs. B.P.

Results of Biomization

The results of the biome reconstruction obtained for the pollen records from Hangai are shown in Fig. 10. In this figure the highest value of the biome score at each level is associated with a greater likelihood of that vegetation type being present near the site. It suggests that steppe was always the main vegetation, while the second and third places were shared between boreal forest (taiga) and tundra. Thus, the results obtained with the objective method of biomization (Fig. 10) support the qualitative interpretation of the pollen data from Hangai. However, some additional information can be obtained by examining the relative values of forest and non-forest biome scores.

Thus, pollen data from Daba-Nur suggests that forest vegetation occupied a smaller area than today between 11,500 and 9500 yrs. BP, when steppe and tundra associations dominated in the surrounding landscape. The scores of steppe are generally close to those of taiga in the middle Holocene part of the records, suggesting greater than modern aforestation in the Hangai

Mountains between 8000 and 5000 yrs. BP. However, the data does not indicate that conditions have been stable during this period or synchronous at all mountain sites. The described broad pattern of vegetation change is more pronounced in the records from high-elevated sites (e.g., Daba-Nur, Huh-Nur and Shiret-Nur), while results from lower-elevated Hudo-Nur, Terkhiin-Tsagan-Nur and Urmiin-Tsagan-Nur sites demonstrate a little difference between the middle and the latest part of the Holocene. Moreover, biomization for lowermost Urmiin-Tsagan-Nur lake shows that the similarity between scores of steppe and those of taiga is stronger sometime after 5200 yrs. BP and today, suggesting that the optimum for forest development in the eastern part of Mongolia probably occurred later than in the mountainous western part. Expansion of boreal forests, including evergreen and summergreen conifers in the region would require wetter and milder conditions than today or during the Late Glacial-early Holocene time.

Available pollen records show that the distance between steppe and taiga scores became greater sometime before 3100–3400 yrs. BP, suggesting deforestation of the area. More precise dating for this event has been obtained only in the Terkhiin-Tsagan-Nur record, suggesting that the shift to the drier (similar to modern) vegetation and climate took place about 4200–4150 yrs. BP.

Discussion and Conclusion

The qualitative interpretation of Hangai vegetation history agrees with that reconstructed by the quantitative method of biomization, suggesting that our interpretation is robust. Reconstructed vegetation changes in the Hangai Mountains are broadly parallel to the vegetation and climate changes derived from pollen, plant macrofossil and diatom records from western Mongolia (Dorofeyuk and Tarasov, 2000; Tarasov

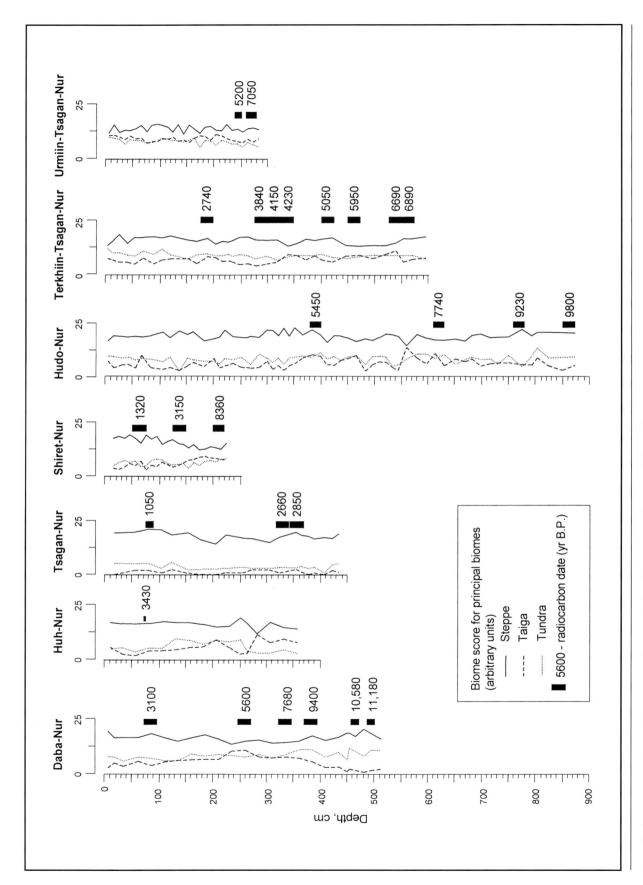

Fig. 10. Biome scores for tundra, steppe and taiga calculated for seven pollen records from Hangai.

et al., 2000). The later study has suggested that changes in available moisture can explain the reconstructed pattern of vegetation in Mongolian Altai between 9000 yrs. BP and the present, while low winter temperatures alone, or combined with dry summers, might have caused the absence of boreal conifers before then. Climate is the major factor causing environmental changes in Mongolia since the Late Glacial, as the area covered by the mountain glaciers was not substantially different from today and human impact was unimportant in the region until recently (Vipper *et al.,* 1989; Gunin *et al.,* 1999; Tarasov *et al.,* 2000).

Drier than present conditions seen in the pollen and lithology record from Daba-Nur sometime between 11,500 and 9500 yrs. BP correspond to the interval when the limit of the summer monsoon reconstructed from Chinese records (Winkler and Wang, 1993) was south of its present position. This caused a moisture deficit in central and northern China and Mongolia. Peat accumulation in the Daba-Nur core began in 11,180±120 yrs. BP. It is synchronous with a phase of very dry climate which occurred between 11,300 and 10,000 yrs. BP in the arid northwestern China as reconstructed by sedimentary record from Yiema lake (39°06'N, 103°40'E) and recently reported by Chen *et al.* (1999).

Wetter than present conditions are reconstructed in Hangai between 9000 and 4000 yrs. BP. A similar change to a wetter climate reported at many sites in eastern Asia has been most probably associated with the strengthening and northward displacement of the Pacific monsoon (Van Campo and Gasse, 1993; Winkler and Wang, 1993; Gasse *et al.,* 1996; Harrison *et al.,* 1996; Tarasov and Harrison, 1998; Gunin *et al.,* 1999; Chen *et al.,* 1999; Tarasov *et al.,* 2000). Pollen data from Hoton-Nur lake in western Mongolia (Fig. 1) suggests that patchy forests dominated by *Picea, Pinus sibirica* and *Larix* occupied a larger area between 9000 and 4000

yrs. BP. Pollen data from Achit-Nur lake (Fig. 1) shows a similar increase in the amount of tree pollen from less than 5% at 10,000 cal. yrs. BP to 50% of total pollen sum between 9397±80 and 6540±100 yrs. BP (Gunin *et al.,* 1999). Wood macrofossils of *Picea* and *Abies* found in the steppe area at Bayan-Sair (Fig. 1) are radiocarbon dated to between 4350 and 3800 yrs. BP. (Dinesman *et al.,* 1989), suggesting that the boreal forest patches have penetrated much further south from their present-day limit during the middle Holocene. However, fluctuations in the pollen assemblages observed in our data from Hangai suggest that the middle Holocene climate was also not very stable. A similar conclusion has been reported by Chen *et al.* (1999), who reconstructed a dry climate event in Yiema lake at 7600 yrs. BP and reported the occurrence of strong sand storms with about 400-yr. periodicity during the middle Holocene.

Boreal forest vegetation in Hangai became less important, being pressed by steppe associations after 4250–4000 yrs. BP. A similar change in vegetation accompanied by a noticeable decrease in the planctonic diatom abundance appeared at Hoton-Nur about 4000 yrs. BP. A decrease in precipitation and a transition to present-day conditions, as a consequence of the insolation-induced attenuation of the monsoon in the late Holocene, have been inferred from biostratigraphic records from China (Van Campo and Gasse, 1993; Winkler and Wang, 1993; Pachur *et al.,* 1995; Gasse *et al.,* 1996) and Mongolia (Dorofeyuk and Tarasov, 1998; Gunin *et al.,* 1999). Chen *et al.,* (1999) reported that dessication processes started in Yiema lake at about 4200 yrs. BP (e.g., at the same as in Hangai), suggesting that the process of aridization has been rather synchronous at the large area of eastern and central Asia. A short-term drought episode that occurred in Yiema lake around 1500 yrs. BP also shows a nice consequence with the deforestation episode reconstructed at two sites in Hangai.

A slight variation in the monsoon would have deep impacts on the lives of people in Mongolia, located at the periphery of the Asian monsoon region. A flock of sheep moving through the grasslands. (Photo by Takeshi Takeda)

The symbol of the Mongolian civilization, which rose at the periphery of the Asian monsoon region, is the horse. The invasion of the agricultural people in the Asian monsoon region by the pastoral people on horseback greatly affected the course of history. A girl on horseback. (Photo by Takeshi Takeda)

Changes in Urmiin-Tsagan-Nur pollen record from Hangai region are slightly different from the other records and do not show clear deforestation during the late Holocene. Similar results have been reported for Yamant-Nur (1000 m a.s.l.) and Gun-Nur (600 m a.s.l.) sites from Northern Mongolia (Fig. 1). At these sites arboreal vegetation is mainly represented today by *Pinus sylvestris* and by tree forms of *Betula* (Gunin *et al.*, 1999), which do not grow in the Hangai Mountains. Thus, it is difficult to provide a reasonable explanation for this phenomena. Recently a new coring project has started in Mongolia with the Japanese-Russian-Mongolian collaboration. In summer 1999 we obtained cores from three lakes situated east of Hangai (Nakagawa *et al.*, 2000). We believe that detailed pollen analysis and precise radiocarbon dating of these cores will help in a better understanding of the vegetation and climate history in central and eastern Mongolia.

Discussed pollen records from Hangai have certain limitations in respect to time resolution and the dating quality. However, they provide clear evidence of changes in vegetation since the Late Glacial to today. Reconstructed patterns are parallel to the millennial-scale environmental changes reconstructed at other sites from northwestern China and western Mongolia and can be explained by a strengthening followed by a decay of the monsoon circulation in the region. However, the data suggests that these gradual changes were interrupted by shorter-term climatic events causing observed changes in the pollen and sedimentary records from Mongolia and China. The underlying mechanism for these short-term climatic oscillations is not yet clear.

Acknowledgements

The paper is dedicated to Pavel B. Vipper, who was the first head and the real leader of the Soviet-Mongolian (now Russian-Mongolian) Biological Expedition and made a great contribution to the studies of vegetation and environmental history in Mongolia and Buriatia. The authors also acknowledge the financial support from Japan Society for the Promotion of Science.

References

Byazrov, L.G., I.A. Gubanov, E. Ganbold, A.N. Dul'geserov, and Ts. Tsegmed, (1983): *Flora Vostochnogo Hangaya* (Flora of Eastern Hangai). Nauka, Moscow, 185 pp.

Chen, F.H., Q. Shi, and J.M. Wang, (1999): Environmental changes documented by sedimentation of Lake Yiema in arid China since the Late Glaciation. *Journal of Palaeolimnology* 22: 159–169.

Dinesman, L.G., N.K. Kiseleva, and A.V. Kniazev, (1989): *Istoriya stepnykh ekosistem Mongol'skoi Narodnoi Respubliki* (The history of the steppe ecosystems of the Mongolian People Republic). Nauka, Moscow, 215 pp.

Dorofeyuk, N.I. (1988): Palaeogeography of Holocene of the Mongolian People's Republic according to results of diatomic analysis of lake bottom sediments. In Gubanov, I.A., Dorofeyuk, N.I. and V.M. Neronov (eds.): *Prirodnye usloviya, rastitel'nyi pokrov i zhivotnyi mir Mongolii* (Natural environments, vegetation and animal communities of Mongolia), pp. 61–83. Nauka, Pushchino.

Dorofeyuk, N.I. (1992): Century changes of lakes tanatosenozes and reconstruction of their history (on Buir-Nur lake example). In: *Ekologiya i prirodopol'zovanie v Mongolii* (Ecology and using of nature in Mongolia). Pushchino, pp. 151–166.

Dorofeyuk, N.I. and P.E. Tarasov, (1998): Vegetation and lake levels of northern Mongolia since 12,500 yrs. BP based on the pollen and diatom records. *Stratigraphy and Geological Correlation*, 6: 70–83.

Dorofeyuk, N.I. and P.E. Tarasov, (2000): Vegetation of western and southern Mongolia in the late Pleistocene and Holocene. Botanicheskii Zhurnal 85(2): 1–17 (in Russian).

Gasse, F., J.Ch. Fontes, Van E. Campo, and K. Wei, (1996): Holocene environmental changes in Bangong Co basin (Western Tibet). Part 4: Discussion and conclusions. *Palaeogeography, Palaeoclimatology, Palaeoecology* 120: 79–92.

Golubeva, L.V. (1976): Rastitel'nost' Severo-Vostochnoi Mongolii v pleistotsene i golotsene. *In* Lavrenko, E.M. and E.I. Rachkovskaya, (eds.): Struktura i dinamika osnovnykh ekosistem MNR (*The structure and dynamics of the main ecosystems in MPR*), pp. 59–71. Nauka, Leningrad.

Golubeva, L.V. (1978): Rastitel'nost' Severnoi Mongolii v pleistotsene i golotsene (basseiny rek Selengi i Orkhona). *Izvestiya AN SSSR, Seriya Geologicheskaya* 3: 68–81.

Grichuk, V.P. and E.D. Zaklinskaya, (1948): *Analiz iskopaemykh pyl'tsy i spor i ego primenenie v palinologii* (Analysis of fossil pollen and spores and its application to palynology). Geografgiz, Moscow, 224 pp.

Grimm, E. (1987): CONISS: A Fortran 77 Program for Stratigraphically Constrained Cluster Analysis by the Method of Incremental Sum of Squares. *Computers and Geosciences* 13: 13–35.

Grimm, E. (1991): Tilia 1.12, Tilia*Graph 1.18. Illinois State Museum, Research and Collection Center. Springfield, Illinois.

Guiot, J. and C. Goeury, (1996): PPPBASE, a software for statistical analysis of palaeoecological and palaeoclimatological data. *Dendrochronologia* 14: 295–300.

Gunin, P.D., E.A. Vostokova, N.I. Dorofeyuk, P.E. Tarasov, and C.C. Black, (eds.) (1999): Vegetation dynamics of Mongolia. *Geobotany 26.* Kluwer Academic Publishers, Dordrecht, 238 pp.

Harrison, S.P., G. Yu, and P.E. Tarasov, (1996): Late Quaternary lake-level record from northern Eurasia. *Quaternary Research* 45: 138–159.

Hilbig, W. (1995): *The vegetation of Mongolia.* SPB Academic Publishing, Amsterdam: 253.

Krengel, M. (2000): Discourse on history of vegetation and climate in Mongolia— palynological report of sediment core Bayan Nuur 1 (NW Mongolia). *Berliner geowissenschaftliche abhandlungen (A)* 205: 80–84.

Lavrenko, E.M. (ed.) (1979): *Karta rastitel'nosti Mongol'skoi Narodnoi Respubliki* (Vegetation map of Mongolian People's Republic). Scale 1:1500,000. GUGK, Moscow (in Russian).

Lavrenko, E.M. (ed.) (1983): *Karta lesov Mongol'skoi Narodnoi Respubliki* (The map of forests of Mongolian People's Republic). Scale 1:1500,000. GUGK, Moscow. (in Russian).

Malaeva, E.M. 1989: The history of Pleistocene and Holocene vegetation in Mongolia and palaeoindicative features of fossil pollen floras. *In* Logatchov, N.A. (ed.): *Pozdnii kainozoi Mongolii* (Late Cainozoic of Mongolia) Nauka, Moscow, pp. 158–177.

Nakagawa, T., P.E. Tarasov, Y. Inoue, and Y. Yasuda, (2000): A preliminary report of the now-running multidisciplinary study on Mongolian-Siberian transect: the overview of the project and a key lacustrine core from lake Gun-Nur. *In* Yasuda, Y. (ed.): *Environmental Change in Eurasia:* The 1st ALDP/ELDP Joint Meeting 20–25 March 2000 Kyoto and Mikata, Japan. *Monsoon* 1: 71–72.

Pachur, H.J., B. Wunnemann, and H. Zhang, (1995): Lake evolution in the Tengger desert, Northwestern China, during the last 40,000 years. *Quaternary Research 44,* 171–180.

Prentice, I.C., W. Cramer, S.P. Harrison, R. Leemans, R.A. Monserud, and A.M. Solomon, (1992): A global biome model based on plant physiology and dominance, soil properties and climate. *Journal of Biogeography* 19: 117–134.

Prentice, I.C., J. Guiot, B. Huntley, D. Jolly, and R. Cheddadi, (1996): Reconstructing biomes from palaeoecological data: a general method and its application to European pollen data at 0 and 6 ka. *Climate Dynamics* 12: 185–194.

Savina, L.N. and T.A. Burenina, (1981): Sokhrannost' pyl'tsy listvennitsy v lesnykh pochvakh i otrazhenie sostava listvennichnykh lesov Mongolii v retsentnykh spektrakh. *In* Savina, L.N. (ed.): *Palaeobotanicheskie issledovaniya v lesakh Severnoi Azii (Palaeobotanical studies in the forests of Northern Asia),* Nauka, Novosibirsk, pp. 62–83.

Savina, L.N., I.A. Korotkov, A.V. Ogorodnikov, E.N. Savin, and T.A. Burenina, (1981): Tendentsii razvitiya lesnoi rastitel'nosti Mongol'skoi Narodnoi Respubliki. *In* Savina, L.N. (ed.): *Palaeobotanicheskie issledovaniya v lesakh Severnoi Azii* (Palaeobotanical studies in the forests of Northern Asia), Nauka, Novosibirsk, pp. 83–158.

Sevastyanov, D.V., Yu.P. Seliverstov, and G.M. Chernova, (1993): K istorii razvitiya landshaftov Ubsunurskoi kotloviny. *Vestnik St.-Peterburgskogo Universuteta, Seriya Geologiya, Geografiya* 28: 71–81.

Sevastyanov, D.V. and N.I. Dorofeyuk, (1992). The history of the water ecosystem of Mongolia. *Izv. Vses. Geogr. Obshch.*, vol. 124, no. 2: 123–138.

Sevastyanov, D.V., N.I. Dorofeyuk, and A.A. Liiva, (1989). The origin and evolution of the volcanic Terkhiin-Tsagan-Nur Lake in Central Hangai (MPR). *Izv. Vses. Geogr. Obshch.*, vol. 121: 223–227.

Tarasov, P.E. and S.P. Harrison, (1998): Lake status records from the Former Soviet Union and Mongolia: a continental-scale synthesis. *Paläoklimaforschung 25,* pp. 115–130.

Tarasov, P., N. Dorofeyuk, and E. Metel'tseva, (2000): Holocene vegetation and climate in Hoton-Nur basin, northwest Mongolia. *Boreas* 29/2: 117–126.

Tarasov, P.E., M.Y. Pushenko, S.P. Harrison, L. Saarse, A.A. Andreev, Z.V. Aleshinskaya, N.N. Davydova, N.I. Dorofeyuk, Y.V. Efremov, G.A. Elina, Y.K. Elovicheva, L.V. Filimonova, V.S. Gunova, V.I. Khomutova, E.V. Kvavadze, I.Y. Neustrueva, V.V. Pisareva, D.V. Sevastyanov, T.S. Shelekhova, D.A. Subetto, O.N. Uspenskaya, and V.P. Zernitskaya (1996): Lake Status Records from the former Soviet Union and Mongolia: Documentation of the Second Version of the Data Base, *NOAA Palaeoclimatology Publications Series Report No. 5.* Boulder, 224 pp.

Tarasov, P.E., T. Webb III, A.A. Andreev, N.B. Afanas'eva, N.A. Berezina, L.G. Bezusko, T.A. Blyakharchuk, N.S. Bolikhovskaya, R. Cheddadi, M.M. Chernavskaya, G.M. Chernova, N.I. Dorofeyuk, V.G. Dirksen, G.A. Elina, L.V. Filimonova, F.Z. Glebov, J. Guiot, V.S. Gunova, S.P. Harrison, D. Jolly, V.I. Khomutova, E.V. Kvavadze, I. M. Osipova, N.K. Panova, I.C. Prentice, L. Saarse, D.V. Sevastyanov, V.S. Volkova, and V.P. Zernitskaya, (1998): Present-day and middle-Holocene Biomes Reconstructed from Pollen and Plant Macrofossil Data from the Former Soviet Union and Mongolia. *Journal of Biogeography* 25: 1029–1054.

Van Campo, E. and F. Gasse, (1993): Pollen- and diatom-inferred climatic and hydrological changes in Sumxi Co basin,

Western Tibet) since 13,000 yrs. BP. *Quaternary Research* 39: 300–313.

Vipper, P.B., N.I. Dorofeyuk, E.P. Metel'tseva, V.T. Sokolovskaya, and K.S. Shulia, (1976): Opyt rekonstruktsii rastitel'nosti zapadnoi i tsentral'noi Mongolii v golocene na osnove izucheniya donnykh osadkov presnovodnykh ozer. *In* Lavrenko, E.M. and E.I. Rachkovskaya, (eds.): Struktura i dinamika osnovnykh ekosistem MNR (*The structure and dynamics of the main ecosystems in MPR*), Nauka, Leningrad, pp. 35–59.

Vipper, P.B., N.I. Dorofeyuk, E.P. Metel'tseva, and V.T. Sokolovskaya, (1989): Landshaftno-klimaticheskie izmeneniya v tsentral'noi Mongolii v golotsene. *In* Khotinskii, N.A. (ed.): *Paleoklimaty pozdnelednikovya i golotsena* (Palaeoclimates of the Late Glacial and Holocene). Nauka, Moscow, pp. 160–167.

Whitlock, C. and P.J. Bartlein (1997): Vegetation and climate change in northwest America during the past 125 kyr. *Nature* 388: 57–61.

Winkler, M.G. and P.K. Wang 1993: The Late-Quaternary Vegetation and Climate of China. *In* Wright, H.E., J.E. Kutzbach, T. Webb III, W.F. Ruddiman, F.A. Street-Perrott and P.J. Bartlein (eds.): *Global Climates since the Last Glacial Maximum,* University of Minnesota Press, Minneapolis, pp. 265–293.

Zhambazhamts, B. and B. Bat (1985): The Atlas of the climate and ground water resources in the Mongolian People's Republic. Goskomgidromet SSSR, GUGMS MNR, GUGK SSSR, Ulaanbaatar: 88 pp.

Chapter **2**

A 78,000 Year Record of Climatic Changes from the South China Coast—the Huguang Maar Lake (Huguangyan)

JENS MINGRAM, NORBERT NOWACZYK, GEORG SCHETTLER, XIANGJUN LUO, HOUYUAN LU, JIAQI LIU AND JÖRG F.W. NEGENDANK

Introduction

Despite the recent progress made especially with the investigation of new marine cores (Wang and Sarnthein, 1999; Wang et al., 1999, Huang et al., 1997, Pelejero et al., 1999, Hanebuth et al., 2000, Sun and Li, 1999), in Southeast Asia there is still a lack of high-resolution, long-reaching palaeoclimate data. With pronounced seasonality in rainfall, temperature, wind directions, tropical typhoons and marine currents, the South China coast offers a great potential for recovering the fluctuation of climatic zones in space and time. Between 19–23°N and 108–112°E volcanic craters and maar lakes of the Lei-Qiong volcanic field (LQVF) provide several potentially long palaeoenvironmental and palaeoclimatic records. The LQVF comprises the central part of the Beibuwan basin which has been an active rift system since the early Tertiary (Huang et al., 1993) and includes the volcanic areas of the northern Hainan Island, the Leizhou Peninsula and some underwater volcanic cones. Within the 7295 km^2 of the LQVF more than 100 volcanic structures have been identified, among them the Tianyang maar and the Huguang maar, the deepest recent lake among the volcanic structures of the LQVF (Fig. 1, Table 1). The Tianyang maar has been extensively drilled by Chinese companies in the course of their oil prospection, and a 400,000 year-long record of

lake sediments, consisting of 8.4 m alluvial silty clay, 34.8 m dark algal gyttja and 178.8 m diatomaceous gyttja (from top to bottom) has been published by Chen et al. (1988) and Zheng and Lei (1999).

Results

The age of formation of the Huguang maar (Huguangyan) is still questionable, but a K/Ar-

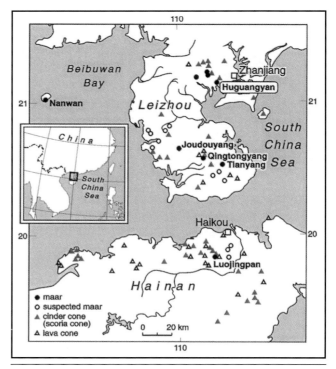

Fig. I: Location and volcano types of the Lei-Qiong volcanic field (LQVF).

51

Table 1. Maar lakes and dry maars of the Lei-Qiong volcanic field.

Name	Locality	Lake Area (km²)	Water Depth (m)	Sediment Thickness (m)	Max. Sediment Age (ka)
Huguangyan	21° 05'N, 110° 05'E	3.6	20	> 56	78
Nanwan	21° 01'N, 109° 02'E	26	< 10	n.a.	n.a.
Joudouyang	20° 37'N, 110° 01'E	5.0	dry	> 80	n.a.
Qingtongyang	20° 35'N, 110° 03'E	10	dry	150	n.a.
Tianyang	20° 30'N, 110° 05'E	7.3	dry	222	400
Luojinpan	19° 57'N, 110° 05'E	3.5	3–4	n.a.	n.a.

dating of a basalt from the volcaniclastic breccia of the crater rim yielded an age of 127 ka (Fong, 1992). A drilling outside the crater of the Huguang maar down to about 80 m revealed a sequence of 2 m of weathering crust, 15 m of pyroclastics, 5 m of basalt and 60 m of sandy clay of the Lower Pleistocene shallow marine Zhanjiang Formation (from top to bottom).

While preparing for a drilling campaign in 1997 we first constructed a detailed isobath map of the Huguang maar from seventeen sonar profiles. From three different positions of the lake we got seven cores by high-precision piston coring (Usinger system) down to a maximum drilling depth of 57.8 m below the lake bottom (Fig. 2).

After core description and photography, measurements of magnetic susceptibility of each core with 2 or 1 mm stepper resolution were performed. From overlapping cores of the two main positions (D/F and B/C) water content and dry density measurements were done with 1 cm-resolution. Geochemical data as total inorganic carbon (TIC), total organic carbon (TOC), total nitrogen (TN) and biogenic silica (BSi) were estimated every 10 cm of the C-core, and a preliminary pollen count was performed of one sample/m of the composed B/C-section. A representative set of 10 cm-long petrographic thin sections was prepared for each lithological unit of the Huguang maar profile.

The age model was established with the help of twelve ^{14}C-AMS-datings of single leafs, one bulk sample and one seed. Beyond the limits of the actual ^{14}C-calibration curve (Stuiver et al., 1998) all conventional ^{14}C ages were transferred into calibrated ages with the magnetic calibration of the radiocarbon timescale from Laj et al. (1996). As there are only two horizons with macroscopically observable layering, the correlation between the cores was made by comparison of the high resolution magnetic susceptibility and dry density measurements. The correlation of the lowermost part of the two main sections D/F and B/C beyond the limits of ^{14}C age determination remains still somewhat questionable. Some thin (single-grain or layers, and layers up to max. 1.8 cm) tephra could be found only in the lower part of core F, and a thick breccia at the bottom of core F and G probably originates from a large slump.

The sedimentary record of the Huguang maar is at first glance composed of a quite monotonous, greenish-black algal gyttja. This rather homogeneous section is interrupted twice by a zone with a lot of mm - to cm - thick brownish layers with numerous reworked woody (bark-like) material. The comparison of the macroscopic and microscopic appearance of the sediments and of all physical and chemical sediment data allowed the subdivision of the Huguang maar profile into eight lithozones as described below (from top to bottom, Fig. 3):

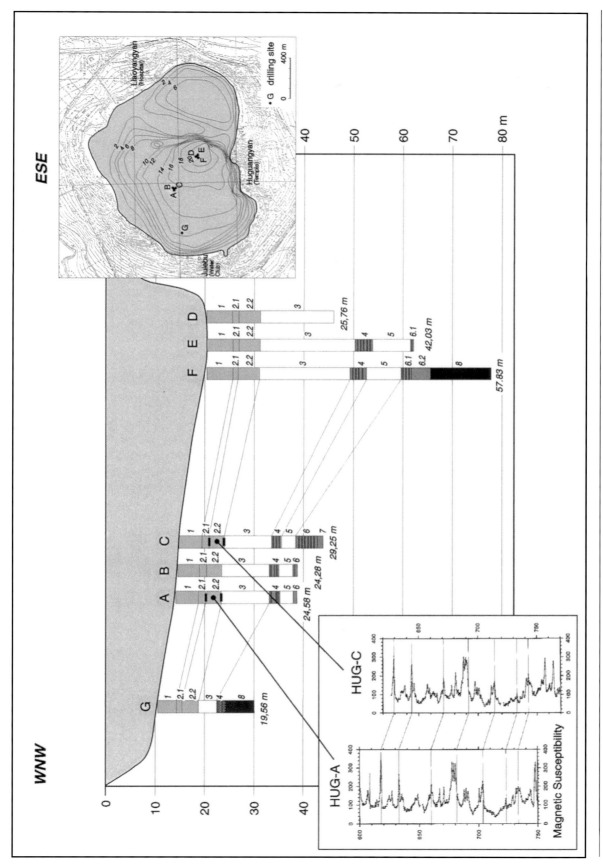

Fig. 2: Huguang maar lake isobath map, drill sites with sediment profiles and lithozones, and an example for inter-core correlation with high-resolution magnetic susceptibility data.

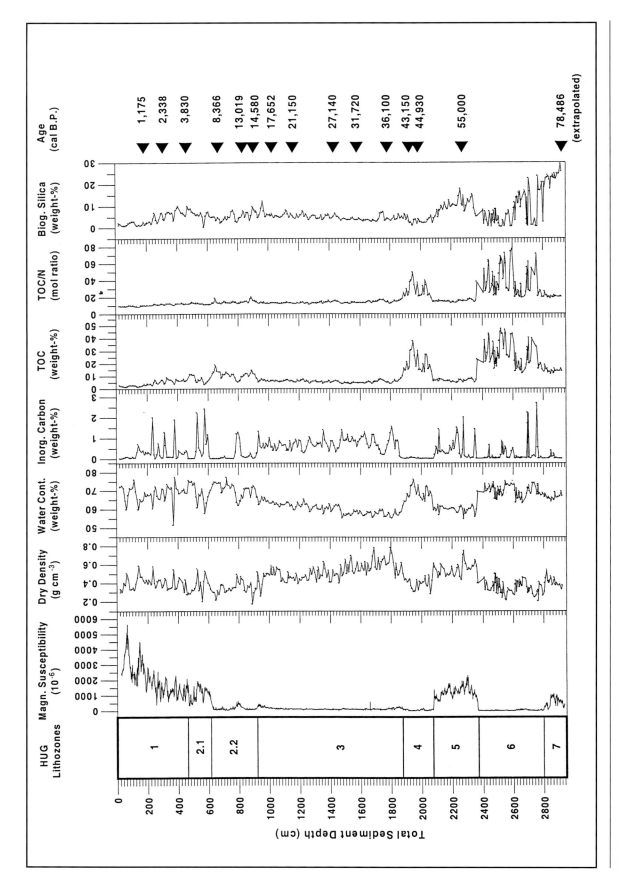

Fig. 3: Lithozones, physical and chemical sediment data of core HUG-C.

54

Lithozone 1 (0–4.41 m in section B/C and 0–5.23 in section D/F): homogeneous algal gyttja, the amount of organic carbon and biogenic silica increases continuously downwards, high - frequency changes of dry density, water content and magnetic susceptibility with large amplitudes.

Lithozone 2.1. (-5.93 m in section B/C and - 6.42 m in section D/F): homogeneous algal gyttja, with a lower frequency of change of dry density and magnetic susceptibility, two large positive peaks of inorganic carbon, dry density and magnetic susceptibility, and corresponding negative ones of water content.

Lithozone 2.2. (-8.95 m in section B/C and - 10.70 m in section D/F): only differs from lithozone 2.1. by much lower values of magnetic susceptibility, and contains again one sharp peak of inorganic carbon on a nearly zero-level.

Lithozone 3 (-18.51 m in section B/C and - 29.17 m in section D/F): homogeneous algal gyttja, with low and nearly invariable level of C/N, with high (and downcore slightly increasing) values of dry density, low values of magnetic susceptibility and moderate changes of inorganic carbon on a general high level.

Lithozone 4 (-20.53 m in section B/C and - 32.43 m in section F): algal gyttja with intercalations of numerous brownish layers with reworked, bark-like plant remains, exceptionally high C/N ratio and scattered minerogenic particles (mainly quartz) up to 2 mm.

Lithozone 5 (-23.33 m in section B/C and - 39.52 in section F): homogeneous algal gyttja, resembles lithozone 3, but with higher amounts of biogenic silica and higher values of magnetic susceptibility.

Lithozone 6 (-28.00 m in section B/C and - 45.46 m in section F): closely resembles lithozone 4, but contains numerous carbonate peaks and some varve-like laminae with layered enrichments of diatoms (*Aulacoseira* spp.).

Lithozone 7 (-29.25 m, only in section B/C): homogeneous algal gyttja, closely resembling those from lithozone 5, but with higher amounts of biogenic silica, and no visible lamination. In thin sections there are observable some scattered woody macro-remains.

Lithozone 8 (> 45.46 m in section F and >11.94 m in section G): volcaniclastic breccia (probably a slump) at the bottom of section D/F and G, however it could not be proved whether they are both of the same age or not.

Although our data set is preliminary and mainly suffers from the lack of a high-resolution pollen section, it is already possible to arrive at the first palaeoenvironmental and palaeoclimatic conclusions.

■ **The Last Glacial Stage**

During lithozone 7 (78–73.5 cal. ka BP), which corresponds to the Marine Isotope Stage (MIS) 5a (timescale of the MIS after Martinson *et al.*, 1987, and Bassinot *et al.*, 1994) a high lake level, high autochthonous productivity of the lake and especially the large amount of tropical and moisture-sensitive faunal elements (as e.g., *Altingia*) point to very humid and warm climatic conditions. The change between lithozone 7 and lithozone 6 at 73.5 cal. ka BP correlates chronologically with the transition between the MIS 5a and MIS 4. The sea level dropped down substantially during MIS 4 (Linsley, 1996), thus exposing vast shelf areas of the South China Sea. Most physical and bulk chemical parameters of the Huguang maar record are influenced mainly by the content of reworked woody material of the lake´s sediments during lithozone 6, but the pollen record (Fig. 4) shows only slight variations. There is still a predominance of tropical elements and only a slight increase of mesic deciduous trees and Fagaceous forest elements. Some varve-like laminations of the F-core from the deepest part of the lake with probably reworked carbonates are hints for the precipitation of carbonates in a shallow shore zone and the existence of an anoxic hypolimnion in the central part of the lake. The existence of a

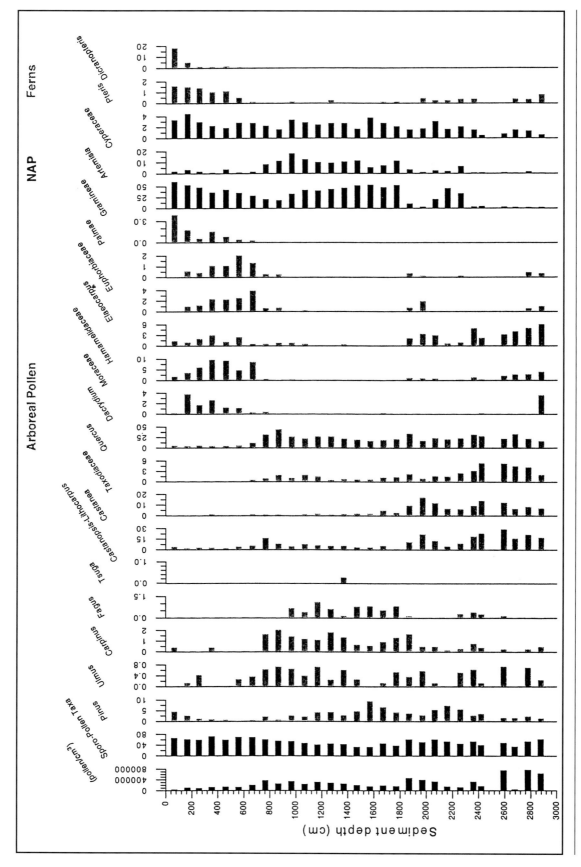

Fig. 4: Pollen diagram of composite section HUG-B/C. (AP and NAP calculated from the total pollen sum, excluding aquatic pollen and fern spores. Pteridophytes calculated from total pollen and spores).

56

humid and warm climate during MIS 4 is in contrast with observations from the Chinese loess area (e.g., Porter and An, 1995, and Chen *et al.*, 1999) and Biwa lake (Xiao *et al.*, 1997; Xiao *et al.*, 1999), where MIS 4 is usually connected with increased winter monsoon activity. But the observations from the Huguang maar are in accordance with the results of Wang *et al.*, (1999) from the South China Sea. They recognized increased clay percentages during MIS 4 in core 17961 in front of the emerged Sunda shelf and concluded a tropical climate with high fluvial runoff and a dense tropical forest cover of the shelf area.

The transition between MIS 4 and MIS 3 fits well with the change between lithozone 6 and 5 of the Huguang maar. But the climatic variations as inferred from the Huguang maar data are not of the same direction as expected e.g., from ice-core temperature and methane records of high (Blunier *et al.*, 1998) and low (Thompson *et al.*, 1997) latitudes. The period between 58 and 48 cal. yrs. BP is much colder and dryer than the previous one at the Huguang maar site. The minerogenic input increases and the pollen record (Fig. 4) is characterized by a first rise in NAP amounts and a drop in tropical pollen percentages. A climatic deterioration during the same period has been described from Taiwan (Tsukada, 1967), and from the Tianyang maar (Southern Leizhou Peninsula) Zheng and Lei (1999) inferred a cooling at 60 cal. ka BP from increased amounts of montane conifers and temperate forest elements. Between 48 and 40 cal. yrs. BP (lithozone 4 of the Huguang maar) the increase of tropical pollen and the decrease of Gramineae and *Artemisia* reflects a climatic amelioration with higher precipitation and temperatures. The lake level should have been higher than during lithozone 5 and the minerogenic input was low. Again the long terrestrial record from Taiwan (Tsukada, 1967) shows the same climatic trend, and Hodell *et al.*, (1999) reported a warm period from Lake Qilu Hu in southwest China between 50 and 41 cal. ka BP.

A major palaeoenvironmental and palaeoclimatic change occurred around 40 cal. ka BP at the Huguang maar site. Lithozone 3 (40 – 15 cal. ka BP) represents a long, stable period with high minerogenic input, low level of TOC, high level of TIC and a C/N-ratio of around thirteen without any major variations. The pollen record shows a remarkable increase of NAP, increased amounts of temperate deciduous tree pollen and a distinct decrease of tropical tree pollen. All these observations point to a pronounced climatic deterioration at 40 cal. ka BP, and cooler and less humid conditions lasted more or less unvariable until the onset of Termination I. The only hint for a further climatic deterioration at the Last Glacial Maximum could be a minor peak in dry density between 21 and 22.5 cal. ka BP and a slight increase in *Artemisia* pollen percentages towards the end of lithozone 3.

■ The Late-glacial (Termination I)

Due to a ^{14}C plateau from 15.3 to 14.4 cal. ka BP (Stuiver *et al.*, 1998) and the lack of an independent chronology for the Huguang maar the exact timing of the beginning of Termination I at the Huguang maar site is not clear. Considering sedimentation rates of our sections, the most probable calibrated age is 15 cal. ka BP, which is three hundred years earlier than the onset of the Greenland Interstadial 1 (Björck *et al.*, 1998). From the South China Sea, Wang and Sarnthein (1999) and Wang *et al.*, (1999) found the earliest intensification of the summer monsoon 3.5–3.8 cal. ka prior to the Post-glacial sea level rise and a weakening of the winter monsoon at about 15 cal. ka BP. Consequently, even if we take into account the uncertainty of the ^{14}C calibration, the palaeoenvironmental shift at the Huguang maar site at the end of the Last Glacial depends more on the sea level rise and/or the weakening of the winter monsoon than the intensification of the summer monsoon. The Younger Dryas (or Greenland Stadial 1 after Björck *et al.*, 1998) is marked by a peak in TIC and smaller amounts of TOC in the Huguang maar record. The low resolution of the preliminary pollen record does not allow the recognition of the Younger Dryas.

■ The Holocene

The Holocene sediments from the Huguang maar are characterized by high-frequency and high-amplitude changes of dry density, water content, TIC percentages and magnetic susceptibility. The palaeoenvironmental conditions seem to be more variable than during the high glacial between 40 and 15 cal. ka BP. The often recorded early Holocene summer monsoon maximum (e.g., Wang *et al.*, 1999 - maximum of fluvial runoff from the Pearl River; Hodell *et al.*, 1999 and Jarvis, 1993 - stable isotopes and pollen records from lakes in Southwest China; Thompson *et al.*, 1997 - oxygen isotope values and CH_4 levels of the Guliya ice core from Western Tibet) correlates with high TOC percentages, low TIC percentages and low values of the magnetic susceptibility from the Huguang maar sediments between 11.5 and 7 cal. ka BP (upper part of lithozone 2.2.). The two peaks of TIC between 7 and 5 cal. ka BP (lithozone 2.1.), which correlate with two palaeosalinity peaks of the South China Sea (Wang *et al.*, 1999), probably represent a period with reduced precipitation. The exact cause for the major change of the magnetic susceptibility record from the Huguang maar at 7 cal. ka BP between low early Holocene and high middle to late Holocene values is still open. But it might be related to palaeoclimatic changes as indicated by a major drop of $\delta^{18}O$ values of the Guliya ice core (Thompson *et al.*, 1997) at the same time.

Large variations of the dry density record with a trend of steady increase and the parallel continuously decreasing TOC percentages indicate the beginning of human influence from 4 cal. ka BP onwards (lithozone 1). The pollen record with increasing *Pinus* percentages (which are considered to be elements of secondary forests) and the gradual spread of open grasslands with *Dicranopteris* support this assumption.

References

Bassinot, F.C., L.D. Labeyrie, E. Vincent, X. Quidelleur, N.J. Shackleton, and Y. Lancelot (1994): The astronomical theory of climate and the age of the Brunhes-Matuyama reversal. *Earth and Planetary Science Letters*, 126: 91–108.

Björck, S., M.J.C. Walker, L.C. Cwynar, S. Johnsen, K.L. Knudsen, J.J. Lowe, B. Wohlfarth and I. Members (1998): An event stratigraphy for the Last Termination in the North Atlantic region based on the Greenland ice-core record: a proposal by the INTIMATE group. *Journal of Quaternary Science*, 13: 283–292.

Blunier, T., J. Chappellaz, J. Schwander, A. Dällenbach, B. Stauffer, T.F. Stocker, D. Raynaud, J. Jouzel, H.B. Clausen, C.U. Hammer and S.J. Johnsen (1998): Asynchrony of Antarctic and Greenland climate change during the last glacial period. *Nature*, 394: 739–743.

Chen, J., Z. An and J. Head (1999): Variation of Rb/Sr Ratios in the Loess-Paleosol sequences of Central China during the last 130,000 years and their implications for monsoon palaeoclimatology. *Quaternary Research*, 51: 215–219.

Chen, J., Huang, C., Lin, M., Jin, Q., and Han, J. (1988): *Quaternary Geology of Tianyang Volcanic Lake, Guangdong Province, China*. Geological Publishing House, Beijing, pp. 256, (In Chinese, with Eng. abstr.).

Fong, G. R. (1992): Cenozoic Basalts in southern China and their relationship with tectonic environment. *Journal of Zhongshan University*, 27: 93–103 (In Chinese, with Engl. abstr.).

Hanebuth, T., K. Stattegger and P.M. Grootes (2000): Rapid flooding of the Sunda shelf: A Late-glacial sea-level Record. *Science*, 288: 1033–1035.

Hodell, D.A., M. Brenner, S.L. Kanfoush, J.H. Curtis, J. Stoner, X. Song, Y. Wu and T.J. Whitmore (1999): Palaeoclimate of Southwestern China for the past 50,000 yrs. inferred from Lake Sediment Records. *Quaternary Research*, 52: 369–380.

Huang, C.-Y., P. M. Liew, M. Zhao, T.-C. Chang, C.-M. Kuo, M.-T. Chen, C.-H. Wang and L.-F. Zheng (1997): Deep sea and lake records of the Southeast Asian palaeomonsoons for the last 25 thousand years. *Earth and Planetary Science Letters*, 146: 59–72.

Huang, Z.G., F.X. Chai, Z.Y. Han, J.H. Chen, Y.Q. Zhong and X.D. Lin (1993): *The Quaternary Lei-Qiong Volcanic Field*. Science Press, Beijing, pp. 281, (In Chinese, with Engl. abstr.).

Jarvis, D.I. (1993): Pollen evidence of changing Holocene monsoon climate in Sichuan Province, China. *Quaternary Research*, 39: 325–337.

Laj, C., A. Mazaud and J.-C. Duplessy, (1996): Geomagnetic intensity and ^{14}C abundance in the atmosphere and ocean during the past 50 cal. yrs. *Geophysical Research Letters*, 23: 2045–2048.

Linsley, B.K. (1996): Oxygen-isotope record of sea level and climate variations in the Sulu Sea over the past 150,000 years. *Nature*, 380: 234–237.

Martinson, D.G., N.G. Pisias, J.D. Hays, J. Imbrie, T.C. Moore and N.J. Shackleton (1987): Age dating and the Orbital Theory of the Ice Ages: development of a high-resolution 0–300,000-year chronostratigraphy. *Quaternary Research*, 27: 1–29

Pelejero, C., J.O. Grimalt, M. Sarnthein, L. Wang and J.-A. Flores (1999): Molecular biomarker record of sea surface temperature and climatic change in the South China Sea during the last 140,000 years. *Marine Geology*, 156: 109–121.

Porter, S.C., and Z. An (1995): Correlation between climate events in the North Atlantic and China during the last glaciation. *Nature*, 375: 305–308.

Stuiver, M., P.J. Reimer, E. Bard, J.W. Beck, G.S. Burr, K.A. Hughen, B. Kromer, G. McCormac, J. Van der Plicht, and M. Spurk, (1998): Intcal 98 radiocarbon age calibration, 24,000–0 cal. yrs. BP. *Radiocarbon*, 40: 1041–1083.

Sun, X., and X. Li (1999): A pollen record of the last 37 ka in deep sea core 17940 from the northern slope of the South China Sea. *Marine Geology*: 156: 227–244.

Thompson, L.G., T. Yao, M.E. Davis, K.A. Henderson, E. Mosley-Thompson, P.N. Lin, J. Beer, H.A. Synal, J. Cole-Dai and J.F. Bolzan (1997): Tropical climate instability: the Last glacial cycle from a Quinghai-Tibetian ice core. *Science*, 276: 1821–1825.

Tsukada, M. (1967) :Vegetation in subtropical Formosa during the Pleistocene glaciations and the Holocene. *Palaeogeography, Palaeoclimatology, Palaeoecology*, 3: 49–64.

Wang, L., and M. Sarnthein (1999): Millennial reoccurrence of century-scale abrupt events of East Asian monsoon: A possible heat conveyor for the global deglaciation. *Palaeo-oceanography*, 14: 725–731.

Wang, L., Sarnthein, M., Erlenkeuser, H., Grimalt, J., Grootes, P., Heilig, S., Ivanova, E., Kienast, M., Pelejero, C., and Pflaumann, U. (1999). East Asian monsoon climate during the late Pleistocene: high-resolution sediment records from the South China Sea. *Marine Geology*, 156: 245–284.

Xiao, J., Y. Inouchi, H. Kumai, S. Yoshikawa, Y. Kondo, T. Liu and Z. An (1997): Biogenic silica record in Lake Biwa of Central Japan over the past 145,000 Years. *Quaternary Research*, 47: 277–283.

Xiao, J.L., Z.S. An, T.S. Liu, Y. Inouchi, H. Kumai, S. Yoshikawa and Y. Kondo (1999): East Asian monsoon variation during the last 130,000 years: evidence from the Loess Plateau of central China and Lake Biwa of Japan. *Quaternary Science Reviews* 18: 147–157.

Zheng, Z., and Z.-Q. Lei (1999): A 400,000 year record of vegetational and climatic changes from a volcanic basin, Leizhou Peninsula, southern China. *Palaeogeography, Palaeoclimatology, Palaeoecology* 145: 339–362.

Net-fishing on the Dongting lake. The net is shaken up and down to scoop up the fish. (Photo by Takeshi Takeda)

The Asian monsoon region has a large and dense population. A village on the shores of Er-Hai lake in Yunnan province, China. (Photo by Takeshi Takeda)

Chapter 3

Maar and Crater Lakes of the Longwan Volcanic Field (Northeast China) and their potential for Palaeoclimatic Studies

JENS MINGRAM, GEORG SCHETTLER, JUDY R.M. ALLEN, CATHRIN BRÜCHMANN, XIANGJUN LUO, JIAQI LIU, NORBERT NOWACZYK AND JÖRG F.W. NEGENDANK

The Longwan maar lakes are situated within the mixed conifer-hardwood forest region of Northeast China, at 600 to 800 m a.s.l. and belong to the Long Gang Volcanic Field (LGVF) near the China-North Korean border (Fig.1).

The climate in the Longwan area is temperate, with an annual mean temperature of 2.9°C (January: -17.8°C; August: +20.2°C) and an annual mean precipitation of 757 mm (Jingyu station, period 1955–1997). There is a strong seasonal contrast between long, dry cool winters and short, wet and hot summers. Annual sunshine is ca 2300 hours.

First scientific geological investigations in the LGVF date back to 1929 (Tokyo Geological Survey, 1929; Ogura, 1969). In the last two decades substantial progress was made with the compilation of geological maps (1:200,000 and 1:20,000, Geological Survey of Jilin, 1994). There are mainly two stratigraphic units in the Longwan area—the Upper Archean Anshan migmatite group and the Quaternary Long Gang Volcanic Group. The migmatites, as a part of the basement of the North China Craton, form a series of northeast-oriented anticlines and synclines, and the east-west striking Cenozoic alkalibasalts cover the older structures forming a plateau-like area at 600 m a.s.l. with 148 volcanic cones on top. At present there are nine water-filled maars and cones with water depths ranging from ca 1 m (swamp) to 127 m (Tab. 1, Fig.1).

The catchment areas of the lakes (which comprise only the inner crater slopes) are mainly forest-covered; there is no lake with a natural tributary or outlet.

The stratification and hydrochemical composition of the lakes have been evaluated during two field campaigns in 1998 and 1999. Chemical water analyses of the surface water and of water profiles of selected lakes (SHL, XIA, ERL, SJL) have been performed and depth profiles of temperature, pH, Eh, conductivity (Cond.), dissolved oxygen (DO_2), and turbidity were measured at the end of the summer stratification using the multiparameter water probe H20 (YSI Inc., Ohio, U.S.A., see Figs. 2 and 3).

All deep lakes show a distinct thermocline at about 10 m. The pH values are increased in the phototrophic zone (~9) due to the CO_2 consumption by phyto-plankton. The water column of ERL, LOW and the eastern (shallower) basin of SJL are slightly oxygenated down to the lake bottom. The other lakes and the western basin of SJL have anoxic deep water. The lake water compositions (Table 2) seem to reflect different contributions by inflow of ground water. The latter might be higher for DAL, LOW and DOLO, than for the other lakes. The silica concentrations in the ground water discharge are between 13 and 14 mg/l, while that of phosphorus are in the order of 0.03 mg/l. The silica concentration of the lakes varies between 0.2 (SJL, ERL, SHL, XIA) and 4.5

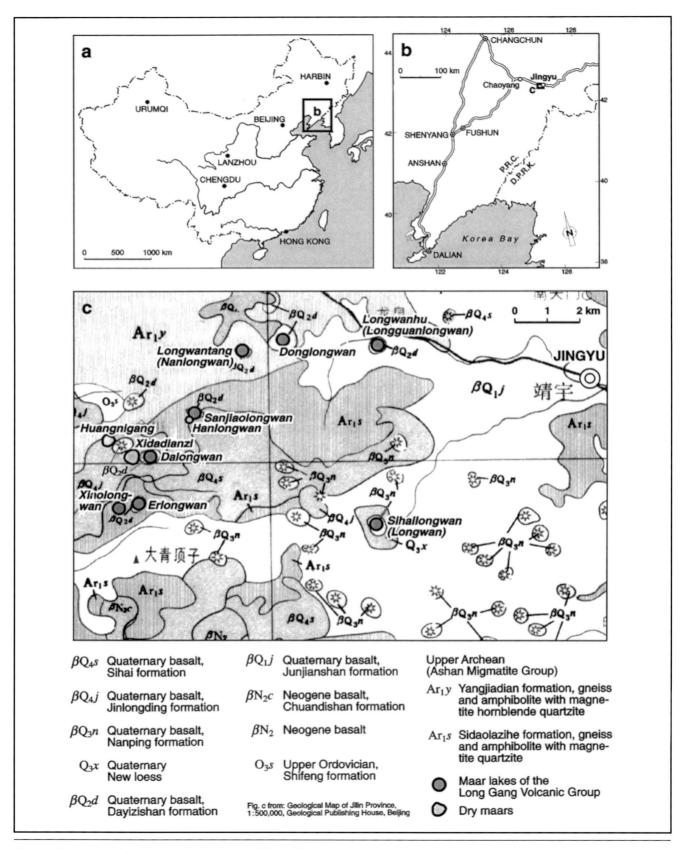

Fig. 1. Location and geology of the Long Gang Volcanic Field.

βQ₄s Quaternary basalt, Sihai formation

βQ₄j Quaternary basalt, Jinlongding formation

βQ₃n Quaternary basalt, Nanping formation

Q₃x Quaternary New loess

βQ₂d Quaternary basalt, Dayizishan formation

βQ₁j Quaternary basalt, Junjianshan formation

βN₂c Neogene basalt, Chuandishan formation

βN₂ Neogene basalt

O₃s Upper Ordovician, Shifeng formation

Fig. c from: Geological Map of Jilin Province, 1:500,000, Geological Publishing House, Beijing

Upper Archean (Ashan Migmatite Group)

Ar₁y Yangjiadian formation, gneiss and amphibolite with magnetite hornblende quartzite

Ar₁s Sidaolazihe formation, gneiss and amphibolite with magnetite quartzite

⬤ Maar lakes of the Long Gang Volcanic Group

⬡ Dry maars

Table 1. Location and morphometric features of maar- and crater lakes of the Long Gang Volcanic Field. Max. water depth as derived by echo sounding in August/September 1998/99 (device: FURUNO FE 6300)

Name	Location	Elevation (m a.s.l.)	Lake Area (km²)	Catchment Area (km²)	Max. Water Depth (m)	Rim Height (m)
Dalongwan (DAL)	42°20'N, 126°22'E	635	0.8	1.0	85	20
Donglongwan (DOLO)	42°26'N, 126°31'E	599	0.4	0.5	127	50
Erlongwan (ERL)	42°18'N, 126°21'E	724	0.3	0.4	36	30
Hanlongwan	42°22'N, 126°25'E	744	0.03	0.2	swamp	15
Huangnigang	42°22'N, 126°21'E	545	-	0.7	dry	60
Longwanhu (LWH) (Longguanlongwan)	42°25'N, 126°36'E	618	0.85	1.0	115	10
Longwantang (LOW) (Nanlongwan)	42°25'N, 126°28'E	655	0.3	2.0	69	20
Sanjiaolongwan (SJL)	42°22'N, 126°25'E	730	0.7	0.9	76	40
Sihailongwan (SHL) (Longwan)	42°17'N, 126°36'E	797	0.5	0.7	50	20
Xiaolongwan (XIA)	42°18'N, 126°19'E	655	0.1	0.15	15	20
Xidadianzi	42°21'N, 126°22'E	618	-	1.3	dry	15

mg/l Si (DOLO). Primary production in XIA, SHL, ERL, and SJL is limited by SRP (soluble reactive phosphorus) which could not be detected in the epilimnion of these lakes during the summer stagnation. Sulphate which decreases in the anoxic deeper water of ERL, SHL and SJL, shows an opposite trend in Lake XIA. The latter indicates subsurface inflow of ground water into the hypolimnion of this lake.

Sediment cores were taken from all eight lakes with a lightweight freefall coring system with core lengths between 32 and 112 cm. For the laminated sediments of Sihailongwan lake the UWITEC freeze-coring system was used to obtain undisturbed sediments of the last years. All cores show macroscopically, at least partly, different types of layering. Cores from five lakes (Sihailongwan, Sanjiaolongwan, Xiaolongwan, Dalongwan and Longwanhu) are mainly composed of layered or finely laminated diatomaceous gyttja with some intercalations of graded layers. Sediments from Donglongwan, Erlongwan

and Longwantang lakes are more influenced by allochthonous minerogenic input.

Magnetic susceptibility was measured with an automated Bartington point sensor. The measurements were performed with a point separation of 1 mm and a spatial sensor resolution of 4 mm. Even cores which are composed mainly of organic gyttja show susceptibilities between 20-100 x 10^{-6}SI. These values are much higher than, for example, susceptibilities of Holocene diatomaceous gyttja from Lago Grande di Monticchio (Brandt *et al.*, 1999), thus indicating a significant content of minerogenic material. Although some cores from different lakes do have a similar susceptibility pattern it is not possible to correlate sediment profiles by means of magnetic susceptibility.

Sihailongwan Lake

The basin of Sihailongwan lake is a single circular, simple U-shaped structure (Figs. 2A, B). The

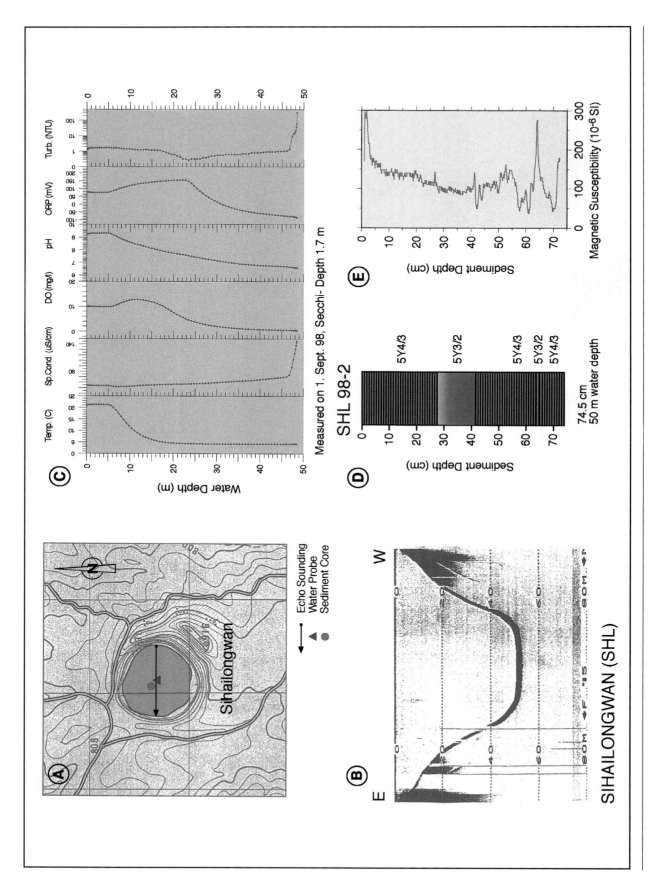

Fig. 2. Lake Sihailongwan (SHL). A. Lake map with trace of echo sounding, positions of water probe measurements and sediment coring; B. Echo sounding profile; C. Hydrological parameters of the water column; D. Sediment section; E. Magnetic susceptibility of the sediment core.

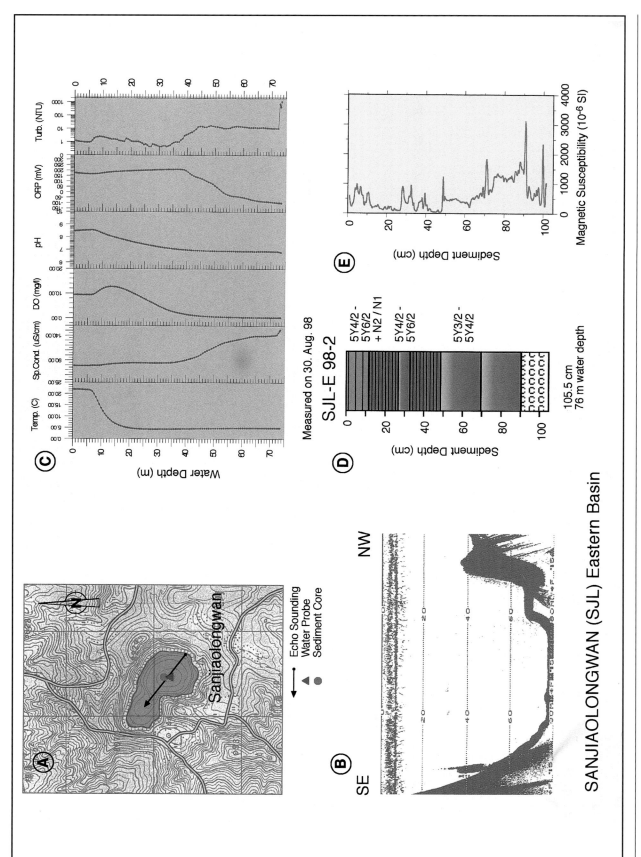

Fig. 3. Sanjiaolongwan lake (SJL), Eastern Basin. A. Lake map with trace of echo sounding, positions of water probe measurements and sediment coring; B. Echo sounding profile; C. Hydrological parameters of the water column; D. Sediment section; E. Magnetic susceptibility of the sediment core.

67

Table 2. Lake and ground water (GW) chemistry, membrane filtrated water samples (0.45 µm), all concentrations in mg/l; n.d. = not detectable; n.a. = not analysed.

Lake (0 m)	date	F	Cl	NO$_3$	SO$_4$	SRP	Si	DIC	Na	K	Ca	Mg
SJL-W	30/08/98	0.08	0.7	n.d.	3.2	n.d.	0.25	7.7	5.1	2.4	5.7	2.4
DOLO	04/09/98	0.13	0.8	n.d.	3.4	n.a.	4.50	15.3	10.3	3.5	13.4	3.2
LOW	03/09/98	0.13	0.8	0.8	5.1	n.a.	0.55	16.3	7.8	3.5	15.8	4.7
DAL	31/08/98	0.11	2.2	n.d.	9.0	n.a.	2.80	31.1	13.6	7.1	13.2	20.5
ERL	06/09/98	0.03	0.5	n.d.	4.8	n.d.	0.18	3.4	0.9	1.0	4.3	1.4
SHL	01/09/98	0.05	0.5	n.d.	3.4	n.d.	0.17	5.9	2.1	1.4	5.6	2.4
XIA	13/09/99	0.08	1.1	0.4	3.7	n.d.	0.13	7.1	0.8	1.4	8.3	1.7
GW(SJL-W)	22/09/99	0.18	1.6	n.d.	10.3	0.044	14.2	16.6	5.9	3.9	14.5	6.3
GW(SHL)	13/09/99	0.13	1.5	0.089	6.4	0.013	13.3	18.6	7.3	3.6	12.5	7.9

laminated sediments of the upper metre are only disturbed by one graded layer which is considered to be the product of an internal slump, and therefore this lake was chosen for detailed investigations.

Bulk Geochemistry

Al$_2$O$_3$ concentrations of around twelve weight-% confirm high allochthonous minerogenic matter concentration for the SHL sediment record (Fig. 4). Increased inwash of local soil particles at 4.5 cm and 28 cm is demonstrated by the profiles of conservative major and trace elements (Al, Zr). The allochthonous detrital input has been increasingly diluted by autochthonous biogenic deposits during the last sixty years (eutrophication). The sulphur flux into the surface sediment has increased synchronously. The latter may reflect increased atmospheric sulphur deposition as well as longer periods of oxygen depletion in the hypolimnion. There is a distinct increase in Pb/Sc from about 1942 which could reflect increased atmospheric deposition of lead contaminated dust particles. An increased release of geogenic lead by acid rain also seems probable for the investigated site.

Lamination and Diatoms

For detailed investigations of the lamination a set of nine 10 cm-long overlapping thin sections was prepared from core SHL 98–2 (74 cm sediment depth). The individual laminae (varves) are composed of three microscopically distinguishable sublaminae—a thin layer enriched in diatoms, immediately followed by an allochthonous minerogenic layer with a sharp lower and gradational upper contact and a layer of mixed composition (mainly mineral—and plant detritus, diatoms, green algae). Near the base of the mixed layer there often occurs an enrichment of *Pinus* pollen, so the allochthonous minerogenic layers have been deposited most likely in spring after lake ice out and snow melt. Varves, composed of these three sublaminae, range in thickness from 0.4 mm to 2.5 mm, with an average of 680 µm. A set of smear slides from SHL 98–2 was investigated for diatoms. Almost 95–99% of the diatom population belong to two different *Cyclotella*-species. The large one, with diameters of 12–20 mm, is *Cyclotella* cf. *radiosa*, a cosmopolitan planktonic species, commonly found in meso- to eutrophic lakes. The smaller species, with diameters of 4–9 mm, is somewhat difficult to estimate due to taxonomic

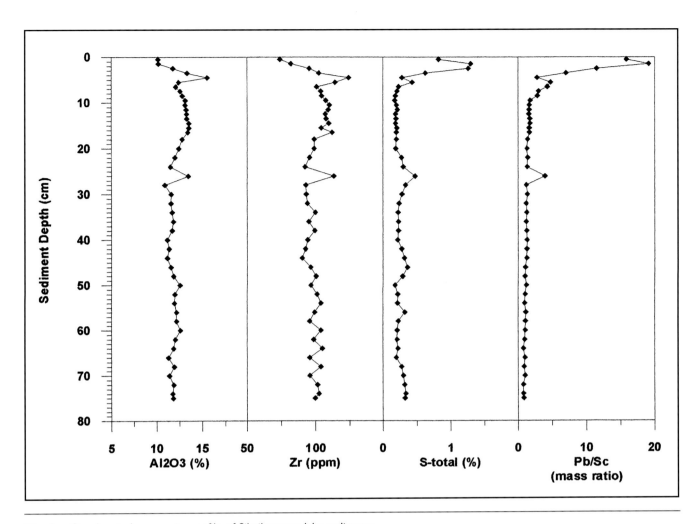

Fig. 4. Geochemical composite profile of Sihailongwan lake sediments.

uncertainties. This species is probably *Cyclotella* cf. *pseudostelligera*, also a planktonic diatom. Another common planktonic species in Sihailongwan lake is *Asterionella formosa*. Some benthic species found in Sihailongwan lake are *Fragilaria capucina*, *Achnanthes lanceolata*, *Epithemia sorex*, *Navicula radiosa*, *Cocconeis* sp., *Cymbella* spp., *Diploneis* spp., *Pinnularia* spp., *Nitzschia* sp., *Tabellaria flocculosa* and *Surirella splendida*. This diatom assemblage is typical for alcaline lakes.

Thin sections do not show distinct seasonal successions of diatoms because there is only a scant number of benthic ones. But there is at least a strong hint to interpret the *cyclotella* layers as seasonal algal blooms and thus the mixed layer—

diatom layer—lamination as varves. The extreme climatic conditions with strong winter seasons (with a regular, long-lasting ice cover) possibly favour diatom blooms in springtime still under the ice cover.

Age Determination

A total of 915 varves could be counted for the 74 cm sediment section of SHL 98–2 (Fig. 5). This varve count is in agreement with a ^{14}C-AMS age determination of a single leaf found at 44 cm sediment depth (which corresponds to 1630 AD from varve counting). The ^{14}C age determination yielded a radiocarbon age of 335 +/- 32 years and (due to a

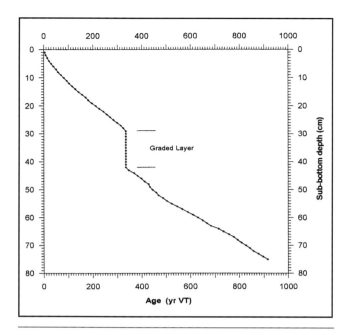

Fig. 5. Age-depth-plot of core SHL 98-2 from Sihailongwan lake.

[14]C plateau) calibrated ages of 1520, 1587 and 1625 AD.

Pollen Analysis

Forty-eight samples were taken from the 81 cm-long SHL 99–2 core using a 7 mm diameter brass sampler. Each pollen sample represents nine calendar years with twenty-five years between the middle points of the samples taken at 2 cm intervals.

The pollen record (Fig. 6) is dominated by pollen of woody taxa (> 80%) throughout; but with a slight decline in the upper 7 cm. The size of the basin is such that the pollen record is a mainly regional signal and thus it can be inferred that forest has dominated the landscape for the interval represented by this sequence.

Pinus is abundant throughout the record; and apart from *Pinus* the main woody taxa present were *Betula, Quercus, Ulmus, Juglans*-type, *Corylus, Ostrya* and *Tilia.* The only herbaceous taxa to make a major contribution to the pollen in the sediments were Chenopodiaceae and

Artemisia, other taxa which are frequently present at low relative abundances are Cyperaceae, Gramineae, and *Thalictrum.*

Total pollen concentration in grains per cc of wet sediment decreases up the stratigraphy. This may be a consequence of decreasing sediment consolidation up the sequence.

For almost the entire profile the biome represented by the pollen assemblage (Prentice *et al.*, 1996), (Tarasov *et al.*, 1998) is that of temperate deciduous forest. The exceptions are samples at 3, 5 and 6 cm which biomise as wooded steppe because the tree sum in the upper 7 cm (pollen zone 1a) are the only occurrences of pollen of cereals (possibly Oryza—rice), and other grasses also are more abundant than previously. There is an increase in unknowns and cryptogams which may be due to opening of the landscape.

Sihailongwan lies within the deciduous broad-leaved forest zone of Wu (1983) and the pollen-based biome reconstruction of temperate deciduous forest/wooded steppe is in agreement with this. The pollen taxa recorded in samples from Sihailongwan correspond closest to the *Quercus mongolica/Betula dahurica* communities of Liu *et al.* (1999) as described in their investigation of surface pollen from southeastern Inner Mongolia.

Throughout pollen zone 1b to 1e the woody taxa sum does not fall below 80% which is in agreement with the isopoll maps by Ren and Zhang (1998) for 800 cal. yrs. BP for northeast China.

The palynological record indicates that for most of the interval represented by the record the landscape has been dominated by mixed boreal/deciduous forest. Only in the last 150 years is there evidence of thinning of the forest; agricultural activity is indicated for the last seventy years.

These preliminary investigations of short cores from Lake Sihailongwan have demonstrated the potential of this site to establish a master varve

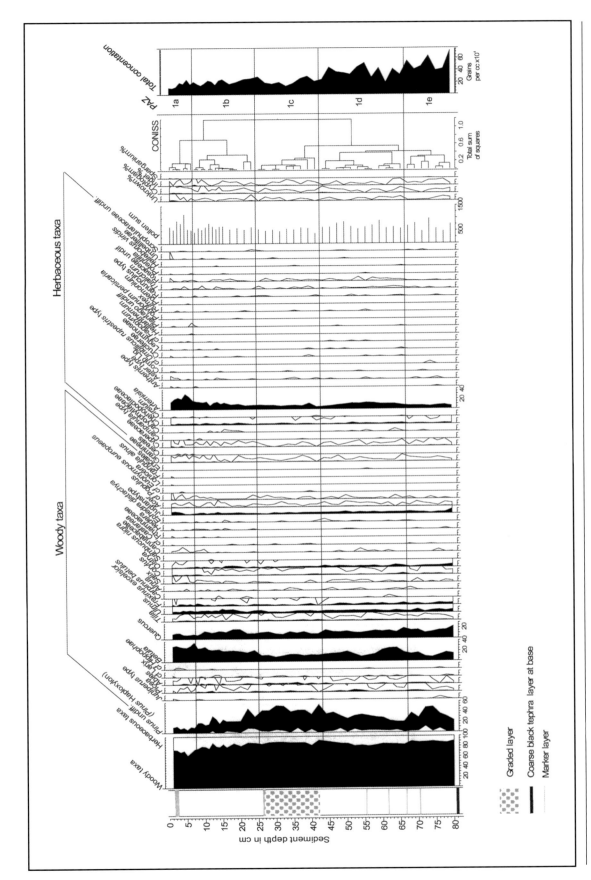

Fig. 6. Pollen percentages of core SHL 99-2 from Sihailongwan lake.

71

chronology for northeast China and to yield an outstanding pollen record for this area of China from which there is little data.

References

Brandt, U., N.R. Nowaczyk, A. Ramrath, A. Brauer, J. Mingram, S. Wulf and J.F.W. Negendank (1999): Palaeomagnetism of Holocene and Late Pleistocene sediments from Lago di Mezzano and Lago Grande di Monticchio (Italy): initial results. *Quaternary Science Reviews*, 18: 961–976.

Geological Survey of Jilin (1994): *Geology of Jilin, China*. Geological Publishing House, Beijing (in Chinese).

Liu, H., H. Cui, R. Pott and M. Speier (1999): The surface pollen of the woodland-steppe ecotone in southeastern Inner Mongolia. *Review of Palaeobotany and Palynology*, 105: 237–250.

Ogura, T. (1969): Volcanoes in Manchuria. *In* Ogura T. (ed.): *Geology and mineralogy of the Far East*. University of Tokyo Press, Tokyo, pp. 373–413.

Prentice, I. C., J. Guiot, B. Huntley, D. Jolly and R. Cheddadi (1996): Reconstructing biomes from palaeoecological data: a general method and its application to European pollen data at 0 and 6 cal. yrs. *Climate Dynamics*, 12: 185–194.

Ren, G., and L. Zhang (1998): A preliminary mapped summary of Holocene pollen data for Northeast China. *Quaternary Science Reviews*, 17: 669–688.

Tarasov, P.E., R. Cheddadi, J. Guiot, S. Bottema, O. Peyron, J. Belmonte, V. Ruiz-Sanchez, F. Saadi and S. Brewer (1998): A method to determine warm and cool steppe biomes from pollen data; application to the Mediterranean and Kazakhstan regions. *Journal of Quaternary Science*, 13: 335–344.

Tokyo Geological Survey (1929): *Geological Atlas of Eastern China*.

Wu, C.-Y. (1983): *Vegetation of China* (in Chinese). Science Press, Beijing.

Chapter 4

Radiocarbon Dating of an Ancient Bridge near Xianyang, Shaangxi Province, China, and Implications for Palaeoclimate Studies

ZHOU WEIJIAN AND JOHN HEAD

Introduction

The Wei river (Fig. 1), a well-documented river in ancient Chinese historical records, originates in Liao Shu Shan, Wei Yuan county, Gan Shu province, then flows in a south-easterly direction through Shaanxi province, running for about 400 km through the Qin Chuan area. More than nine rivers flow into the Wei river, which has been called the 'radiated river'. The Wei river played a very important role in the development of Shaanxi province's economy, providing resources for agricultural development, and the facility for water transportation. During Chinese history, important dynasties such as Zhou, Qin, Han, and Tang, (Zhou/Chou, 1121 BC–249 BC; Qin/Ch'in, 248 BC–207 BC; Han 206 BC–AD 220; Tang/T'ang, AD 618–AD 907) established their capitals in ancient Guan Zhong, which was the political, economic, and cultural centre of China. Xianyang was the capital during the Qin dynasty, and Chang An (now Xian) was the capital during the Han and Tang dynasties (Fig. 1).

Large logs, about 0.4–0.5 m in diameter have been found in a vertical position, extending about 1 m above the surface of a dry, former river bed, near Xianyang, Shaanxi province. Similar logs, about 2–3.5 m in length, and 0.7–1.2 m in diameter were found alongside. These logs were obviously part of a bridge, which would

have been about 15 m in width, and about 50 m in length. Historical records indicate the construction of three famous bridges across the Wei river. So far only two bridges of East Wei and Middle Wei have been discovered. The wooden bridges were periodically destroyed during wars and suffered periods of disuse during climatic and geomorphological changes, when the river bed modified its position within the landscape. Archaeological evidence as to the position, structure, time of construction, period of use, and shape of these three bridges is scarce. The aim of our study is to date the wooden bridge and try to determine the West Wei bridge

Archaeology

Literature records supplied by the Xianyang Management Committee of Cultural Relics have provided the following evidence for the three bridges.

Middle Wei River Bridge: This bridge was constructed during the reign of Qin Zhao Wang, about 2,300 years ago, and provided a crossing point over the Wei river to the south of Yaodian district, which is now Xianyang City (Fig. 1). According to the book *Shan Fu Jiou Shi* (Qin Zhao Wang used the Yu Gong (Yu Palace) south of the Wei river, and the Xianyang Gong north of the river. In order to provide direct access from one palace to the other, he constructed a

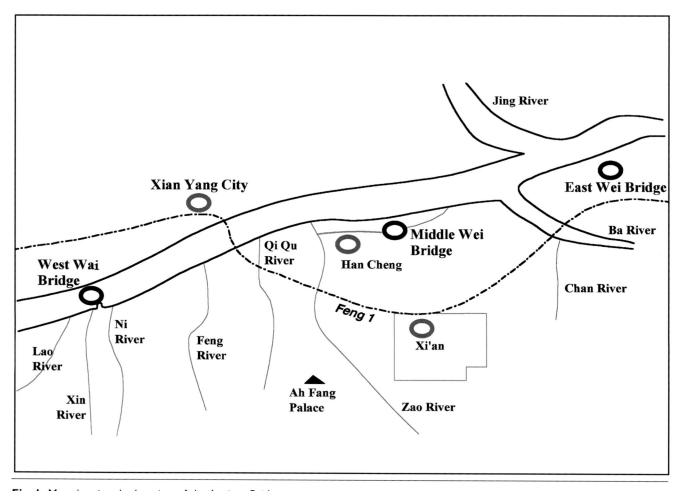

Fig. 1. Map showing the location of the Ancient Bridge.

horizontal bridge of length c. 380 steps, and evidence of the position and structure of this bridge can be easily found. After Qin Shihuan united six countries in 221 BC, he extended and reinforced this bridge in order to improve the access between the two palaces.

During the Han Dynasty, the capital was transferred to the south of the Wei river, to Chang An (now Xian), and it was not necessary to connect the two palaces. With the development of a feudalistic economy, and hence an increase in transport within the area, the bridge was repaired for further use, and was renamed Wei Qiao (Wei Bridge). At the end of the East Han, Dong Zhuo had the bridge destroyed. Cao Cao had the bridge rebuilt, but the width was only about 8 m.

During the North and South dynasties, Liu Yu (Song Wudi) had the bridge burned, and it was rebuilt after the North dynasty. During the Tang dynasty, the Middle Wei Bridge was shifted about 10 Li (about 4–5 km) to the east, but was later destroyed during the Qing dynasty (AD 1616–AD 1911). The main bridge structure was sunk, and now most probably lies within the area southeast of Yaodian, Xianyang city.

East Wei Bridge: This bridge was located near the town of Wei Qiao, now Gao Ling county, and was constructed by Han Gao Zu, one of the emperors of the Han Dynasty, around 152 BC. This bridge was constructed of wood, and played a very important role in the transportation of agricultural crops (Fig. 1). Tang Gao Zong constructed a storehouse for grain near the bridge, and grain from crops grown to the east of the Wei river was stored here before it was transported to Chang An (now Xian), when needed. Relics of the East Wei Bridge, and a stone recording the bridge construction were found near the town of Gen Zheng, about 3 km south of the bridge on the Xian highway.

West Wei Bridge: This bridge was located at Li Xu, southeast of Xianyang (Fig. 1), and was constructed during the reign of Liu Qie, an emperor during the Han Dynasty, around 138 BC. The bridge provided a connection between Chang An and Mao Ling, and was also known as the 'Convenient Bridge'. This bridge was made of wood in a fashion similar to the East Wei Bridge, and was not as large as the Middle Wei Bridge. Since this bridge was located near Xianyang, it was known as the Xianyang Bridge during the Tang Dynasty. After the Tang Dynasty came to an end, the bridge was destroyed by flood waters, then was repeatedly repaired and destroyed. The book *Chang An Zhi* (Records of Chang An), written during the Song Dynasty, indicates that the so-called Convenient Bridge was rebuilt by the first emperor of this dynasty over a period of four years, to be destroyed again by flood waters later on.

In the winter of 1985, peasants digging sand from the dry river bed close to Xi Dun village, Diao Tai district, found wooden logs, buried in a north-south direction. The northern area of this site has been excavated, and consists of six rows of wooden supports. The distance between each support is about 2.3 to 2.6 m, and the supports form a rectangular array. Horizontal poles were found near the end of the fifth row of wooden supports together with some charcoal fragments. Each support had been cut to form a cone-like shape at the bottom in which knives had been embedded to aid penetration into the river sediments. The excavated wooden poles are 2–3.5 m in length, and about 0.67–1.23 m in diameter. The wood is mainly pine. Near the bridge, relics such as bronze weapons, ceramic utensils and tile fragments were found.

It was thought that the remnants of this bridge could possibly belong to the West Wei Bridge, since no previous remnants of this bridge had been found and the bridge supports seemed to be in the right area, even though the Wei river now flows a few kilometres away.

A portion of the outer wood from one of the excavated poles was dated in the State Key Laboratory of Loess and Quaternary Geology and subsequently, three sections from vertical support posts, still *in situ*, were cut for further examination in the laboratory.

Experimental

Two of the three sections cut from vertical support poles of the bridge were polished on one side, and the rings were counted and accurately measured using a magnifying lens and a specially constructed measuring device. The first section had a diameter of 0.5 m, and the second section had a diameter of 0.48 m. A large slice of each section was separated into portions weighing 20–25 g. In most cases this meant that the samples contained four to five rings. In some cases, where the rings were very narrow, up to ten rings were used.

The wood was cut into small fragments (< 1 mm), and placed into a Soxhlet Extractor. A series of solvents was used to extract resins, turpenes, and other extractable materials from the wood. The first solvent used was a mixture of benzene and ethanol (2:1), then ethanol followed by water. Each extraction was carried out for a time period ranging from six to twenty-four hours (or until the solvent in contact with the wood was colourless). The wood was then placed into a 2-litre Erlenmeyer flask with about 800 ml distilled water. The flask was placed on a steam bath, and the contents were heated to 75°C. Approximately 3 ml hydrochloric acid was added to the water to obtain a pH of about 3, and 7.5 g sodium chlorite was added very slowly to the flask. The reaction was allowed to proceed at 75°C for about three hours, with the pH being checked periodically, and a few drops of hydrochloric acid added if the pH became > 3. If after three hours, the wood did not appreciably lighten in colour, another 7.5 g of sodium chlorite could be added, and the reaction continued for another three hours at 75°C. After the sodium chlorite treatment had been completed, the flask was allowed to cool, and the liquid was filtered off. The crude cellulose (holocellulose) that remained was rinsed thoroughly with distilled water, and air-dried to minimize decomposition of the carbohydrate component (Head, 1979; Gupta and Polach, 1985). This fraction was fluffy in texture and white in colour with a slight yellowish tinge.

The sodium chlorite treatment dissolves most of the lignin component of the wood, leaving the carbohydrate component (cellulose and hemicelluloses) intact. Usually, depending on the species and structure of the wood, up to 4% by weight of lignin can remain. It is thought that this lignin is bound to either portion of the hemicelluloses, or to a portion of the cellulose, and further treatment with sodium chlorite would cause the break-up and dissolution of some of the carbohydrate component, resulting in a lower yield. Pure cellulose can be isolated from the holocellulose fraction by treatment with alkaline solutions in a nitrogen atmosphere. In this case, purification of the holocellulose fraction was not attempted since the wood samples for treatment were a species of pine, and serious contamination of the samples was not expected.

Benzene synthesis from the holocellulose samples was carried out as described in two handbooks published by the Guiyang Institute of Geochemistry, Academia Sinica, in 1973 and 1977, by Tamers (1975), and by Zhou Mingfu et al., (1983). ^{14}C age determinations were carried out using a Wallac Quantulus^{TM} liquid scintillation counter, using 3 ml benzene samples with butyl PBD as the scintillant (Polach et al., 1983; Kojola et al., 1985). The counting vials used were 3 ml Teflon^{TM} copper vials (Gupta and Polach, 1985). Chinese sugar carbon was used as the Modern Standard, with benzene synthesized from anthracite collected from Mentougou, Beijing, being used to determine equipment background characteristics. Conventional ages were calculated using the criteria defined by Stuiver and Polach (1977).

Results and Discussion

The ^{14}C ages determined for wood from the outer rings of one of the horizontal poles excavated from the bridge site, and from the first section obtained from one of the vertical supports, together with the most probable Christian Calendar ages obtained from the Tree Ring Calibration Curves of Stuiver and Pearson (1986), are shown in Table 1. Sample XLLQ-75, with a ^{14}C age of 2120 ± 80 cal. yrs. BP has a Christian calendar age of 167 BC years. This result supports the archaeological evidence that this bridge was the West Wei Bridge. A section will be cut from the log supplying this sample, and this age should be verified by a series of samples that should provide a chronology of about 150 years.

Table 1. Tree ring ^{14}C ages and Christian Calendar ages for samples from Section 1 and one of the horizontal beams.

Sample	Lab Code	^{14}C Age (BP)	Calib Age
Vertical support Section 1			
Rings 1–12	XLLQ-132	2050 ± 80	1995
Rings 78–82	XLLQ-147	1790 ± 75	1710
Rings 113–117	XLLQ -131	1780 ± 50	1710
Rings 133–137	XLLQ -148	1940 ± 75	1707
Rings 156–169	XLLQ -130	1730 ± 55	1663
Outer rings of horizontal beam Section 2			
Outer rings	XLLQ -75	2120 ± 80	2116

So far only five from a possible twenty-six samples have been dated from the first section, though a trend seems to be present. A linear regression was plotted using the most probable Christian calendar age values against the central ring number, and the ages obtained from the regression line are shown in Table 1. The coefficient of correlation for the regression line was 0.76. The data indicates that the tree started growing, 1,995 years ago, and was probably cut down at 1558 cal. yrs. BP. This data is not precise enough at this stage, since a time range of about 362 years is indicated for a section containing 170 rings.

A plot of tree ring number against ring width for the first section can be seen in Fig. 2. The variability in ring width most probably indicates changing climatic conditions though tree ring width time series plots contain a physiological component indicating individual tree growth characteristics (Wigley *et al.*, 1987). The first section most probably covers the early AD time range, which is at least 150 years after the construction of the bridge. Archaeological evidence indicates that the bridge was periodically destroyed and rebuilt, so the horizontal beam from which sample XLLQ-75 was taken would have been one of the original beams (2120 cal. yrs. BP), while the support from which this section was taken would have been a later addition, added during one of the rebuilding phases, at about 1995 cal. yrs. BP. The support from which the second section was taken could cover an age range older or younger than the age range of the first section.

If this plot is compared against the plot for the

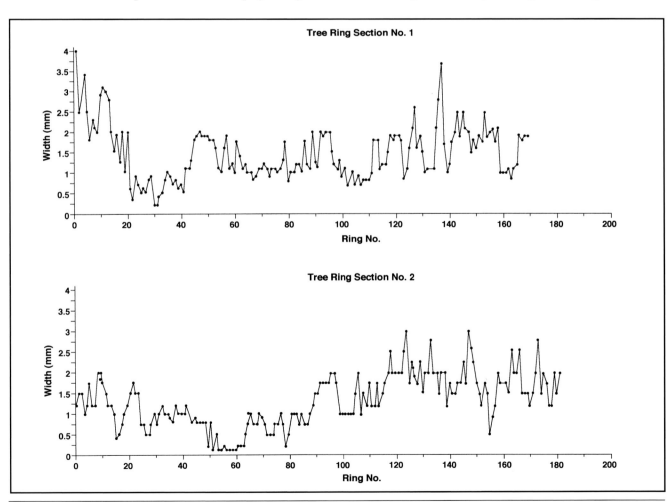

Fig. 2.   Tree-ring results of the Ancient Bridge log.

second section shown below it, no distinct comparative pattern emerges. Hence, these sections seem to cover quite different age ranges.

In Fig. 3, plots of the deviation of 10-year ring width means from the total ring width mean are shown for both sections. This statistical filtering technique tends to minimize the physiological component of the data (La Marche, 1974). Relatively short-term fluctuations from possible cold, dry (or hot, dry) to warm, wet periods are indicated. The first section indicates quite sharp fluctuations, while the second section indicates much more gradual changes. Fig. 3 also indicates that the trees represented by these sections most probably grew within different time periods.

The data presented in this paper forms a preliminary summary of this project though the potential indicated so far for a detailed palaeoclimatic picture of the area in the vicinity of Xian and Xianyang during occupation by many emperors, is extremely great. Continuation of this study will be time-consuming though the possible information to be obtained should provide an extremely valuable addition to the historic information already available.

The evidence obtained so far strongly indicates that the wood examined in this study originally formed part of the West Wei Bridge, which completes the archaeological and historic record indicating the presence of the three bridges over the Wei river.

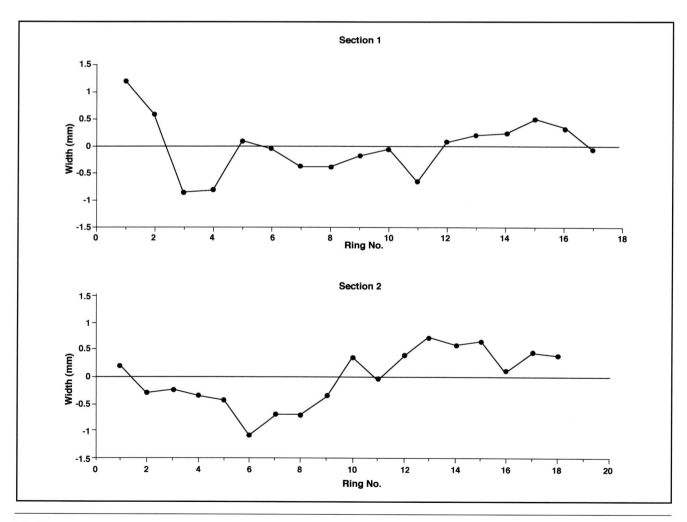

Fig. 3. Tree-ring Smoothing Curve.

Acknowledgements

Mr Wu Fa Rong and the staff of the Xianyang Museum supplied the samples, and provided valuable information with respect to historical and archaeological records. This study was supported by KZCX3-SW-120, KZCX2-SW-118, NSFC40121303, NKPBR-2001CCB-00100, and NKBRSFG199043400.

References

Gupta, S.K., and H.A. Polach (1985): *Radiocarbon Practices at ANU, Monograph*, ANU Radiocarbon Laboratory, 171 pp.

Head, M.J. (1979): *Structure and Chemical Properties of Fresh and Degraded Wood: Their Effects on Radiocarbon Dating Measurements*. Thesis for Master of Science, ANU: 130, Unpublished Manuscript.

Kojola, H., H. Polach, J. Nurmi, A. Heinonen, T. Oikari, and E. Soini (1985): Low-Level Liquid Scintillation Spectrometer for Beta-Counting. In *Proceedings, Nordic Archaeometry Conference, Marienham, Finland, 1984*. Archaeological Society of Finland, ISKOS, pp. 539–543.

La Marche Jr., V.C. (1974): Palaeoclimatic Inferences from Long Tree-Ring Records. *Science* 183: 1043–1048.

Polach, H.A., J. Gower, H. Kojola and A. Heinonen, A. (1983): An Ideal Vial and Cocktail for Low-Level Scintillation Counting. In S.A. McQuarrie, C. Ediss, and L.I. Wiebe (eds.): *Advances in Scintillation Counting*, University of Alberta Press, pp. 508–525.

Radiocarbon Laboratory, Guiyang Institute of Geochemistry, Academia Sinica. (1973): Radiocarbon Dating of Archaeological Samples. In *Geochimica*: 135–137 (in Chinese).

Radiocarbon Laboratory, Guiyang Institute of Geochemistry, Academia Sinica (1977): *Method of Radiocarbon Dating and its Applications*, Science Press, pp. 5–12 (in Chinese).

Stuiver, M. and G. W. Pearson (1986): High Precision Calibration of the Radiocarbon Time Scale, AD 1950–500 BC. *Radiocarbon* 28 (2B), pp. 805–838.

Stuiver M. and H. G. Polach (1997): Discussion: Reporting of ^{14}C data, *Radiocarbon* 19: 355–363.

Tamers, M.A. (1975): Chemical Yield Optimisation of the Benzene Synthesis for Radiocarbon Dating. *Int., Jour. Appl. Radiation and Isotopes* 26: 676–682.

Wigley, T.M.L., P. D. Jones and K. R. Briffa (1987): Cross-Dating Methods in Dendrochronology. *Journal of Archaeological Science* 14: 51–64.

Wu Fa Rong and Staff of the Xianyang Museum, personal communication.

Zhou Mingfu and Sheng Chende (1983): An Investigation of Li_2C_2 Method for Radiocarbon Dating. In *Geochimica*, pp. 411–416 (in Chinese).

Song Mingqiou (1931): *Chang An Zhi*, The Press of Chang An County Zhi Bureau, Vol. 12: 2.

In Rajasthan, located on the northern periphery of the Indian monsoon, securing water supply is a matter of life and death. A boy herding sheep in Rajasthan, India. (Photo by Takeshi Takeda)

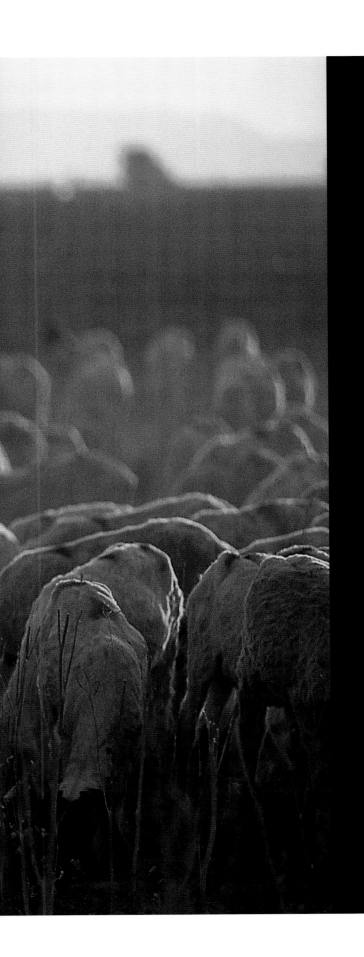

Part II

Indian Monsoon Variability
and Human Adaptations

A woman carrying a water
pot on her head.
(Photo by Takeshi Takeda)

Chapter 5

Palaeomonsoons and Palaeoclimatic Background to the Prehistoric Cultures of the Western and Central Thar Desert, Rajasthan, Northwestern India

MUKUND D. KAJALE, BHASKAR C. DEOTARE AND SHARAD N. RAJAGURU

Introduction

The Thar desert is an easternmost extension of the middle latitude desert belts of Africa and western Asia. It is primarily affected by the vagaries of the southwest summer monsoon, with an annual rainfall varying from 500 mm in the eastern margin to about 100 mm in the western margin of the Thar. Winter rainfall is caused by the incursion of westerly disturbances. It is less than 10% of the mean annual rainfall (Sikka, 1997). The high rate of evaporation (PET or Potential of Evapo Transpiration exceeding 2000 mm/annum) and high temperatures all year round considerably reduces available moisture in the Thar. Wind speed reaches upto around 25–30 km per hour during May and June while PET reduces to about 500–600 mm during the monsoon months, i.e., July to September. Droughts are frequent in the Thar due to high variability of rainfall. The Perso-Arabian dry vegetational belt of arid to semi-arid western Rajasthan gradually merges into the Indo-Malayan elements of the eastern and southeastern flanks of semi-arid to semi-humid belts respectively because of gaps in the Aravalli hill ranges which run from northeast to southwest encompassing Delhi in the north and Kheda in north Gujarat in the south (Individual observations and Bhandari: Personal Communication).

On the eastern side of the Thar (24° 30'N 30°N, 69° 30'E, 76°E) the Aravalli hill ranges act as a major water divide separating the well-integrated drainage of the Indo Gangetic plain on the east from the poorly developed ephemeral drainage (inclusive of the Luni river) on the west. The western margin of the Thar in India is predominated by dunes which pass into the Indus alluvial plain, further west in Pakistan. In the south, the dunes of the Thar abruptly terminate against the Great Rann of Kutch, while in the north, the desert grades into the Indo-Gangetic alluvial plain.

Geologically speaking, a variety of rocks of the Proterozoic age dominate the eastern side while the Permo-Carboniferous and Jurrasic sediments occur in the west. Tertiary sediments rich in bentonitic clay and lignite occur as pockets in the southern and northern parts of the Thar. Quaternary alluvial and aeolian deposits cover these older geological formations. Quaternary sediments are thick upto 300 m in the Luni basin since it is affected by graben and horst structures controlled by faults and lineaments, particularly in north-northeast to south-southwest directions (Dhir, 1992; Wadhawan and Sharma, 1997; Bajpai et al., 2001), leading to exceptionally thick deposition during the Late Neogene-Early Pleistocene period.

Geomorphologic features such as pediments

with inselbergs, ephemeral streams, dry rock-cut gorges and gravel veneers dominate the rocky upland of the Thar, while varieties of dunes, interdunal sandy plains, playas and calcritized regoliths occur in non-rocky lower plains of the Thar. Detailed geomorphologic studies based mainly on the study of satellite imageries have made it possible to map a number of palaeochannels in the area under discussion (Kar, 1999). In the present study, however, the northern and northwestern aggraded alluvial plain, a part of the sub-Himalayan drainage system, has been omitted in view of our constraints in field studies in that area. The following framework, tentatively suggested for Quaternary environmental and cultural study covers Jodhpur, Jaisalmer, Nagaur and part of the Pali, Bikaner and Barmer districts of the Thar (Fig. 1).

An attempt has been made here to focus on the geomorphological, palynological and stable isotopic ($\delta^{13}C$ and $\delta^{18}O$) composition of soil carbonates and clay mineralogical signatures which have been found to be useful while

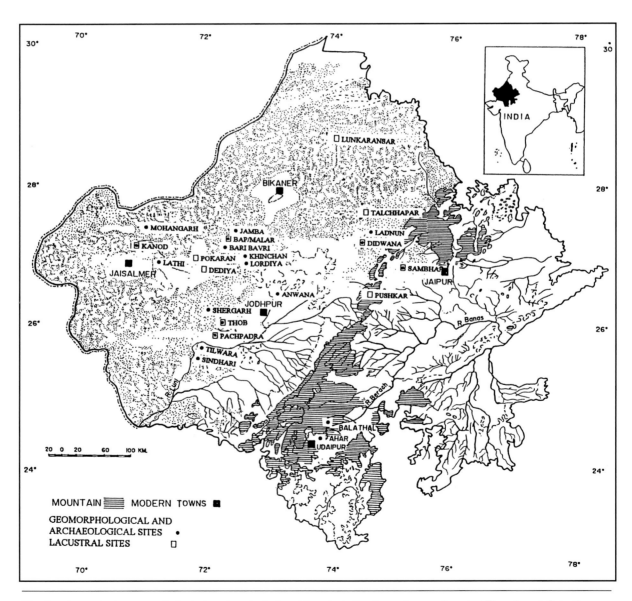

Fig. I. Important geomorphological, archaeological and lacustral sites in western and central Thar desert, Rajasthan.

investigating the antiquity of the Indian summer monsoon in the Thar in general and its changing pattern of intensity, duration, etc., since the Neogene. Development of prehistoric cultures in the Thar is also discussed against data obtained by various scholars working in the recently concluded multi-institutional research projects supported by the Department of Science and Technology, New Delhi.

In this paper, we have tried to integrate these results, taking a resume of the works and avoiding details of field and laboratory tests. This is essentially to develop a conceptualized tentative framework of the palaeoenvironment of the Thar against the backdrop of which the archaeological studies may be visualized by the archaeologists and the Quaternary scientific community. It is hoped that the results summarized below will accelerate further multi-disciplinary studies not only in the Thar but also in relatively less studied deserts of Asia and well studied parts of central southern Africa. The theme of the paper is organized as follows:

1. Neogene and Early Pleistocene environment,
2. Middle Pleistocene environment and evidences for prehistoric sites,
3. Late Pleistocene environment and palaeolithic sites,
4. Holocene environment and archaeological sites belonging to later cultural periods (Mesolithic to Late Historic) and
5. Conclusions.

Neogene and Early Pleistocene Environment

The main driving force for the summer monsoon of India is a sharp thermal contrast between the glaciated Himalayas and the warm Indian Ocean (Kutzback et al., 1989). The monsoon circulation appears to have begun whilst the Himalayas and the Tibetan plateau attained a reasonable height during the middle Miocene (–21 to 17), (Xu Quin

Qi, 1993) and reached close to the present elevation during the Pliocene or the early Pleistocene (Fort et al., 1996). It is thus obvious that the summer monsoon circulation got well established during the Neogene. The stable isotope studies by Quaide et al., (1989) on pedogenic carbonates developed within Siwalik sediments of the Neogene age in Pakistan support the view that the seasonal (monsoonic) rainfall pattern got established during the Neogene. Hoom et al., (2001) carried out systematic palynological studies on lower Siwalik sediments exposed in central Nepal and suggested that during the middle Miocene, the Himalayan foothills and the Gangetic plain of Nepal were mainly forested with subtropical to temperate broad-leaved taxa. The grasslands appeared during the late Miocene (8–6.5 Ma), a change related to uplift and intensified monsoonal rainfall. C4 plant-dominated biomass existed during the early Pleistocene period. The recent studies by Dhir et al., (1998) on well-developed deep calcretes, mostly hardpan type, on rocky pediments of sandstone and schists of Proterozoic age in the eastern margin of the Thar desert (e.g., Anwana, Didwana) (Fig. 1) and on Jurrasic sediments in the western margin indicate seasonality of rainfall. These calcretes have recently been dated by the ESR method (Kailath et al., 2000) and cover a time range from the Neogene to early Pleistocene. Stable isotope values ($\delta^{18}O$ and $\delta^{13}C$ on carbonates) show dominance of C3 type of plant biomass thriving in a warm humid climate (Dhir et al., 1998). This calcretized and duricrusted landscape of central and western Thar was also drained by braided stream channels as indicated by calcretized as well as ferricretized pebbly gravels in the southwestern part of Jaisalmer district. These have been designated as Shumar Formation and are assignable to the Neogene-Pleistocene (Achyuthan, 1998). Around Bikaner, one observes thick gravel and sandstones, partially calcretized which have yielded a large number of fossil woods of *Lagerstroemia speciosa, L. parviflora, Ougeinia, oojeinensis and Dialium* (Guleria, 1990). Guleria (1990) and Ganjoo *et al.*, (1984) have also ascribed these fluvial sediments to the Neogene. An occurrence of such fossil, free species in the present arid part of Bikaner, definitely speaks in favour of a warm humid climate during the Neogene (Guleria, 1990).

Recently, Tandon *et al.*, (1999) and Jain *et al.*, (1999) have studied a 10 m thick exposed fluvial section near Sindhari village (Fig. 1) in the Luni valley, about 30 km downstream of the town of Balotra. The sequence is dominated by braided bar gravels and sandstones and near channel flood plain silts and clays affected by pedogenesis leading to calcisol and calcic vertisol development. The luminescence dating was carried out on fluvial sediments, and tentatively the Sindhari sequence has now been ascribed to >400 ka or even to the Pliocene period. Presence of abundant smectite clays and stable isotope values ($\delta^{18}O$ 6.8% and $\delta^{13}C$ 6.16%) of pedogenic carbonates associated with these sediments indicate mixed C3 and C4 plant biomass within a semi-arid climate, which was wetter than the present one, probably because of the longer duration of the summer monsoon (Jain *et al.*, 1999).

On the whole, the development of thick calcretized regolith over rocky pediments, occurrence of ferricretized and calcretized braided channels with well-developed flood plains, dominance of C3 plant biomass and the presence of deciduous trees in a grass-dominated vegetation cover in the central and western part of the Thar suggest a relatively strong summer yet semi-arid monsoonic type climate during the Neogene and the early Pleistocene.

This suggests that the Neogene/early Pleistocene environment was significantly different from the present water-deficient landscape showing poor regolithic cover, shrubby degraded thorny vegetation and ephemeral streams. So far, there is no evidence of the existence of Early Man during this period.

Middle Pleistocene Environment and Evidences for Prehistoric Sites

Well-developed nodular calcretes and loamy and clayey sediments of fluvial and fluviolacustral origin represent the middle Pleistocene formations. The aeolian sands, (particularly as fossil obstruction dunes) also represent the terminal phase of the middle Pleistocene period. A good number of Palaeolithic sites—Early Acheulian to Late Acheulian and the middle Palaeolithic have been excavated, particularly around Didwana in the eastern margin of the Thar desert (Fig. 1) (Gaillard *et al.*, 1983; Raghavan *et al.*, 1989; Misra and Rajaguru, 1989).

An Early Acheulian site of Singi Talav near Didwana is associated with ancient playa, while Late Acheulian and middle Palaeolithic artefacts have been recovered from the mottled grayish clayey loams which grade into 2–3 m nodular calcrete in the upper post of the profile. None of these sites could be precisely dated by absolute dating methods. On the otherhand, middle Palaeolithic and Acheulian artefacts found on a moderately pedogenised surface on fossil dunes at an 16R location around Didwana have been assigned ages by Th-Ur series dating on calcrete nodules. Uncorrected Th-Ur series dates indicate a date of about 400 ka for Acheulian artefacts and a date of around 150 ka for middle Palaeolithic artefacts (Raghavan *et al.*, 1989). Additionally, the luminescence dates on Aeolian sands indicate ages of around 200 ka and 125 ka for Acheulian and middle Palaeolithic artefacts respectively (Singhvi *et al.*, 2001). Kailath *et al.*, (2000) have dated some of the well-developed nodular calcretes by ESR method and these tentative dated range in ages from early middle Pleistocene (~600 ka) to around late middle Pleistocene (~200 ka).

Dhir (1995) and Dhir *et al.*, (1998) conducted fairly detailed field, petrographic (including cathod luminescence and electron microprobe and XRD) and stable isotope studies on nodular calcretes in the Thar. These studies show that the nodular calcretes are associated with sheet floods and with hydromorphic soils. Further they have also been affected by seasonal ground water fluctuations. The presence of clay minerals like palygorskite (Singhvi *et al.*, 2001) and attapulgita (Achyuthan and Rajaguru, 1998) point towards the prevalence of an arid to semi-arid environment. The stable isotope studies also corroborate this view, as they show dominance of C4 plant biomass. In comparison with the Neogene and early Pleistocene environment, the middle Pleistocene environment was distinctly arid to semi-arid, with the dominance of a low energy drainage system, favouring occasional sheet floods. Pools and playas could have existed in deflated depressions and groundwater remained close to the valley pediment surface, with considerable fluctuations. Dunes might have developed in geomorphic niches, particularly in the footslopes of rocky hills. Early Man camped on the shores of playa, on stabilized dunes and also on the surface of gravel ridges of Neogene age (Misra, 1995).

The Neogene landscape, as represented by gravel ridges and calcrete plains with relief inversion, is grossly a relict one and hence a misfit in the arid and semi-arid environmental setup of the Thar. The middle Pleistocene regolithic flat is inset into the Neogence landscape and to a certain extent, it is in conformity with the arid environment of the Thar.

Late Pleistocene Environment and Palaeolithic Sites

As compared to the middle Pleistocene chronology, the late Pleistocene lithological units like aeolian sand sheet, dunes and associated pedogenic carbonates from the Thar have been well dated by varieties of luminescence dating techniques in the Physical Research laboratory, Ahmedabad.

One of the best-studied dune profiles (20 m deep) is located in the vicinity of Didwana (Misra and Rajaguru, 1989; Raghavan, 1987 Raghavan *et al.*, 1989).

Systematic excavations over three seasons have yielded Upper Palaeolithic, Middle Palaeolithic and Lower Palaeolithic artefacts in excellent stratigraphic contexts. The entire profile has been thoroughly investigated in the laboratory including techniques of micromorphology, XRD, SEM and stable isotope on soil carbonates (Raghavan, 1987). As mentioned in the previous section, aeolian accumulation occurred at 200 ka, and it continued with interruptions around 125–100 ka, 60 ka and 35 ka, as indicated by calcisol development over sand dunes during these phases of weaker sand accumulation. A non-diagnostic blade flake on quartzite was found within hard crystalline, calcrete nodules (5–7 cm diameter) at a depth of 18.5 m below the surface. TL date on aeolian sand is ~200 ka, while Th-Ur series uncorrected date on associated calcrete nodules is 400 ka. Such discrepancies in absolute dates/ages by two different methods caution us against uncritical acceptance of isolated absolute dates. Further detailed geomorphological and high resolution geochronological studies are necessary to get a sound time frame for aeolian sand accumulation in the eastern margin and human occupation on dune surfaces.

The luminescence ages on the Didwana profile, subsequent to 100 ka very well match with other profiles, particularly at Amarsar (Fig. 2) near Jaipur (Chaugaonkar *et al.*, 1999), Chamu near Jodhpur (Singhvi, *et al.*, 2001) and Dhori Mana near Barmer. All these dates/ages indicate extensive aeolian sand accumulation around 100 ka in favourable geomorphic inches like foothill slopes and pediments, far off the effective stream activity. The latter is indicated by lenticular calcrete rich pebbly sandy gravels commonly

Fig. 2. Late Quaternary formations exposed around Amarsar.

phase in the Thar. The climate was, however, semi-arid as indicated by the presence of palygorskite in calcrete nodules and of easily weatherable minerals like amphiboles in palaeosols developed over dune sands at Didwana (Raghavan *et al.*, 1989). Interesting new data as based on stable isotopic studies of soil carbonate nodules developed in dune sands at Shergarh trijunction in district Jodhpur (Fig. 1) has thrown light on climatic conditions between 70 ka and 45 ka (Andrews *et al.*, 1998). Such studies indicate that the summer monsoon was fairly active, similar to the present one, between 60 ka and 40 ka. The palaeosol (reddish calcisol) developed over reworked aeolian sands located in a cliff section exposed on the right bank of Luni near Kudala also indicates a relatively wetter monsoonic climate during this time slice as compared to the preceding and succeeding phase of the late Quaternary in the Thar (Kar *et al.*, 2001). The Luni river near the town of Balotra has also preserved similar short palaeoclimatic events. During this regional wet climatic phase spread over Rajasthan and Gujarat in western India, there is, however, no convincing evidence of human occupation in the region, save Didwana. This point needs to be investigated in future.

Another wet climatic phase is indicated by the development of calcisols on the aeolian sands at Didwana (around 35 ka) and by fluvial gravel capping the Aeolian sand at Manwara near Balotra. At Didwana, a rich assemblage of Upper Palaeolithic artefacts along with a few miniature handaxes have been recovered. At Manwara, a good number of flakes have been collected from the gravel, which has been TL dated to around 26 ka (S. Mishra: Personal Communication). These two Palaeolithic sites appear to be the isolated examples of human occupation during the Last-glacial period, not only in the Thar but also in arid and, semi-arid parts of Gujarat.

The climate during the Last Glacial Maximum in the Thar was extremely dry with a weak

observed within the Luni basin. A few TL dates and occurrences of reworked middle Palaeolithic and Late Acheulian artefacts tend to suggest early late Pleilstocene age for these gravels.

The transitional climatic phase from the relatively warm and wet climate of the Last Interglacial (as indicated by palaeosol developed over dune sands in the fossil dune profile of Didwana) to a dry climate at the beginning of the Last Glaciation (~80 ka approximately) was geomorphologically sensitive. An epiosodic and largely ephemeral stream activity is indicated by sheet gravels.

Contemporary to the fluvial activity, sand dunes were growing in size in foothill regions. Palaeolithic Man was certainly present during this Interglacial-Glacial transitional climatic summer monsoon (Allchin *et al.,* 1978; Wasson *et al.,* 1983; Andrews *et al.,* 1998). Geomorphologically, there is some corroborating evidence for the deposition of aeolian sand, particularly in the southern margin of the Thar (Juyal *et al.,* 2000). The aeolian dynamism was weak due to the absence of strong winds related to the summer monsoon. The presence of hyper saline playas at Didwana (dated to around 18 ka) and of a saline playa at Bap-Malar (dated to around >15 ka during the later phase of LGM indicate that there were some winter rains, at least in the northwest part of the Thar (Singh *et al.,* 1990; Deotare *et al.,* 1999; Kajale *et al.,* 2000; Deotare *et al.,* 2001). The cloud cover in the Thar during the LGM might have been for a longer duration, covering summer and winter monsoonal months. The PET may have been considerably reduced and the playa could therefore retain some water most probably saline to hyper saline. So far, no archaeological sites have been detected either around the playas or in the southern margin where evidence for active dunes has been noted. The vegetational cover was primarily steppe as indicated by the presence of pollen such as those recovered from playa sediments at Didwana (Singh *et al.,* 1990). During the end phase of the late Pleistocene (spanning 16 ka to 12–11 ka,) there is evidence of climatic change due to the gradual strengthening of the summer monsoonal winds, aeolian accretion was rapid for a limited time slice (1000–2000 yrs) and extensive dune development took place all over the Thar, particularly in a time bracket of 14 ka – 12 ka (Chawla *et al.,* 1992; Thomas *et al.,* 1999; Juyal *et al.,* 2000). The Luni river also emerged out of a nearly defunct phase of the LGM and started depositing gravels during episodic strong flood (Tandon *et al.,* 1999). Owing to an overall rise in groundwater and the surface flow of water, the playas like Didwana in the eastern margin and Bap-Malar and Kanod in the western margin possibly turned into shallow fluctuating saline water bodies. The pollen analysis of gypsum rich sediments of Bap-Malar and Kanod shows that the vegetation continued to be dominated by Gramineae, Cyperaceae, Chenopodiaceae/Amranthaceae, Salvadoraceae, etc., not substantially different from what is observed today (Kajale *et al.,* 1999; Kajale *et al.,* 2000; Deotare *et al.,* 2001).

In summary, the late Pleistocene in the Thar shows overall aridity in the area with short spells of wet periods, yet within the bracket of semi-aridity. The overall fluvial system became ineffective and aeolian activity remained an important environmental parameter throughout the late Pleistocene. Pedogenesis was also weak and C4 plant biomass became an important aspect of the vegetational environment. The Thar, however, experienced a drastic environmental change since the LGM. Winter rains appeared in the climatic system and the summer monsoons became stronger around 15–16 ka. Southwest winds generated a dynamic aeolian system, which blanketed the major part of the Thar with a variety of dunes and sandsheets. Though playas appeared on the landscape, they remained essentially saline. The Thar landscape remained inhospitable and harsh for human activity during the end of the Pleistocene. On the contratry, the same inhospitable landscape developed into a slightly better vegetated one with adequate semi-saline/freshwater and stream channels with water pools in braided courses. This change in the environment occurred during the Holocene, around 8–5 ka.

Holocene Environment and Archaeological Sites Belonging to Later Cultural Periods (Mesolithic to Late Historic)

In the Kashmir valley, the process of deglaciation started somewhat early around 18 ka as compared to the rest of India (Agrawal and Yadav, 1998). The Thar, however, has belatedly registered changes in the behaviour of the summer monsoon in response to global glacial and Inter-glacial climatic changes. It shows marked changes in landscape and lacustral environment only during the end Pleistocene and the early Holocene.

The playas that existed intermittently over the entire Thar, have been investigated by various scholars (Singh *et al.*, 1974; Kajale and Deotare, 1995, 1997; Kajale *et al.*, 1999 & 2000., Deotare *et al.*, 1998 and 2001; Wadhawan and Sharma, 1997; Deshmukh and Rai, 1991; Rai, 1990; Rai and Sinha, 1990; Rai and Absar, 1996; Wasson *et al.*, 1983 & 1984) in the last thirty years. However, only four playas, like Didwana (in the eastern semi-arid margin with a present annual rainfall of about 350 mm), Lunkaransar (in the north central part on the boundary between semi-arid and arid with a mean annual rainfall of about 250 mm), Bap-Malar and Kanod (Figs. 3-5) (in the western arid core with a mean annual rainfall around 200–150 mm) have been investigated geochronologically, geomorphologically, palynologically and archaeologically (Singh *et al.*, 1974, 1990; Deotare and Kajale, 1996; Kajale and Deotare, 1997; Enzel *et al.*, 1999; Kajale *et al.*, 1999 & 2000; Deotare *et al.*, 2001). These relatively well investigated playas reveal a marked change in the interplay of summer and winter monsoon though individual lakes vary in their response due to local factors. The arid core playas remained lakefull (perennial) and yet semi-saline between 8 ka–7 ka and dried up around 5.5 ka. The Lunkaransar playa was lakefull (perennial) with a fresh water phase between 6 ka and 5 ka and dried up completely around 4.8 ka. Didwana lake remained perennially full with a fresh water phase between 6 ka and 4 ka and suffered dessication around 3.5 ka

It appears that these playas remained lakefull so long as both summer and winter rains were effective, particularly during the early Holocene and part of early-middle Holocene (i.e., approximately between 7 and 4 ka). For reasons not yet understood properly, the winter rains became ineffective and the summer rains marginally weak after 4 ka. As a result, most of the playas became ephemeral and dry with saltish encrustation. The vegetation in the eastern and central part show predominance of salt tolerant species, grasscover and intermittent trees belonging to *Acacia, Prosopis, Zizyphus,* etc.

In the arid western part of the Thar, grass and sedge dominated perhaps even during the early Holocene. Stable isotopic analysis of organic remains from Bap-Malar and Kanod show dominance of C4 plant biomass during the early-middle Holocene (Kajale *et al.,* 1999 & 2000; Deotare *et al.*, 2001).

Fig. 3. General view of Trench W9 (Bap-Malar Playa) showing predominantly sandy layers.

Fig. 4. General view of the Kanod Playa showing pediment surface in the foreground and interspersed vegetational thickets of Salvadora oleoides.

Though playas in the Thar have registered Holocene climatic changes fairly well, the dunes show a complex response to climatic fluctuations. The aeolian activity almost came to an end around 10 ka in the extreme southern margin of the Thar (Juyal *et al.*, 2000) while it continued with lesser intensity than the earlier one until 5 ka in the other part of the Thar. Moderate aeolian activity has been registered around 3 ka, 2 ka, 0.7 ka and 0.3 ka (Thomas *et al.*, 1999). In fact, the Thar has no significant natural aeolian accretion in the late Holocene, despite relative increase in the aridity. The fluvial response to the Holocene climatic changes is not well registered. The Luni river shows gravel aggradation around 12–13 ka, incision in the gorge during 2.8 to 1 ka and alluvial fill aggradation with slack water flood deposits laid down during floods, at times catastrophic in the last 1000 years as seen from palaeohydrological records (Kale *et al.*, 2000). Ephemeral streams, which intermittently acted as feeders to playas in arid western parts have also preserved archaeological records of Early Historic and Medieval periods in gravels and flood silts respectively as observed at Bari Bavri near Phalodi (Deotare *et al.*, 1999). These feeder streams also remained ephemeral and largely

Fig. 5. General view of the Trench K4 in the Kanod Playa showing the gypsum-rich nature of the laminated predominantly clayey sediments.

inactive during middle and the late Holocene (Deotare *et al.*, 1999).

The environmental history as preserved in playa sediments, dune sands and ephemeral stream deposits provide a complex picture of the intricate environmental system. The details need to be unravelled by future multi-disciplinary high-resolution and quantitative palaeoclimatic investigations.

Early Man's response to Holocene environmental changes is equally complex. One observes rich scatters of microlithic tools (Fig. 6) throughout the Thar indicating high population

Fig. 6. Close view of the microlithic tools and cores on sand dunes at the site of Jamba, in the vicinity of Bap-Malar Playa.

density during the early-middle Holocene (6 ka–3 ka) (Misra, 1995) and ceramic tradition appears around 3.5 ka (Mishra *et al.*, 1999). The ceramics and a few isolated large-sized bricks appear around 1.5 ka in arid core of the Thar and the pastoral way of life continued all throughout the Holocene.

Conclusion

A brief reconnaissance of literature on the palaeoclimatic history of deserts in Asia, Africa, America, Australia, etc., suggests that the Thar environmental and cultural history is to some extent, comparable to that of the Sahara and Arabian regions. Hence discussions are made here with reference to some recent Egyptian studies.

A good deal of new geoarchaeological data has recently emerged on the Egyptian Sahara which has been one of the driest deserts (with rainfall < 10 mm/annum) in the world (Haynes, 2001). It appears that the Egyptian desert had changed from the present lifeless, hyper-arid environment to arid-semi-arid (pluvial) during the middle Pleistocene (~300 ka), the Last Interglacial (~125 ka) and the Early-middle Holocene (9–8 ka to 5 ka). In fact, the middle Pleistocene wet phase was stronger than the 'Neolithic Pluvial' (early to middle Holocene). The Acheulian and middle Palaeolithic hunter-gatherers and the Neolithic agro-pastoralists made optimum use of available surface waters in braided stream channels, playas, pans, oasis and springs. In contrast, the hyper aridity of the LGM seriously affected Early Man's activity in this part of Sahara.

Compared to the Egyptian Sahara, the semi-arid Thar did not experience any drastic climatic changes at least since the middle Pleistocene (~600 ka) as summarized in the accompanying Table 1. Inspite of relatively mild changes in climate, landscape response was fairly sensitive and it controlled Early Man's settlements in the Thar. Like in the Sahara, human activity was intense during the middle Pleistocene, the Last Interglacial and during the early to middle Holocene. It was minimal during the LGM. Since the early Holocene the Thar has been occupied by man uninterruptedly till today while the Saharan desert region lost its major significance for human settlement after the end of the Neolithic Pluvial.

Table 1: Resume of Landforms, Palaeomonsoon, Palaeoclimate and Human Cultures in central and western Thar desert.

Geological Period and Approx. Age	Geomorphic Features/ Parameters	Inferred Monsoon & Palaeoclimate	Culture
Neogene-Early Pleistocene (8 Ma–0.7 Ma)	Ferricretes, Hardpan sands, Calcretes, Fluvial gravels and sands	Strong sub-humid to semi-arid	Probably Man not present in the area
Middle Pleistocene (600 ka To ~ 200 ka)	Nodular calcretes, fluvial clays, silts, and sands obstacle dunes	Modest to weak, semi-arid to arid	Early to late Acheulian
Early late Pleistocene (125 ka to ~ 75 ka)	Fluvial gravels, calc pans, pedogenic calcretes reworked aeolian sands	Modest to weak, semi-arid to arid	Middle Palaeolithic
Late Pleistocene (~ 60 ka to ~ 25 ka)	Palaeosols, aeolian sands and pedogenic calcretes	Modest and semi-arid	Middle and upper Palaeolithic (Rare)
Latest Pleistocene including LGM (~ 22 ka to ~ 15 ka)	Weak aeolian activity, playas with evaporites, gravels, reworked aeolian sands	Weak summer monsoon, moderate winter rains, arid	Stray upper Palaeolithic
Pleistocene-Holocene transition (~ 14 ka to ~ 9 ka)	Strong aeolian sands, playas, wind activity strong	Moderate, semi-arid	Microlithic
Early to middle Holocene (~ 8 ka to ~5 ka)	Playas and moderate wind activity	Moderate summer and winter monsoons, semi-arid	Prolific Microlithic cultural spread
Late Holocene (~ 3 ka to 1.4 ka)	Ephemeral streams, dwindling playas, moderate to weak wind activity	Weak summer monsoon, declining winter monsoon	Microlithic and Early Historic

The peripheral regions of Thar desert are rich in wildlife.
(Photo by Takeshi Takeda)

Acknowledgements

The present brief review of research has been feasible because of the initial research support provided by the Deccan College Postgraduate and Research Institute and subsequently by the Department of Science and Technology, New Delhi through research project Nos. ESS/CA/A3-12/94 and ESS/CA/A3-8/92. We are, therefore, indebted to our institutional authorities as well as to DST, New Delhi for their invaluable support.

We are also grateful to our inter-institutional colleagues from Central Arid Zone Research Institute, Jodhpur; Physical Research Laboratory, Ahmedabad; Delhi University, Delhi whose published and unpublished work has been duly referred to in this summary paper. We regret any errors that may have occurred.

References

Achyuthan, H. (1998): Neogene Quaternary duricrusts of Jaisalmer basin, Rajasthan – Palaeoenvironmental significance, *DST Project completion Report* (ESS/CA/A3-10/93).

Achyuthan, H. and S.N. Rajaguru (1998): Micromorphology of Quaternary calcretes around Didwana in the Thar Desert, Rajasthan, *Annals of Arid Zone*, 37(1): 25–35.

Agrawal, D.P. and M.G. Yadav (1998): A comparison of proxy climatic data from Tibet and Kashmir, *Himalayan Geology*, 19(2): 55–59.

Allchin, B., A.S. Gaudie, and K.T.M. Hegde (1978): *The Prehistory and Palaeogeography of the Great Indian Desert*, Academic Press, London.

Andrew, J.E., A.K. Singhvi, A.J. Kailath, R. Kuhn, P.E. Denis, S.K. Tandon, and R.P. Dhir (1998): Do stable isotope data from calcrete record Late Pleistocene Monsoonal climate variation in the Thar desert of India? *Quaternary Research*, 50: 240–251.

Bajpai, V.N., T.K. Saha Roy and S.K. Tandon (2001): Subsurface sediment accumulation patterns and their relationship with tectonic lineaments in the semi-arid Luni river basin, Rajasthan, Western India. *Journal of Arid Environment*, 48: 603-621.

Chaugaonkar, M.P., K.S. Raghav, S.N. Rajaguru, A. Kar, A.K. Singhvi and K.V.S. Nambi (1999): Luminescence dating results of dune profiles from the margins of the Thar desert and their implications. *Man and Environment*, XXIV (1): 21-26.

Chawla S., R.P. Dhir, and A.K. Singhvi (1992): Thermoluminiscence chronology in sand profiles in the Thar Desert and their implications, *Quaternary Science Review*, 11: 25–32.

Deotare, B.C. and M.D. Kajale (1996): Quaternary pollen analysis and palaeoenvironmental studies on the salt basins at Pachpadra and Thob, western Rajasthan, India: Preliminary observations, *Man and Environment*, 21(1): 24–31.

Deotare, B.C., M.D. Kajale, S. Kshirsagar, A.A. and S.N. Rajaguru (1998): Geoarchaeological and palaeoenvironmental studies around Bap–Malar playa, District Jodhpur, Rajasthan, *Current Science*, 75 (3): 316–320.

Deotare, B.C., M.D. Kajale, S. Kusumgar, and S.N. Rajaguru (1999): Late Holocene environment and culture at Bari Bavri, western Rajasthan, India, *Man and Environment*, 24(1): 27–38.

Deotare, B.C., M.D. Kajale, S. Kusumgar, V.D. Gogte, R. Ramesh, S.N. Rajaguru, J.D. Donahue, and A.J.T. Jull (2001): Pleistocene-Holocene Climatic Changes in Arid Core of the Thar Desert: A Case Study of Kanod Playa, District Jaisalmer. Rajasthan. In Simon Kaner, Akira Matsui, Liliana Janik and P. Rowley-Conwy (Eds.): From *The Jomon to Star-Carr*, Oxford: Hadrian Books (British Archaeological Reports International Series.

Deshumukh, G.P. and V. Rai (1991): Some Observations on the Origin of saline lakes/Ranns of Rajasthan, India, *Proceedings of Quaternary Landscape of Indian Subcontinent,* Geology Department, M.S. University of Baroda, Baroda, pp. 41–47.

Dhir, R.P. (1992): Calcretes, In Singhvi, A.K. and Amal Kar (eds.), *Thar Desert: Man and Environment* 55–60, Bangalore: Geological Society of India: Vol. 43: 435–447.

Dhir, R.P. (1995): The genesis and distribution of arid zone calcretes, *Memoir Geological Society of India,* 32: 191–209.

Dhir, R.P., S.K. Tandon, S.N. Rajaguru, and R. Ramesh (1998): Calcretes: Their genesis and significance in palaeo-environmental reconstruction in arid Rajasthan, India. In Klauss Heine, A. Faure and A.K. Singhvi (eds.): *Palaeoecology of Africa and the surrounding Islands*, Balkema, pp. 223–230.

Enzel, Y., L. Ely, S. Mishra, R. Ramesh, R. Amit, B. Lazar, S.N. Rajaguru, V.R. Baker, and A. Sandier (1999): High resolution Holocene environmental changes in the Thar desert, Northwestern India, *Science*, 284: 125–128.

Fort, M. (1996): Late Cenozoic environmental changes and uplift on the northern side of the Central Himalaya: a reappraisal from field data, *Palaeogeography, Palaeoclimatology and Palaeoecology*, 120: 123–145.

Gaillard, C., D.R. Raju and S.N. Rajaguru (1983): Acheulian occupation at Singi Talav in the Thar desert: A preliminary report on 1982 excavation, *Man and Environment*, 7: 112–130.

Ganjoo, R.K., H. Raghavan, S.N. Rajaguru, and C. Gaillard (1984): Late Neogene fossil wood from the Bikaner gravel bed, *Current Science*, 53 (22): 1207–1208.

Guleria, J.S. (1989, issued 1990): Fossil Dicotyledonous Woods from Bikaner, Rajasthan, India, *Geophytology*, 19 (2): 182–188.

Haynes, V.C. (2001): Geochronology and climatic change of the Pleistocene-Holocene Transition in Darb el Arba in desert, *Geoarchaeology*, 16 (1): 119–141.

Hoom, C., T. Ohja, and J. Quade (2001): Palynological evidence for vegetation development and climatic change in the sub-Himalayan zone (Neogene-Central Himalaya), *Palaeogeography, Palaeoclimatology and Palaeoecology*, 16 (1): 133–161.

Jain, M., S.K. Tandon, S.C. Bhatt, A.K. Singhvi, and S. Mishra (1999): Alluvial and Aeolian sequences along the river Luni, Barmer district: Physical stratigraphy and feasibility of Luminiscence chronology methods, *Memoirs of Geological Society of India*, No. 42: 273–295.

Juyal, N., A. Kar, S.N. Rajaguru and A.K. Singhvi (2000): Chronostratigraphic evidence for episodes of desertification in the southern margin of Thar Desert, India, International Conference on *Desertification*, Dubai, 12–16 Feb. 2000: 195–197.

Kailath, A.J., T.K. G. Rao, R.P. Dhir, K.V.S. Nambi, V.D. Gogte, and A.K. Singhvi (2000): ESR characterization of calcrete from the Thar desert for dating application, *Radiation Measurements*, 32: 371–383.

Kajale, M.D. and B.C. Deotare (1995): Field observations and lithostratigraphy of three salt lake deposits in Indian desert of western Rajasthan, *Bulletin of Deccan College Postgraduate and Research Institute*, 53: 117–134.

Kajale, M.D. and B.C. Deotare (1997): Late Quaternary environmental studies on saline lakes in western Rajasthan: A summarized view, *Journal of Quaternary Science*, 12 (5): 405–412.

Kajale, M.D., B.C. Deotare, S. Kusumgar, A.J. Jull, T. Donahue, V.D. Gogte, S.N. Rajaguru, R. Ramesh, and A.K. Singhvi (1999): Quaternary palaeoenvironmental History of Bap-Malar saline Playa in the Thar Desert of western Rajasthan, Abstract in *Quaternary of India* (eds. Tiwar, M.P. and D.M. Mohabey), Gondwana Geological Magazine, Vol. 4.

Kajale, M.D., S.N. Rajaguru, B.C. Deotare, S. Kusumgar, A.K. Singhvi, R. Ramesh, V.D. Gogte, and N. Basavaiah (2000): Late Quaternary Climatic Changes from Indian Thar Desert: Evidence from Kanod and Bap-Malar Playas, Dubai International Conference on Desertification (Abstracts), 12–16 February 2000: 198–199.

Kale, V.S., A.K. Singhvi, P.K. Mishra, and D. Banerjee (2000): Sedimentary records and luminescence chronology of Late Holocene palaeoflood in the Luni river, Thar Desert, NW India, *Catena*, 40: 337–358.

Kar, A. (1999): Climatic and neotectonics controls on drainage evolution in the Thar. In Paliwal, B.S. (ed.): *Geological evolution of Northwestern India*, Scientific Publishers, Jodhpur, pp. 246–259.

Kar, A., A.K. Singhvi, S.N. Rajaguru, N. Juyal, J.V. Thomas, D. Banerjee, and R.P. Dhir (2001): Reconstruction of the Late Quaternary environment in the lower Luni plains, Thar Desert, *Journal of Quaternary Science*, 16 (1): 61–68.

Kutzbach, J.E., P.J. Guetter, W.F. Ruddiman, and W.L. Prell (1989): Sensitivity of climate to Late Cenozoic uplift in Southern Asia and the American West: Numerical experiments, *Journal of Geophysical Research*, 94, D15: 18: 393–18,407.

Mishra, S., M. Jain, S.K. Tandon, A.K. Singhvi, P.P. Joglekar, S.C. Bhatt, A.A. Kshirsagar, S. Naik and A. Deshpande-Mushergee (1999): Prehistoric cultures and Late Quaternary environment in the Luni basin around Balotra, *Man and Environment*, 24 (1): 39–50.

Misra, V.N. (1995): Geoarchaeology of the Thar Desert, northwest India, In Statira Wadia, Ravi Korisetter and Vishwas Kale (eds.): *Memoirs of Geological Society of India*, No.32, Bangalore, pp. 210–230.

Misra, V.N. and S.N. Rajaguru (1989): Palaeoenvironments and Prehistory of the Thar desert, Rajasthan, India. In Frifelt K. and R. Sorensen (eds.): *South Asian Archaeology* 1985, Scandinavian Institute of Asian Studies Occasional papers no. 4, Copenhagen.

Quaide, J., T.E. Cerling, and J.R. Bownan (1989): Development of Asian monsoon revealed by marked ecological shift during the latest Miocene in Northern Pakistan, *Nature*, 342: 163–166.

Raghavan, H. (1987): *Quaternary Geology of Nagaur district, Rajasthan*, Unpublished Ph. D. thesis, Pune University.

Raghavan, H., S.N. Rajaguru, and V.N. Misra (1989): Radiometric dating of Quaternary dune section, *Man and Environment*, 13: 19–22.

Rai, V., (1990): Facies Analysis and Depositional Environment of Pokran saline Rann, District Jaisalmer, Rajasthan, India, *Journal of Geological Society of India*, vol. 36: 317–322.

Rai V. and A.K. Sinha (1990): Geological Evolution of Kuchaman lake, district Nagaur, Rajasthan, *Journal of the Palaeontological Society of India*, 35: 137–142.

Rai, V. and A. Absar (1996): Sub-Lacustrine Hydrothermal activity in Kuchaman and Sargot saline lakes, district Nagaur, Rajasthan. In *Geothermal Energy in India*, (eds.): U.L. Pitale and R.N. Padhi GSI special Publication, No. 45, 361–366.

Sikka, D.R. (1997): Desert climate and dynamics, *Current Science*, 72 (1).

Singh, G., R.D. Joshi, S.K. Chopra, and A.B. Singh (1974): Late Quaternary History of Vegetation and Climate of the Rajasthan Desert, India, *Philosophical Transactions of the Royal Society of London*, B: Biological Sciences, Vol. 267 (889): 467–501.

Singh, G., R.J. Wasson, and D.P. Agrawal (1990): Vegetational and Seasonal Climatic Changes since the Last Full Glacial in the

Thar Desert, northwestern India, *Review of Palaeobotany and Palynology*, 64: 351–358.

Singhvi, A.K., S.K. Tandon and R.P. Dhir (2001): Quaternary Stratigraphy and Palaeo-environmental History of the Thar desert, *DST Completion Report (ESS/CA/A3-08/92), Physical Research Laboratory, Ahmedabad.

Tandon, S.K., M. Jain, and A.K. Singhvi (1999): Comparative development of mid to late Quaternary fluvial and fluvio-aeolian stratigraphy in the Luni, Sabarmati and Mahi river basins of Western India, *Gondwana Geological Magazine* special volume 4: 1–16.

Thomas, J.V., A. Kar, A.J. Kailath, N. Juyal, S.N. Rajaguru, and A.K. Singhvi (1999): Late Pleistocene-Holocene history of Aeolian accumulation in the Thar desert, *Zeitschrift fur Geomorphologie*, 166: 1–4.

Wadhawan, S.K. and H.S. Sharma (1997): Quaternary stratigraphy and morphology of desert ranns and evaporite pans in central Rajasthan, India, *Man and Environment*, 22 (2): 1–10.

Wasson, R.J., S.N. Rajaguru, V.N. Misra, D.P. Agrawal, R.P. Dhir, A.K. Singhvi, and K.K. Rao (1983): Geomorphology, late Quaternary stratigraphy and palaeoclimatology of the dune field, *Zeitschrift Geomorphologie*, suppl. 45: 117–151.

Wasson, R.J., G. Singh, and D.P. Agrawal (1984): Late Quaternary sediments, minerals and inferred geochemical history of Didwana lake, Thar Desert, India, *Palaeogeography, Palaeoclimatology and Palaeoecology,* 46: 345–372.

Xu Qunqi Qi (1993): The first appearance of the Himalayas and its relation to global climatic events. *In Evolving landscapes and evolving biotas of East Asia since the Mid-Tertiary*, University of Hongkong, Hongkong.

Chapter 6

Human Occupation and Environmental Changes since the last 130,000 years in Mainland Gujarat, Western India

ANIRUDDHA S. KHADKIKAR AND K. KRISHNAN

Introduction

Human societies have always relied on their natural environment for subsistence, although the degree of dependence has changed with time as man has learned to adapt more to his surroundings. The effect of climate on society is more clearly seen in studies dealing with the past 5000 years (e.g., Wigley *et al.,* 1981). This relationship between climate and society is aptly illustrated in the case of the Egyptian and Andean Civilizations (Hassan, 1997; Binford *et al.,* 1997). A similar climate-culture dependency was explored as a cause for the rise and demise of the Harappan Civilization, which was reviewed by Possehl (1997). Recent work in Rajasthan (Enzel *et al.,* 1999) has renewed this debate with their results indicating no apparent connection between climate change and the 'eclipse' of the Harappan Civilization. Very few earlier studies have demonstrated a definitive climate-culture relationship for longer time scales in the Indian context although attempts have been made (Zeuner, 1950; Allchin *et al.,* 1978). One of the main obstacles for making such comparisons is a lack of adequate chronology. Recently, there have been an effort to date the older levels of the stratigraphic record of Gujarat (Khadkikar *et al.,* 1999; Juyal *et al.,* 2000), which has resulted in a robust stratigraphic and chronological framework which permits comparisons with the archaeological record of the region. This is the exercise that has been attempted here, the results of which demonstrate a clear relationship between the periods of an intensified monsoon and human occupation in Mainland Gujarat.

Regional Setting

Mainland Gujarat forms a part of the state of Gujarat, western India and is bound towards the north and northeast by the Aravalli mountains. Its western boundary is marked by the saline marshes of the Little Rann of Kutch while towards the south it is delimited by the Gulf of Cambay (Fig. 1). Throughout Mainland Gujarat are several westward and southward bound rivers of which two major ones, the Mahi and the Sabarmati transect this region in a roughly north-south direction to flow into the Gulf of Cambay. These rivers have been cut into during the Holocene resulting in vertical cliff-like cutbanks, which allow a glimpse into the archaeological and environmental record of this region.

Today, Mainland Gujarat receives a rainfall of 500–700 mm with the drier regions lying towards the north and northwest. A progressive decrease in the rainfall isohyets is observed from the Mahi river region towards the northern reaches of the Sabarmati river. This rainfall distribution pattern is also seen in the regional occurrence of modern vertisols which generally form in the sub-humid to humid climatic zones (Fig. 2) (Coulombe *et al.,* 1996).

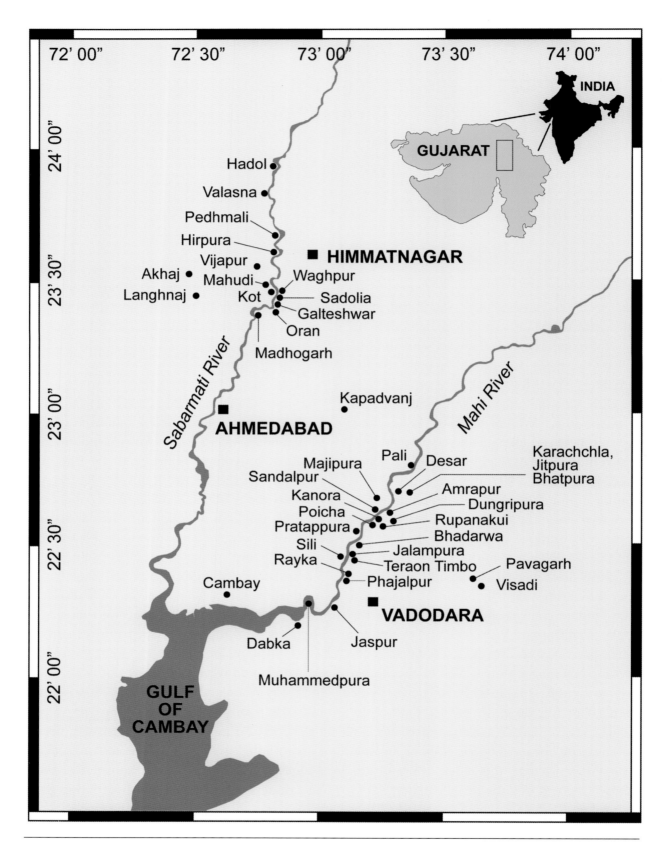

Fig. 1. Location map of the study area showing the region commonly referred to as Mainland Gujarat. All the sites mentioned in the text are marked on the map.

100

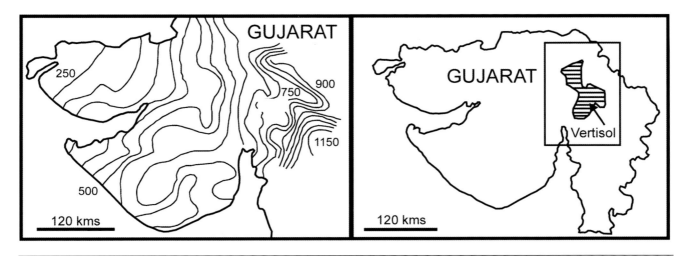

Fig. 2. Comparison with modern rainfall isohyet of Gujarat and the spatial distribution of vertisol to illustrate the close correspondence with regions experiencing around 700 mm mean annual rainfall.

All sedimentological processes in Mainland Gujarat are governed by the vagaries of the Southwest Indian Monsoon. The Indian summer monsoon's origin lies in the land-sea temperature gradient that is generated by differential heating of the ocean and land masses (particularly the Tibetan Plateau). The summer monsoon spans a period of two to three months and sets in usually by the middle of June each year. The winter monsoon does not yield any rain in this region. The Mahi and Sabarmati river basins are adjacent to each other and experience similar flood discharge variations inter-annually (Khadkikar, 1999a). Hence during peak discharges (which are geomorphologically and sedimentologically significant events) both the Mahi and Sabarmati are flooded. This implies that in principle both river basins may be viewed as one geomorphic entity responding similarly to the changes of the Southwest Indian Monsoon on longer time scales. Consequently it is correct to relate the stratigraphic record of the Mahi to the Sabarmati river region.

This region was first studied in the late 19th century (Foote, 1898) which demonstrated the richness of the archaeological record. Subsequently in the 1940s a detailed archaeological survey was undertaken (Sankalia, 1946) in the Sabarmati region that documented the presence of Lower Palaeolithic and Microlithic implements in this region. A similar survey for the Mahi region (Subbarao, 1952) also yielded Lower Palaeolithic and Microlithic records albeit not as profuse as those in the Sabarmati region. For a prolonged period this region remained unattractive to geologists until a joint study was undertaken by England and India to understand its environmental and archaeological history (Allchin & Goudie, 1971; Allchin et al., 1978). Apart from this seminal work other studies have resulted in a crisper archaeological history of Mainland Gujarat (Mehta & Sonawane, 1970; Momin, 1982; Misra & Pandya, 1989; Pandya et al., 1990). The simultaneous sustained interest of geologists has resulted in numerous stratigraphic, sedimentological and geochronological studies that continue till today (Wasson et al., 1983; Chamyal & Merh, 1992; Merh & Chamyal, 1997; Tandon et al., 1997; Malik et al., 1999; Khadkikar et al., 1999; Juyal et al., 2000; Khadkikar et al., 2000).

Lithostratigraphy and Archaeology

The lithostratigraphic record of Mainland Gujarat (Fig. 3) is represented by a sediment

101

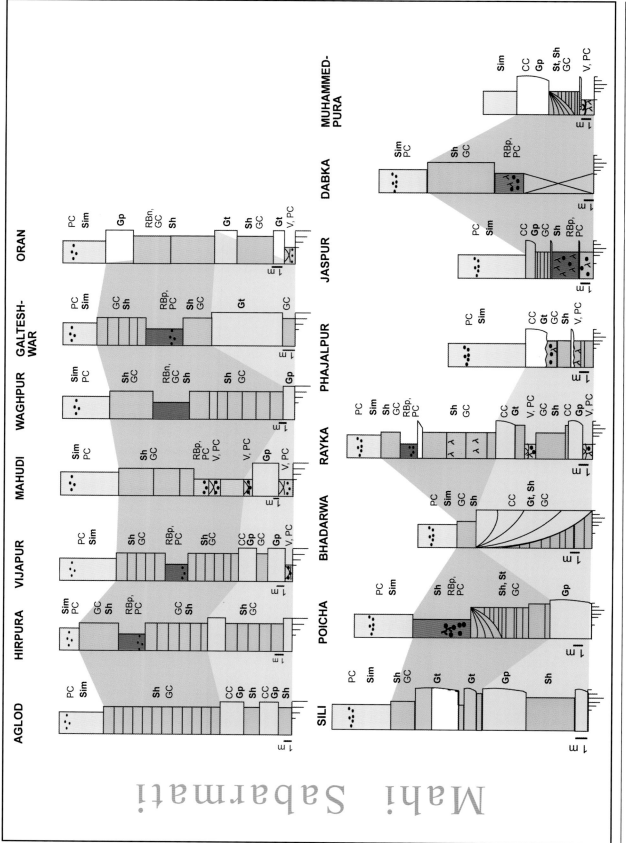

Fig. 3. Lithostratigraphy of the Mahi and Sabarmati regions (modified after Khadkikar et al., 1999). GC = Groundwater calcrete, CC = Calcrete conglomerate, V = Vertisol, RBp = Red Bed pedogenic, RBn = Red Bed non-pedogenic. PC = Pedogenic calcrete, CC = Calcrete conglomerate. For explanations for clastic facies codes and their interpretation, refer Table I.

succession 20–35 m in thickness. Very broadly, the lithostratigraphy has a three-tiered structure comprising three suites of facies representing discrete depositional environments. These have been termed Aggradation Phase 1, 2 and 3 in chronologically and stratigraphically ascending order (Khadkikar *et al.*, 1999; 2000). A comparison with archaeological records shows a correspondence between various phases of cultural development from the prehistoric period. Hence we have chosen to compartmentalize and describe the lithostratigraphy and archaeology under the same three-tiered framework. All the sites discussed herein are derived from published sources of information. For this initial assessment of the archaeological record we have not delved into the typology of the reported finds but have taken the typological and cultural classifications and interpretations by the respective investigators at face value.

Another limitation of this exercise is that archaeologists and geologists have seldom worked together with a common objective in this region. Due to this, the sites that archaeologists have studied in greater detail have not been assessed with similar rigour by geologists and vice versa. Hence comparisons between the two records (viz. lithostratigraphic and archaeological) can only be made at the broadest level i.e., presence/absence of anthropogenic activity in the region through time.

Aggradation Phase 1/Lower Palaeolithic

This phase of aggradation covers the stratigraphically lowermost deposits as exposed along the river cutbanks. The deposits consist of both sand and gravel/conglomerate (Table 1), the latter showing planar cross-bedding and trough cross-bedding (Fig. 4A). This coarser sediment is

Table 1: Facies codes and their interpretation (Khadkikar *et al.,* 1999)

Facies Code	Description	Interpretation
Gp.	Single as well as multi storey, normal graded, 1 to 4–5 m thick, planar cross-stratified with dips of 15° to 24°, clast size up to 15 cm in maximum diameter.	Aggradation along the avalanching slip-faces of a mid-channel bar during bankfull flows.
Gt	Multi-storey normal graded, 0.5–3 m thick, trough cross-stratified with dips of 18°, clast size up to 14 cm in maximum diameter. Calcrete clasts dominant with clasts of quartzite, basalt and laterite.	Downstream migrating trains of sinuous-crested dunes, bedload sheets in channels.
St	Trough stratified sands, width up to 8 m and thickness between 0.45 to 1.5 m.	Shallow channel forming through avulsions within the river valley.
Sh	Horizontally stratified sands, laterally extensive, usually 3 m - 4.5 m thick.	Sheetflood deposits that formed through episodic infrequent flash-flood.
Sim	Massive, unstratified silts, texturally homogenous with calcareous nodules. Geomorphic expression as vertical bluffs.	Aeolian sandy loess deposits.

Fig. 4. Trough cross-bedded calcrete rich gravels associated which overlie calcic-vertisols of Aggradation Phase 1. The location is Rayka in the Mahi river basin (Fig. 4A).

Calcic-vertisols showing pseudo-anticlines at the Rayka site in the Mahi basin. Such soils are widespread across the region at similar stratigraphic levels (Fig. 4B).

Distinctive Red-Bed marker horizon that is prominent across both the Sabarmati and Mahi basins and is seen here at Dabka (Mahi basin) bisecting the section. Also seen overlying the Red-Bed are horizontally bedded sands of Sh facies which belongs to Aggradation Phase 2 (Fig. 4C).

Capping sandy loesses of Aggradation Phase 3, which are seen as steep vertical bluffs throughout the region. Here at Mahudi, Sabarmati basin, the thickness of the deposits reaches up to 8 m (Fig. 4D).

seen in both the Sabarmati and Mahi river regions but is dissimilar in composition, the former showing the presence of laterite whereas the latter has a predominance of calcrete and quartzite. The sandier facies contain groundwater

calcretes (Khadkikar *et al.,* 2000) that impart a laterally extensive banding to the deposits. Stratigraphically these deposits are up to 5–6 m thick. Associated with these clastic sediments are dark-coloured calcic vertisols (Fig. 4B) that display a range of features including oppositely directed concave-up curviplanes, fissuring, pedogenic slickensides, rhizoliths and a profusion of pedogenic calcretes. Mineralogically the vertisols are dominated by montomorillonite

(Malik *et al.,* 1999) but also contain illite and kaolinite. The calcrete nodules are larger in size and reach up to 10 cm in diameter and are rich in calcium carbonate.

The facies suite (Table 1) suggests that the river that laid these sediments was braided in morphology owing to the presence of planar cross-bedded conglomerates which are remnants of mid-channel bar accretion. Evidence of peak discharges is seen in the distribution of trough cross-bedded conglomerates/gravels that led to the formation of subaqueous gravel dunes (Khadkikar, 1999b). Associated calcic-vertisols testify to seasonality in the hydrological cycle and infrequent inundation of riverbanks.

This phase of aggradation has been dated using luminescence-based and electron spin resonance techniques (Khadkikar *et al.,* 1999; Juyal *et al.,* 2000). These results suggest an age bracket between 130,000 to 80,000 BP.

The gravels/conglomerates of this phase have yielded Lower Palaeolithic tools from both the Mahi and Sabarmati regions. In the Sabarmati a total of six sites have been discovered (Sankalia, 1946) whereas in the Mahi four sites are reported (Subbarao, 1952). The different types of Palaeolithic tools recovered from these regions are handaxes, cleavers, scrapers, choppers, flakes, pebble tools and chipped pebbles, the largest number being found in the Sabarmati region. From its typological description it appears that the tool kit along Mainland Gujarat consisted of inferior and superior types (Sankalia, 1946). Pedhmali is the most important Palaeolithic site on the Sabarmati. This is located almost midway on the central reaches. Hadol lies on its north and Ghadhara, Hirpura and Warsora in the south. Despite the presence of a good number of handaxes in different localities on the Sabarmati, split pebbles and pebble tools are most common. Typo-technological analysis of the stone tool assemblage indicated that they represent a mixture of Abbevillian and Acheulian techniques. At Pedhmali, which has yielded a large number of tools (188) this feature is very evident. As the inferior and superior types of tools could not be linked with any specific stratigraphic levels, no technological evolution of tools was inferred by Sankalia (1946, 1974). The Palaeolithic industry in Mahi was similar in character to that of the Sabarmati region as it also consisted of tools belonging entirely to the Abbevillio-Acheulian industry, mixed with a few pebble choppers and flakes. By taking into account the presence of this variety of tools, Subbarao (1952) proposed that this area could have been a meeting ground of the Sohan and Bi-face industry.

Aggradation Phase 2/Middle-Upper Palaeolithic

In this phase, the stratigraphic succession becomes more sand-dominated throughout the entire region. These sandy deposits show internal planar bedding at places whereas in areas proximal to the mountains (e.g., in the Sabarmati region) the deposits show planar cross-bedded facies. Within the sediments, laterally extensive groundwater calcretes are seen along bedding planes (Khadkikar *et al.,* 2000), which impart rigidity to these sediments. Based on the distribution of sedimentary facies, these deposits were interpreted by Khadkikar *et al.,* (1999) to be remnants of ephemeral streams such as the Luni river, which exists north of Mainland Gujarat. The streams deposited sediments primarily during infrequent flash floods that produced the laterally extensive horizontal bedded sand facies.

Within these sediments, a thick well-developed to moderately developed red-bed is usually seen that is both pedogenic and non-pedogenic in origin (Fig. 4C). The pedogenic variant of red-beds shows formation of sub-centimetre sized calcretes and rhizoliths (Khadkikar *et al.,* 2000). The red-bed when pedogenic in origin shows affinity to calcic-alfisols (Khadkikar *et al.,* 2000).

Since these sediments are stratigraphically concordant with the gravels of Aggradation Phase 1 which have been dated to 80 ka BP (Juyal et al., 2000), we use this as the lower age limit of Aggradation Phase 2. The red-bed event has been bracketed between 50–23 ka BP based on luminescence ages of the reddened sediments (Tandon et al., 1997; Juyal et al., 2000; Prasad & Gupta, 2000) and radiocarbon ages (Allchin et al., 1978) on the calcrete nodules. Consequently Aggradation Phase 2 is bracketed between ~80 ka BP to 23–20 ka BP.

The archaeological record of these sediments is patchy if not non-existent. There have been no reports of implements from the red-bed horizon till date. The Middle and Upper Palaeolithic are poorly represented with one site each; Kapadvanj for the former and Visadi for the latter (Fig. 5). Both these sites present problems in relating to the stratigraphic succession discussed in this paper as they are either surface finds (middle Palaeolithic of Kapadvanj) of disputed nature or from obstacle dunes (Visadi) whose stratigraphic position is uncertain. Nonetheless, the record from the Middle to Upper Palaeolithic is seen as a blatant omission, a lacuna in the archaeological history of this region.

Aggradation Phase 3/Microlithic

The uppermost part of the deposits in this phase is remarkably different from the underlying sediments (Fig. 4D). These deposits are massive i.e., they contain no bedding that is seen at the meso-scale. This massive disposition imparts a geomorphic tendency to occur as vertical bluffs. Granulometric analyses on several samples across the region (Malik et al., 1999; unpublished data) have shown them to be sandy silts. Groundwater calcretes are absent from these deposits (Khadkikar et al., 2000). Pedogenic calcretes occur at some locations where they are sub-centimetre in size and have diffuse margins.

The sandy silts are similar to sandy loesses that are commonly observed at desert margins (Pye, 1989). These deposits form due to sediment entrainment and suspension fallout from desert winds known as 'haboobs' or 'dust devils'. Such winds reworked the underlying flash-flood sediments of Aggradation Phase 2 to form a subdued landscape. A regional water deficit is seen in the absence of groundwater calcretes in these deposits. Available luminescence dates on these sediments suggest that wind activity commenced at 22000–20000 yrs. BP (Wasson et al., 1983) and continued till 8000 yrs. BP (Tandon et al., 1997).

The Microlithic is well represented in the region across both the Mahi and Sabarmati (Fig. 5) river regions (Sankalia 1946; Subbarao 1952; Mehta & Sonawane 1970; Sankalia 1974; Allchin et al., 1978; Momin 1982; Mishra & Pandya 1989; Pandya et al., 1990). Most of the information available on the microliths is from surface surveys except that of Langhnaj, Hirpura, Kanewal, Dhansura and Loteshwar where excavations have been carried out (Sankalia 1946; Sankalia 1974; Momin 1982, Sonawane & Ajithprasad 1994). A large variety of tools have been reported from these different excavations, which indicates that there was considerable activity in this region during this phase. The presence or absence of bones along with these microliths is also important. The microliths include lunates or crescents, triangles, semi-triangles, trapezes, long or short blades, roundish, rectangular or square scrapers, tiny disc-like pieces or core trimmings, and cores either roundish or cylindrical. These lithic assemblages could not be divided into geometric and non-geometric to provide information about the evolution of this period as is done in European Prehistory (Sankalia, 1946: p. 133). Many of the sites have also yielded potsherds. These are neither numerous or distinctive in any way. A large number of bone splinters have also been recovered from the

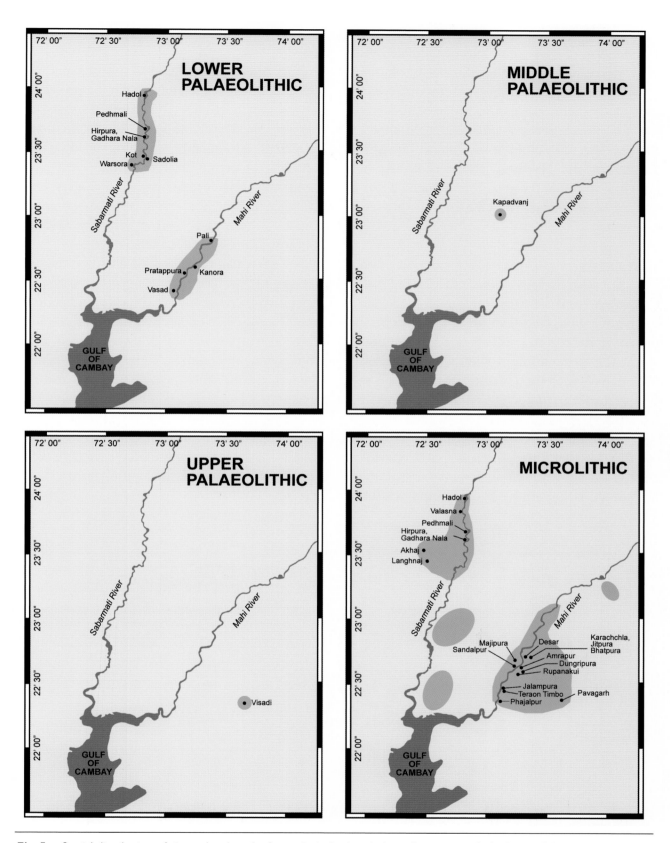

Fig. 5. Spatial distribution of sites related to the four principal cultural phases beginning with the Lower Palaeolithic and ending with the Microlithic in Mainland Gujarat. Oval regions demarcate clusters of Microlithic sites as compiled by Momin (1982).

107

excavations at Hirpura and Langhnaj. Examination of these bone pieces indicated that these were deliberately chipped by human agencies for various purposes. Different types of gastropod shells have also been discovered. When all the evidence is put together it appears that the Microlithic phase of Gujarat has a few features of the Neolithic and Mesolithic, but applying any of these terms which have a cultural connotation to refer to this phase is incorrect.

Discussion

A comparison of the archaeological and sedimentological record suggests the presence of humans in the region during Aggradation Phase 1 and at the termination of Aggradation Phase 3. No activity is seen for a prospective horizon—the red-bed of Aggradation Phase 2. This could be due to the fact that this interval has not been closely examined by archaeologists. Here we examine whether there were any environmental controls on the temporal and spatial distribution of human occupation in Mainland Gujarat (Fig. 6).

■ Temporal Distribution

During Aggradation Phase 1 the climates were sub-humid which is manifested in the presence of calcic-vertisols. Rivers were deeply incised and the banks were stable. Floodplains were inundated infrequently which resulted in deep soil development. Calcic-vertisols form under woodlands and co-existent minor grasslands. Evidence for such a mixed vegetation biomass comes from the stable carbon isotopic composition of calcretes (Khadkikar et al., 2000) which shows a restricted distribution between $-7‰$ and $-5‰$. These values point to a 65% C3 vegetation biomass (which consists dominantly of trees). It may then be inferred that food resources were in surplus although there is no direct evidence for the faunal composition of Mainland Gujarat during this phase. But the evidence suggests that overall the physical and ecological environment was conducive to human habitation. A similar evidence of an intensified monsoon is seen in the high resolution Guliya ice core record (Thompson et al., 1997) from the Tibetan Plateau (Fig. 6).

During Aggradation Phase 2 the climate was

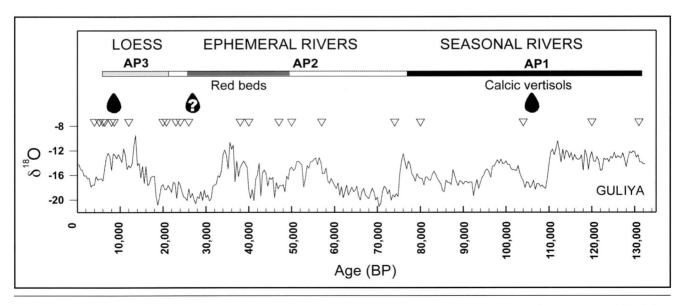

Fig. 6. The history of the Southwest Indian Monsoon over the past 130,000 years as seen in the high-resolution record of the Guliya ice core with which are shown the three depositional phases and the distribution of archaeological tools through time.

semi-arid. Ephemerality of rivers is not conducive to human habitation in two ways: the deficiency in the water budget of such regions and the flooding intensity of ephemeral channels. Both conditions are extreme. Little soil development is observed which may well represent a drastic reduction in biomass during this phase. The red-bed event represents an event of climatic amelioration with the pedogenic variety forming under a tropical savanna kind of landscape (Khadkikar *et al.*, 2000). Stable isotopic values of pedogenic calcretes show a range between – 7‰ to +2‰ reflecting a more spatially variable biomass dominated by grasslands with pockets of wooded vegetation. We think that this stratigraphic level has the prospect of yielding implements, which can only be verified by future detailed studies.

During Aggradation Phase 3 the climate became very arid which is also reflected in the absence of archaeological tools in these sediments. At the upper levels of this phase, evidence of human occupation is seen all across the region in the distribution of surface microlithic sites as well as in the sediments at some sites which have been excavated (e.g., Langhnaj). This is seen to be in tandem with climatic amelioration at the beginning of the Holocene around 10–8 ka BP.

■ **Spatial Distribution**

There is a broad similarity in the zones of occupation during the Microlithic and Lower Palaeolithic phase. The middle reaches of the Sabarmati along with the upper reaches of the Mahi show a population of Lower Palaeolithic sites although the number of sites is less in both the basins. This may be directly related to the human population or could be an artefact resulting due to non-homogenous distribution of excavation sites. The Microlithic is very well represented in this region with the entire Mainland Gujarat region showing

pockets/clusters of sites. One population is seen along the Sabarmati river in the upper reaches, another in the lower reaches between the Sabarmati and Mahi rivers and the third along the lower reaches of the Mahi river. The increase in the number of sites seems to be directly related to an increase in population with respect to the Lower Palaeolithic period. The spatial distribution during the Lower Palaeolithic seems to be governed by the 700-mm mean annual rainfall isohyet (manifested in the distribution of calcic-vertisols), a behaviour mimicked by Microlithic site distribution.

■ **Long-term Human Response to Climate Change**

An increasing number of studies have documented the response of human societies to climate change (Wigley *et.al.*, 1981). This has been demonstrated in the case of Andean Civilizations (Binford *et al.*, 1997), and is debated for the Harappan Civilization (Possehl 1997, Enzel *et al.*, 1999). However very few studies have examined the response of human populations to the last Inter-glacial cycle. Our comparisons between the archaeological and environmental record of Mainland Gujarat suggest the presence of a distinctive control of climate on human migratory behaviour over the past 130,000 years. This is modulated through a control on the biomass and the nature of the physical environment. Through time there is also a tendency towards climatic adaptation which is seen in the distribution of Microlithic sites in the more semi-arid regions, as opposed to the limited distribution during the Lower Palaeolithic period.

Conclusion

The environmental and archaeological record of Mainland Gujarat shows a close interrelationship between human occupation and climate. Climate influenced both the physical and ecological environment. This would mean changes in water resources, natural hazards, vegetation biomass and food resource availability. The next logical step is to examine whether environmental variables had any role whatsoever to play in the technological evolution of tools, a facet of tool evolution contemplated by Zeuner (1950). Hence, it is imperative that an integrated approach be adopted to gain a better understanding of the prehistory of this region and a crisper view of the relationship between human activity and environmental change. Such an approach would essentially involve mapping of the palaeogeographic distribution of man during different time periods, understanding the reasons for barren stratigraphic intervals that too in a congenial environmental setting and appraising the nature of human adaptation to the vagaries of the Indian Monsoon in the Holocene.

References

Allchin, B. & A. Goudie (1971): Dunes, aridity and early man in Gujarat, western India. *Man*, 6: 248–265.

Allchin, B., A. Goudie, & K.T.M. Hegde (1978): *The Prehistory and Palaeogeography of the Great Indian Desert*. Academic Press, London, 370 pp.

Binford, M.W., A.L. Kolata, M. Brenner, J.W. Janusek, M.T. Seddon, M. Abbot & J.H. Curtis (1997): Climate variation and the rise and fall of an Andean Civilization. *Quaternary Research*, 47: 235–248.

Chamyal, L.S. & S.S. Merh (1992): Sequence stratigraphy of the surface Quaternary deposits in the semi-arid basins of Gujarat. *Man and Environment*, 17: 33–40.

Coulombe, C.E., C.P. Wilding & J.B. Dixon (1996): Overview of vertisols: characteristics and impacts on society. *Advances in Agronomy*, 57: 289–375.

Enzel, Y., L.L. Ely, S. Mishra, R. Ramesh, R. Amit, B. Lazar, S.N. Rajaguru, V.R. Baker, & A. Sandler (1999): High resolution Holocene environmental changes in the Thar Desert, northwestern India. *Science*, 284: 25–128.

Foote, B.R. (1898): *Geology of Baroda State*. Government Press, Baroda.

Hassan, F.A. (1997): The dynamics of a riverine civilization: a geoarchaeological perspective on the Nile Valley, Egypt. *World Archaeology*, 29(1): 51–74.

Juyal, N., R. Rachna, D.M. Maurya, L.S. Chamyal & A.K. Singhvi (2000): Chronology of Late Pleistocene environmental changes in the lower Mahi basin, western India. *Journal of Quaternary Science*: 15: 501–508.

Khadkikar, A.S. (1999a): *Sedimentation-calcrete cycles in the late Quaternary deposits of Mainland Gujarat, western India*. Unpublished Ph.D. Thesis, M. S. University of Baroda.

Khadkikar, A.S. (1999b): Trough cross bedded conglomerate facies. *Sedimentary Geology*, 128: 39–49.

Khadkikar, A.S., George Mathew, J.N. Malik, T.K. Gundu Rao M.P. Chowgaonkar & S.S. Merh (1999): The influence of Southwest Indian monsoon on continental deposition over the past 130 kyr. Gujarat, western India. *Terra Nova*, 11(6): 273–277.

Khadkikar, A.S., L.S. Chamyal & R. Ramesh (2000): The character and genesis of calcrete in Late Quaternary alluvial deposits, Gujarat, western India, and its bearing on the interpretation of ancient climates. *Palaeogeography, Palaeoclimatology, Palaeoecology* 162: 239–261.

Malik, J.N., A.S. Khadkikar & S.S. Merh (1999) Allogenic control on late Quaternary continental sedimentation in the Mahi river basin, western India. *Journal of the Geological Society of India*, 53: 299–314.

Mehta, R.N. & V.H. Sonawane (1970): Explorations in the Daskroi Taluka, Dist. Ahmedabad. *The Journal of The M.S. University of Baroda*, 19 (1): 7–14.

Merh. S.S. & L.S. Chamyal (1997): *The Quaternary Geology of Gujarat Alluvial Plains*. Indian National Science Academy, New Delhi, 98 pp.

Misra, V.N. & S. Pandya (1989): Mesolithic occupation around Dhansuri, district Sabarkantha, Gujarat: A preliminary study. *Man and Environment*, 14(1): 123–127.

Momin, K.N. (1982): Mesolithic settlements in Central Gujarat. *Prachya Pratibha*, 10: 91–97.

Pandya, S.H., M.V. Desai & S.B. Makwana (1990): Excavations at Dhansura I: A Preliminary Announcement. *Man and Environment*, 15(2): 39–44.

Possehl, G.L. (1997): Climate and the eclipse of the ancient Indian cities of the Indus. In H. Nüzhet Dalfes, G. Kukla & H. Weiss (Eds.): *Third Millennium BC Climate Change and Old World Collapse*, Series 1, vol. 49, NATO ASI Series, Springer Verlag, Berlin, pp. 193–243.

Prasad, S. & S.K. Gupta (2000): Role of eustasy, climate and tectonics in Late Quaternary evolution of the Nal-Cambay region, NW India. *Zeitschrift für Geomorphologie*, 43: 438–504.

Pye, K. (1989): *Aeolian dust and dust deposits*. Academic Press, London, 334 pp.

Sankalia, H.D. (1946): *Investigations into the Prehistoric Archaeology of Gujarat*. Sri-Pratapasimha maharaja Rajyabhisheka Grantha-mala, Memoir No. 4, Baroda: Baroda State Press.

Sankalia, H.D. (1974): *The Prehistory and Protohistory of India and*

Pakistan. Deccan College Postgraduate and Research Institute, Poona.

Sonawane, V.H. & P. Ajithprasad (1994): Harappa culture and Gujarat. *Man and Environment*, 19(1/2): 129–139.

Subbarao, B. (1952): Archaeological explorations in the Mahi valley. *Journal of the M.S. University of Baroda*, 1: 33–74.

Tandon, S.K., B.K. Sareen, M. Someshwar Rao, & A.K. Singhvi (1997): Aggradation history and luminescence chronology of Late Quaternary semi-arid sequences of the Sabarmati basin, Gujarat, Western India. *Palaeogeography, Palaeoclimatology, Palaeoecology*, 125: 230–253.

Thompson, L.G., T. Yao, M.E. Davis, K.A. Henderson, E.M. Thompson, P.N. Lin, J. Beer, H.A. Synal, J. Cole-Dai & J.F. Bolzan (1997): Tropical climate instability: The last glacial cycle from Qinghai-Tibetan ice core, *Science*, 276: 1821–1825.

Wasson, R.J., S.N. Rajaguru, V.N. Misra, D.P. Agrawal, R.P. Dhir, A.K. Singhvi, & K.K. Rao (1983): Geomorphology, late Quaternary stratigraphy and palaeoclimatology of the Thar dunefield. *Zeitschrift für Geomorphologie*, 45: 117–151.

Wigley, T.M.L., M.J. Ingram & G. Farmer (1981): *Climate and History: Studies in past climates and their impact on man*. Cambridge University Press, Cambridge.

Zeuner, F.E. (1950): Stone age Pleistocene chronology of Gujarat. *Deccan College Monograph Series*, 6: 46.

Beautiful flowers cover the land during the monsoon season of the Indian monsoon region. (Photo by Takeshi Takeda)

A woman carrying water in Rajasthan, India, located on the northern periphery of the Indian monsoon region. (Photo by Takeshi Takeda)

Chapter 7

Holocene Adaptations of the Mesolithic and Chalcolithic Settlements in North Gujarat

P. AJITHPRASAD

Introduction

Northwest India comprising the states of Gujarat and Rajasthan shows several climatic zones, arid to sub-humid, as one moves progressively southward. North Gujarat falls in the arid and semi-arid climatic zones as it lies between the 310 mm and 690 mm isohyets (Singh *et al.,* 1991). Located at the southern fringe of the Thar Desert, the region is particularly sensitive to monsoon induced climatic changes (Fig. 1). The present desert margin, which coincides with the 250 mm isohyet, had extended several kilometres further south up to Baroda in the past, resulting in the formation of several active sand dunes (Goudie *et al.,* 1973), the latest of which occurred during the Last Glacial Maximum, around 18 ka. Most of these sand dunes became stabilized during the Post-glacial period, in the Holocene climate. With the onset of increased monsoonal precipitation in the Holocene, north Gujarat became more hospitable for prehistoric communities to settle down in. Recent archaeological investigations in north Gujarat have revealed several sites of the Mesolithic hunter-gatherers and Chalcolithic food producers belonging to the middle and late Holocene. The discovery of a number of Mesolithic sites on the stabilized surface of the sand dunes clearly indicates that the Mesolithic occupation in north Gujarat post-dates dune stabilization (Goudie *et al.,* 1973). As many of the inter-dunal depressions retained water from monsoon runoff for most of the year, it was now possible not only for Mesolithic hunter-gatherers but also Early Chalcolithic farming communities to make use of the resources available in the monsoonal environment.

One of the important features of the monsoon is its cyclic nature. In response to this, environmental resources also undergo a cyclic decline and rejuvenation. The cyclic nature of environmental reproduction had a profound impact on past human settlements irrespective of their economic status. In fact, the prehistoric economic activities as well as patterns of settlement in north Gujarat during the middle and late Holocene reflect adaptive strategies of past human societies in response to a challenging monsoonal environment. Although the overall climate does not register major changes during the Holocene, variations in the monsoonal precipitation influencing human habitation at different times during this period have been recorded in northwest India (Singh *et al.,* 1990). Therefore, understanding past and present environmental settings in a micro-level is essential for a better understanding of the middle and late Holocene human adaptations in north Gujarat.

Physiography and Environmental Settings

Gujarat comprises three major traditional geo-cultural regions: Kutch, Saurashtra or Kathiawad and Mainland Gujarat (Majumdar, 1960). The

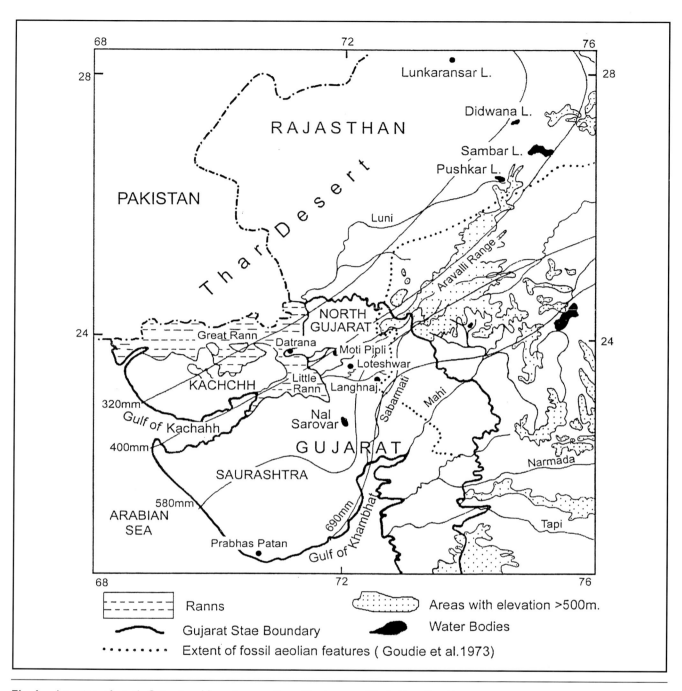

Fig. I. Location of north Gujarat and key sites mentioned in the text. (Kachchh = Kutch)

vast rocky plateau surrounded in the north and east by the Great and the Little Ranns of Kutch and bordering Pakistan in the north is Kutch (Fig. 1). The peninsular region lying south of Kutch and surrounded by the Arabian Sea is Saurashtra/Kathiawad. The fertile alluvial plain lying south of the Sabarmati river and flanked in the east by the Aravalli and the Satpura ranges and drained by the Mahi, the Narmada and the Tapi rivers is Mainland Gujarat. Within this division, north Gujarat is a narrow strip of land that connects Kutch with the mainland of Gujarat. This narrow region is traditionally known by the name 'Anarta' (Majumdar, 1960).

116

It is bounded in the west by the Great and the Little Ranns of Kutch. The two Ranns are vast low-lying flat salt-waste, which retain a thin sheet of water during the monsoon and for a few months thereafter. Today, for most of the year, it is a dissolute flatland with a blistering encrustation of salt on the surface. The Ranns play a significant role in the environmental setting of both present and past human settlements in this region.

North Gujarat presents a relatively flat but slightly undulating landscape with a series of fossil sand dunes extending from the foot of the Aravalli ranges in the northeast and gradually merging with the plains of Saurashtra and central Gujarat in the southwest. Here, at its southwestern margin, is located a large lake, the Nal Sarovar (Fig. 1). The Nal Sarovar (27°45'N, 72°6'E) is a shallow, brackish water lake which is situated about 60 km southwest of Ahmedabad. It is 30 km long and 6 km wide, extending over an area of 180 km². It is an important sanctuary for migratory birds and lies in the low alluvial tract separating north Gujarat from Saurashtra and connects the southeast extremity of the Little Rann of Kutch with the head of the Gulf of Khambat. Major rivers that drain this region are the Rupen, the Saraswati, the Banas and the Sabarmati and their tributaries. All these rivers flow down into the Little Rann except the Sabarmati, which drains into the Gulf of Khambat. The Banas and the Saraswati, which flow in the northern part, are heavily silted and have a broad and shallow channel. The Rupen, on the other hand, is more deeply entrenched in the alluvium and shows little evidence of silting. Although they carry a considerable volume of water during the monsoon, none of them are perennial rivers.

The climate in north Gujarat is characterized by hot summers and cold winters. Located in the arid and semi-arid climatic zones the region today receives very low rainfall ranging from 310 mm to 690 mm (Fig. 1). Since most of this is received during the monsoon, the climate is more or less dry in the greater part of the year. The rains are not only sparse but also erratic and therefore recurrent draughts are very common in this region. During the summer, water in the rivers is confined to a few depressions and puddles in the channel. However, an important feature of the region as far as the water sources are concerned is a number of inter-dunal depressions which, with a little augmentation, today, are used as village tanks. Many of these shallow depressions support open pastoral land in their vicinity soon after the monsoon and are the mainstay for livestock, and pastoral communities.

Soils in north Gujarat are predominantly sandy. Nevertheless, a few stretches of fertile black cotton soil are also found in the alluvial plain. These are extensively used today for agricultural purposes. In many places the soil is poor and saline and the subsoil water is brackish. Such areas have in the course of time afforded good pastoral land. All these environmental factors have a considerable influence on the subsistence practices of the present-day population in north Gujarat, especially in the cultivation of millet—*Bajra*—(Pearl millet; *Pennisetum typhoides*) and cotton, as these two do not need much moisture (Patel, 1977). Vast pastoral grassland and fallowland support a large number of cattle and sheep and therefore pastoralism and agriculture are presently the two important components of the rural economy of north Gujarat.

Palaeoenvironments

The reconstruction of the Holocene environment of prehistoric settlements in north Gujarat has, until very recently, been based on palaeoclimatic records retrieved from the lake sediments of Rajasthan. Palynological and geochemical studies of sediments from Lunkaransar, Didwana, Sambar and Pushkar lakes in Rajasthan have shown that there have been several instances of significant variation in the relative abundance of rainfall at different times during the Holocene. For instance, the Didwana lake showed oscillating fresh and saline conditions from 9.3 ka to 7.5 ka. The lake level culminated in the interval 7.5–6.2 ka indicating a wetter condition, which was followed by moderately fresh water conditions between 6.2–4.2 ka (Wasson *et al.,* 1984; Singh *et al.,* 1990). Between 10.5 and 3.5 ka the estimated precipitation was approximately 40% higher than the present (Swain *et al.,* 1983). From 4.2 ka to the present only ephemeral lakes have persisted in Rajasthan (Wasson *et al.,* 1984; Singh *et al.,* 1990). Palynological data indicated increased salinity at 3.7 ka (Singh 1971; Singh *et al.,* 1972). The above climatic interpretations have broadly been used in the past for understanding the environmental background of prehistoric cultural developments in western India including Gujarat (Singh, 1971; Brayson and Swain, 1981).

However, very recently, sediments from Nal Sarovar at the southwestern margin of north Gujarat have also been studied for palaeoclimatic reconstruction (Sharma and Chauhan, 1991;

Prasad *et al.,* 1997; Pandarinath *et al.,* 1999). A high-resolution record extending back to 6.6 ka BP could be reconstructed by Prasad *et al.,* in 1997 from the core samples from Nal Sarovar showing palaeoenvironmental data slightly different from that of the Rajasthan lake cores. This is significant because Nal Sarovar is located at the southwest margin of north Gujarat, whereas the Rajasthan lakes are located hundreds of kilometres north of Gujarat. Nal lies in a region where the southwest monsoon is the most important factor that influences climatic records. Hence, climatic interpretations from Nal sediments are more relevant in understanding the late Holocene environments of north Gujarat. A brief summary of palaeoclimatic reconstruction of the above study in conjunction with similar studies from Rajasthan lake sediments is reproduced here in Fig. 2.

The palaeoclimatic records for the period 6.6–4.8 ka from both the lakes in Rajasthan showed a higher annual rainfall. Nal Sarovar data for the same period shows a shallow lake level, with periodic drying and short wet spells. The

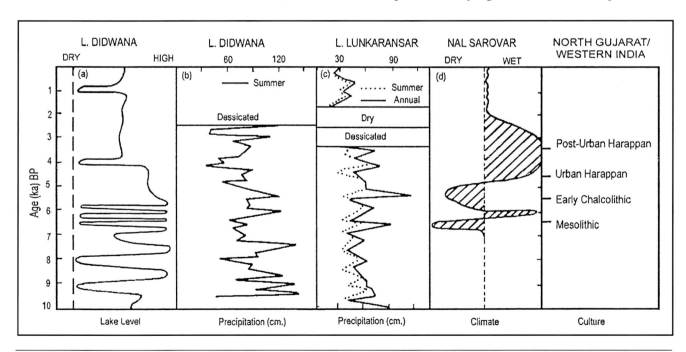

Fig. 2. Palaeoclimatic data from Rajasthan lakes and Nal Sarovar, and prehistoric cultural developments (Adapted from Prasad *et al.,* 1997). (a) Wasson *et al.,* (1984); (b and c) Brayson and Swain (1981).

Mesolithic culture in north Gujarat has been dated to this period. Nal Sarovar data indicates a wetter climate than the present in the succeeding period, from 4.8 ka to 3 ka. This is comparable with the Rajasthan precipitation data, which also indicated a higher than present rainfall. The period of Early Chalcolithic as well as the Harappan culture in north Gujarat coincides with this wet period. Termination of this wet phase in the Rajasthan lakes has been dated to about 3.5 ka to 4 ka. Whereas the Nal Sarovar data indicates the beginning of aridity about 3 ka, the deterioration of the climate may have set in a couple of centuries earlier. Data from both the regions shows the onset of present-day conditions around 2 ka (Prasad *et al.,* 1997).

Middle and Late Holocene Prehistoric Settlements

Having discussed the environmental setting and the late Holocene climatic features of north Gujarat, we may now look at the archaeological records of prehistoric settlements. As has been mentioned earlier, prehistoric settlements of the Holocene period in this region begin with the Mesolithic culture and are followed by the Chalcolithic (Bhan, 1994; Sonawane, 2000). The beginning of the Mesolithic period in this region has been dated to 4700 BC, whereas the Chalcolithic period is dated from 3500 BC to 1500 BC (Table 1). The Chalcolithic settlements are generally confined to the estuaries of the Rupen, the Banas and the Saraswati rivers and the narrow creek-like depression that connects the Little and the Great Ranns. The Mesolithic sites are more or less evenly distributed with a slightly higher concentration along the river channels towards the rugged northeastern part.

The Mesolithic Sites

More than twenty-six sites belonging to this cultural period are found in north Gujarat (Fig. 3). The number of sites may increase considerably if the poorly explored north and northeastern part is explored thoroughly. The sites are located on the stabilized surface of fossil sand dunes indicating that the Mesolithic occupation post-dates stabilizations of the dunes. Besides, microlithic artefacts have been reported along with Chalcolithic assemblage from twelve sites in this region. Some confusion still remains whether

Table 1. Prehistoric cultural sequence and chronology of north Gujarat in the Holocene

Culture	Date	Site	^{14}C dates (cal. yrs. BC)
Harappan			
Post-Urban Harappan	1900–1500 BC		—
Urban Harappan	2500–1900 BC	Nagwada	2133
Early Chalcolithic			
Early Harappan (Sindh related)	<2800 BC		—
Pre-Prabhas	2900 BC	Prabhas Patan	2892, 2911
Anarta	3500 BC	Loteshwar	2921, 3698
Mesolithic	4700 BC	Loteshwar	4750, 4744
		Langhnaj	2452

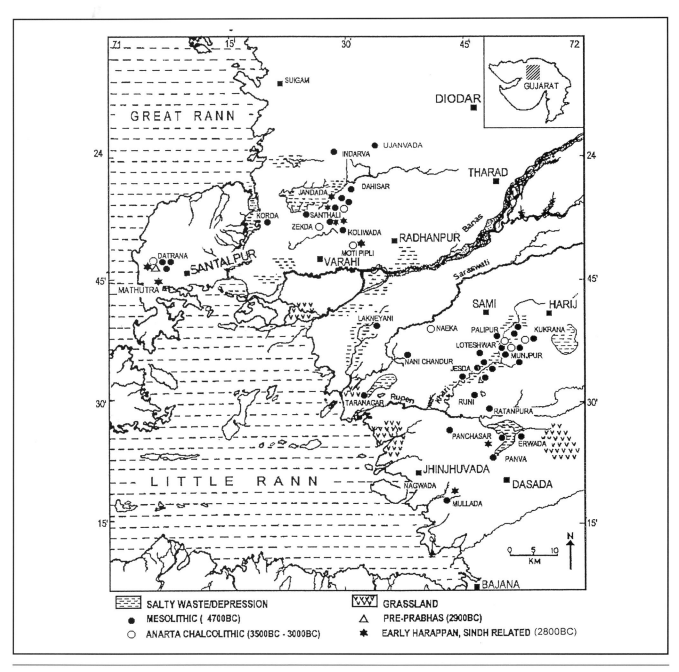

Fig. 3. Mesolithic and Early Chalcolithic settlements in north Gujarat.

these microliths are an integral part of the Chalcolithic assemblage or belong to an earlier Mesolithic period occupation. Stratigraphically, Mesolithic deposits are found lying below the Chalcolithic habitation deposit at Loteshwar, Datrana, Moti Pipli and Ratanpura (Table 2). Radiocarbon dates of the Mesolithic settlements in north Gujarat show a wide range, the earliest

being 4750 ka BC from Loteshwar and the latest, 2452 ka BC (Possehl, 1993), from Langhnaj in Mehsana district (Table 1). The Mesolithic hunting and gathering way of life, therefore, appears to have continued to flourish for several centuries even after the introduction of the Chalcolithic way of life in the region. Hence, some of the sites may be either contemporary

120

Table 2. Chronology and stratigraphy of Mesolithic and Chalcolithic assemblages
in some of the important sites in north Gujarat.

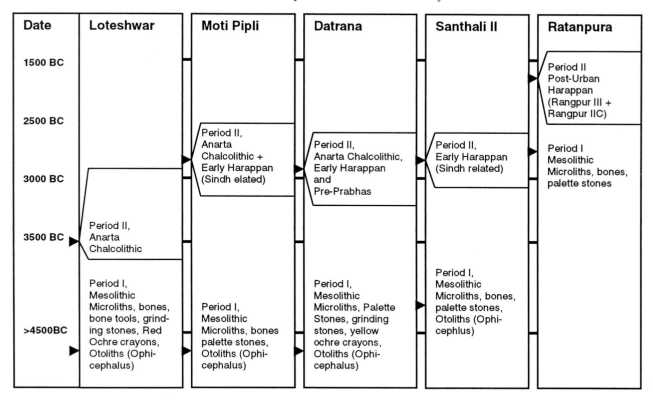

with or even later to the Harappans. The presence of pottery and copper implements in some of the late Mesolithic sites, for instance at Langhnaj (Sankalia, 1965) may be due to this cultural dynamism.

Mesolithic sites are generally located on the top of stabilized fossil sand dunes situated either along the course of river channels or close to natural depressions that retain water. The sites are identified by surface spread of lithic artefacts and skeletal remains of animals exploited by the community. Barring a few, most of the Mesolithic sites are small in size falling in the range of 0.25 to 0.3 ha (Table 3).

In many of the large sites, there is more than one distinct cluster of artefacts spread on the surface, probably suggesting the seasonal occupation history of the site. More often, especially in the monsoonal climate, seasonal migration from or occupation at a particular site or region by hunter-gatherers is generally an articulation of adaptive ways for exploiting seasonally available environmental resources (Barnard, 1979). It becomes clear from the

Table 3. Distribution of prehistoric settlements in north Gujarat based on their size

Sites	Size (ha)								Total
	0.01-< 0.25	0.25-< 0.50	0.5-< 1.0	1.00-< 1.50	1.5-< 2.0	2.0-< 2.50	2.5-< 3.0	> 3	
Chalcolithic	28	23	19	12	6	2	3	8	101
Mesolithic	13	7	3	1	-	-	-	2	26

palaeoclimatic records of the Nal Sarovar that precipitation was not particularly higher when Mesolithic settlements flourished in north Gujarat during the middle and late Holocene. It was, in fact, predominantly dry, punctuated with regular short wet spells. The microlithic assemblages representing Mesolithic culture should, therefore, be viewed in the backdrop of a monsoonal environment swinging periodically between wet and dry phases for a period of about 2000 years.

Apart from the usual microlithic artefacts of both geometric and non-geometric type, the Mesolithic assemblage included several 'palette-stones', hammers and anvils (Fig. 4). The assemblage included bone points and a few ground and faceted red and yellow ochre crayons. Faunal remains collected from many of the sites are predominated by skeletal remains of wild mammalian fauna with a few fish remains and mollusc shells. Loteshwar and a few sites in its

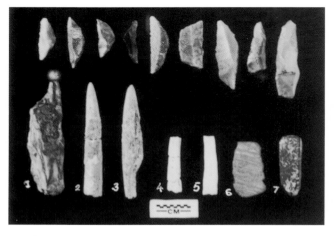

Fig. 4a. I row: Geometric and non-geometric microliths. II row: 1 borer, 2–3 bone points, 4–5 dentalium shell 6–7 yellow and red ochre crayons.

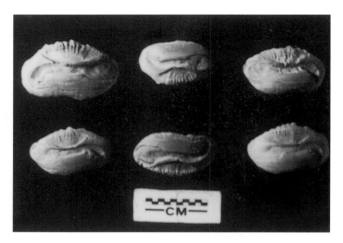

Fig. 4b. Otoliths of ophicephalus fish.

Fig. 4c. Ground and faceted siltstone artefact.

Fig. 4d. Grinding stones, palette stones and hammers stones.

Fig. 4. Mesolithic artefacts from Loteshwar.

122

vicinity in the Rupen estuary are exceptionally rich in lithic artefacts and faunal remains. A preliminary study of the fauna collected from Loteshwar shows a predominance of antelope (*Antelope cervicapra*) and gazelle (*Gazelle dorcas*) in the Mesolithic assemblage[1]. This may indicate a preference towards hunting these two animals by the Mesolithic hunter-gatherers. A large number of grinding stones found in the assemblage are evidence that grinding was a regular practice, probably as a part of food processing as well as tool production. The recovery of a single cylindrical stone artefact showing grinding and polishing (Fig. 4c) from the top layer of the Mesolithic is also significant in this context. Ground and polished stone artefacts of a similar nature have been reported from the Mesolithic assemblage at Langhnaj too (Sankalia, 1965).

Early Chalcolithic Settlements

More than a hundred Chalcolithic sites have so far been reported from north Gujarat (Ajithprasad and Sonawane, 1993; Bhan, 1994). They fall primarily into two categories: (a) The Early Chalcolithic settlements which have been dated earlier than 2500 BC and (b) sites belonging to different phases of Harappan culture dated from 2500 BC to 1500 BC (Table 1). In the first category, settlements of three distinct Early Chalcolithic traditions are found in north Gujarat (Fig. 3). The most important among these is a Chalcolithic tradition largely confined to north Gujarat. Due to its regional character it has been named 'Anarta' Chalcolithic after the traditional name of north Gujarat (Ajithprasad and Sonawane, 1993; Sonawane and Ajithprasad, 1994). It had a wide distribution encompassing the entire north Gujarat and extending to the adjoining parts of Kutch. The origin of this tradition has been dated at 3500 to 3000 BC by radiocarbon determinations at Loteshwar. Pottery vessels, mostly bowls, basins and pots and jars of

varying sizes, associated with this Early Chalcolithic site are mostly hand-made. A fine Red Ware, Gritty Red Ware, Burnished Red Ware and Burnished Grey Ware with characteristic surface treatment and decorated with shades of white and dark brown bichrome designs are the standard pottery of this regional tradition. Loteshwar, Kukrana, Mujpur, Santhali and Nayeka are the important sites where the above-mentioned cultural assemblage is found more or less independently. In many other sites, the Anarta pottery is found associated with the Urban as well as the Post-Urban phases of Harappan culture, indicating its long time span.

The second category of Early Chalcolithic sites reported from this region is the assemblage incorporating cultural traits comparable to the Early Harappan settlements of the Sindh and adjoining regions (Ajithprasad, 2000, Majumdar and Sonawane, 1996–97). The hallmark of this assemblage is a set of well-made wheel-thrown pottery showing distinct vessel forms. They show a technical competence comparable to the Harappans not only in clay preparation but also in firing. Long blades of 'Rohri chert/flint' and semi-precious stone beads are also a part of this assemblage. Eight sites of this category have been reported from north Gujarat (Fig. 3). Important among these are the burials at Nagwada and Santhali and the regular habitation site at Moti Pipli. At most of the sites the above assemblage is associated with the Anarta pottery and their spatial distribution is coterminous with the spread of Anarta sites in north Gujarat. Although not dated radiometrically, it can be confidently dated between 2800 BC and 2500 BC through stratigraphic interpolation.

The third distinct Early Chalcolithic assemblage in north Gujarat is that of Pre-Prabhas, first reported from Prabhas Patan/Somnath (Fig. 1) in the Junagadh district of the Saurashtra coast (Dhavalikar and Possehl, 1992). It is distinguished by a unique set of mostly hand-made pottery incorporating a

Burnished Red Ware with a corrugated or ribbed surface, and a Burnished Grey, Black and Red Ware. All this pottery, mostly represented by large open-mouthed bowls, pots and jars, shows distinct incised surface decoration. It is represented in north Gujarat by a single settlement measuring 0.3 ha at Datrana and has been dated to 2900-2800 BC (Ajithprasad, 2002). In fact, the 30 cm Chalcolithic deposit at Datrana incorporates the above-mentioned early Harappan Sindh related pottery as well as the Anarta pottery in the upper layers indicating the chronological horizon of the above three Early Chalcolithic assemblages at the site.

Harappan Affiliated Chalcolithic Sites

The Early Chalcolithic is followed by sites belonging to the Urban as well as the Post-Urban phases of Harappan culture dating from 2500 BC to 1500 BC. Their distribution in north Gujarat is mapped in Fig. 5. Only very few sites belonging to the Urban Phase which has been dated from 2500 BC to 1900 BC have been reported from north Gujarat. Nagwada and Zekda are prominent among these. These are small rural settlements but incorporating most of the Urban Harappan traits in the artefact assemblage including inscribed terracotta sealing, several distinct pottery vessels, long 'Rohri chert' blades, characteristic beads of semi-precious stones and gold, triangular terracotta cakes, shell objects, etc. (Hegde *et al.,* 1988). However, their indebtedness to the Anarta Chalcolithic tradition is apparent from the large collection of Anarta pottery found from these sites (Ajithprasad and Sonawane, 1993). Excluding the above few, all other sites belong to different phases of Late Harappan/Post–Urban Harappan dating from 1900 BC to 1500 BC. At these later sites, it is the Harappan cultural elements from Saurashtra that predominate and therefore are comparable to the Harappan cultural sequence at Rangpur proposed by Rao (1963; 1979).

A majority of the Chalcolithic sites, irrespective of whether they belong to the Early Chalcolithic period or are affiliated to Harappan culture, are small settlements with shallow habitation debris. In the case of Anarta settlements, several pits with varying diameter and depth (2 m to 0.5 m) are very common at the site. They are generally filled with habitation debris. This feature is not found in the sites of two other Early Chalcolithic cultures. While 70% of the sites measure less than 0.5 ha, 50% are even smaller measuring less than 0.3 ha only (Table. 3). The sites generally have a habitation deposit around 1.0 m indicating short duration of occupation. Pieces of clay lumps with reed impression suggest that their dwelling structures were of wattle and daub. However, the structures at Nagwada belonging to the Urban Phase, were built of mud-bricks as well as stones. The small size and flimsy nature of the habitation devoid of any permanent structures indicates the meagre resources available in a harsh climate with low and irregular rainfall. Moreover, several of the Post-Urban settlements show a tendency to cluster close to shallow natural depressions, which retain fresh water during the monsoon and become brackish only with the onset of summer. These are therefore mostly pastoral settlements, probably indulging in marginal agricultural practices, which become active upon the availability of water. This reflects the optimal exploitation of the cyclic nature of natural resources.

Fluorine-Phosphate Ratio and Mesolithic-Chalcolithic Relationship

An important factor that we looked at in the rich Mesolithic and Early Chalcolithic assemblages in north Gujarat, especially in Loteshwar, was the role of Mesolithic culture in the emergence of the Early Chalcolithic way of life by the second half of the 4th Millennium BC. A very elementary attempt has been made to address this issue through a preliminary study of faunal collection, lithic

Fig. 5. Harappan affiliated Chalcolithic settlements in north Gujarat.

artefacts and the technology of their production and settlement features from well-stratified Mesolithic and Chalcolithic assemblages at Loteshwar, Datrana and Moti Pipli. The radiocarbon dates for the beginning of Mesolithic and Chalcolithic occupations at Loteshwar (Table 1) provided the necessary chronological control for the above study.

Loteshwar revealed large quantities of faunal remains both in the Mesolithic and the Chalcolithic deposits. However, faunal remains from the Mesolithic levels at the site do not include bones of any domesticated animals. Nor do they incorporate bones of wild cattle, sheep or goat: the primary domesticated fauna found in the Early Chalcolithic assemblages in Gujarat as well

125

as in other regions. The Anarta Chalcolithic that follows the Mesolithic occupation by around 3500 BC incorporates bones of fully domesticated cattle and sheep/goat. It is, therefore, clear that there is little evidence for events leading to independent domestication either in the Mesolithic or in the Early Chalcolithic assemblages at Loteshwar.

However, a clear preference for exploiting antelopes and gazelles is obvious from the preponderance of the skeletal remains of these two animals in the Mesolithic assemblage at Loteshwar. This is one of the important steps in the process of intense exploitation leading towards herding. Similarly, extensive use of grinding stones and the presence of some ground and polished artefacts in addition to the microlithic artefacts may indicate technological adaptation for different modes of food processing. The use of microlithic implements and grinding tools continues without any change in the Early Chalcolithic/Anarta levels suggesting a continuation of the technique. Moreover, the Chalcolithic assemblage does not show the use of the crested ridge technique for blade production. Contrary to this, the crested ridge technique was widely used in the Early Chalcolithic context in the Indus valley for producing lithic blades. Consequently, the Early Harappan Sindh-related assemblage at Moti Pipli and the Pre-Prabhas deposit at Datrana show extensive application of the crested ridge technique for blade production. Crested ridge blades have also been reported from the Early Chalcolithic levels at Prabhas Patan (Subbarao, 1958). The technique was apparently unknown to the Early Chalcolithic Anarta community at Loteshwar indicating that the Anarta Chalcolithic settlement at Loteshwar may have preceded the other two categories of Early Chalcolithic sites in this region.

This is substantiated by a study of fluorine-phosphate ratio in the animal bones collected from the Mesolithic and Early Chalcolithic levels at Loteshwar, Datrana and Moti Pipli, carried out in order to understand the temporal relationship between the two.[2] The amount of fluorine in ancient bones found in a particular region gives a clue to the relative age of samples belonging to different periods. Thus, the fluorine value, which is expressed as a percentage ratio between fluorine and phosphate ($100F/P_2O_5$), varies with different cultural periods depending on their date. A higher $100F/P_2O_5$ ratio indicates more time depth. The analysis is based on the fact that the accumulation of fluor-apatite in bone depends on the time elapsed after its burial. The actual accumulation is also influenced by the soil environment in which the bones are buried. However, a value of one and above is generally associated with the samples dating back to the first half of Holocene and a value of less than one is associated with samples belonging to the late Holocene in the semi-arid regions of western India. In the present study we are more interested in knowing the time gap between the Mesolithic and Chalcolithic occupation than the actual age and therefore the influence of burial environment in the calculation need not be accounted for.

Distribution of $100F/P_2O_5$ in different layers belonging to the Early Chalcolithic and the Mesolithic period are shown in Fig. 6. The values range from 0.2 to 0.55 in the case of Early Chalcolithic bone and are generally close to one and above one in the case of Mesolithic samples. Samples from Datrana and Moti Pipli demonstrate this very well, indicating a time gap between the Mesolithic and Chalcolithic deposits. However, in the case of Loteshwar, values for the Mesolithic samples vary from 0.8 to 0.55 and that of the Early Chalcolithic vary from 0.6 to 0.26[3]. Overlapping of the values at the end of the Mesolithic and in the beginning of the Chalcolithic period is therefore very clear from the histogram. It implies that the Mesolithic at Loteshwar may have continued for a prolonged period of time starting from 4700 BC, and there appears to be little time lag between the end of the Mesolithic and the beginning of the Chalcolithic period at the site. This is interesting

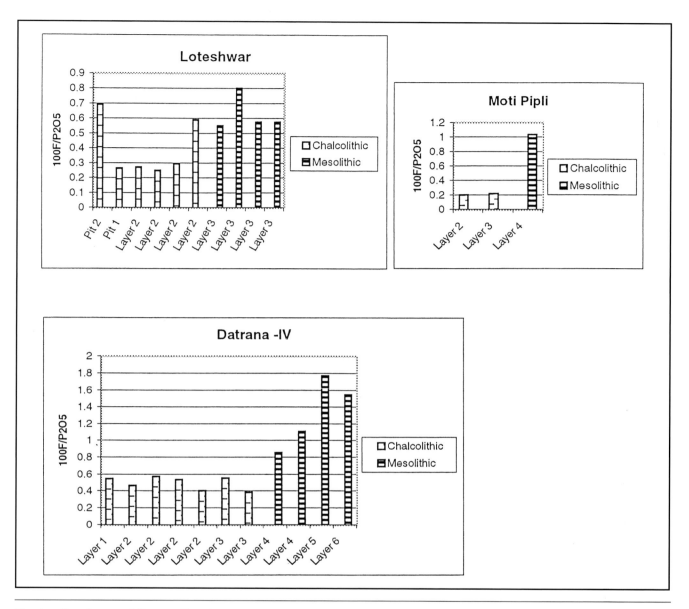

Fig. 6. Distribution of Fluorine-Phosphate ratio ($100F/P_2O_5$) in the Mesolithic and the Early Chalcolithic bone samples from north Gujarat.

for understanding the beginning of the Anarta Chalcolithic tradition in north Gujarat and the dynamics of its interaction with the Mesolithic hunter-gatherers. Considering the fact that the Mesolithic way of life continued to flourish even after the emergence of full-fledged Chalcolithic urbanism in this region, the symbiotic relationship that may have existed between the two needs thorough understanding.

It is also important that the development of Mesolithic culture and the beginning of the

Anarta Chalcolithic culture in north Gujarat according to the Nal Sarovar sedimentary data is associated with an environment which was predominantly dry with regular spells of wet phase. This is hardly surprising because it is the environmental challenges that bring in innovative methods of exploitation and management of resources rather than an environment of plenty. The new trends in the intense exploitation of faunal resources and diversification of technology in the Mesolithic and the beginning of

The scant remains of a giant tree at the Gilund site, India.
(Photo by Takeshi Takeda)

Early Chalcolithic food production could be viewed as adaptive responses to a challenged environmental resource base which persisted for about 2000 years from 6.6 ka to 4.8 ka. The ways and means of the beginning of this Early Chalcolithic food production/pastoralism in north Gujarat need to be understood in the light of several early food producing communities in Sindh and Baluchistan existing earlier than 3500 BC, and the cultural processes involved in the spread of better technologies. Subsequently, when the environment became more wet by around 4.8 ka we come across the spread of Early Chalcolithic farming communities not only of the Anarta, but also of the Early Harappan cultures of Sindh and Pre-Prabhas in this region. This trend culminates in the emergence of urban centres of the Harappan culture in western India including Gujarat, and its repercussion could be seen in north Gujarat settlements too. The wet phase in Gujarat continued, according to the Nal palaeoclimatic data, upto 3 ka and subsequently became progressively dry. However, the Urban elements of the Harappan culture started declining by about 1900 BC (~3.9 ka) and what followed was a predominantly rural set up with pastoralism and marginal farming as the main economic activity for a period of four to five hundred years up to 1500 BC (~3.5ka). These developments cannot be linked with the deterioration of the climate in Gujarat, as the Nal data do not show any decline in precipitation during this period. Hence, climatic change appears to have had little role in the overall decline of the Harappan culture in Gujarat.

Otoliths and the Late Holocene Environment of North Gujarat

Some more interesting information for understanding the environment during the late Holocene is provided by fish otoliths recovered from Mesolithic sites located close to the Little Rann of Kutch (Fig. 4b). Otoliths or ear-stones are hard calcium concretions formed within the inner ear of bony fish (Casteel, 1976:17; Irie et al., 1967). The size of an otolith varies according to the species and size of the fish. Moreover, it is possible to identify the fish up to species level from the surface features of otoliths (Casteel, 1976:24). Mesolithic assemblages from Loteshwar, Moti Pipli, Datrana, and Santhli which are located along the eastern margin of the Little Rann of Kutch have yielded otoliths of a fish tentatively identified as belonging to the family Channidae and the genus *Channa/Ophicephalus*[4], commonly known by the name 'snakeheads'. This is a kind of fish that aestivates in the mud during the dry season. It is found in a muddy estuarine environment as well as in freshwater lakes and is sensitive to a saline environment. Its presence in the Early Chalcolithic level at Loteshwar could not be confirmed due to the disturbed nature of the deposit. However, otoliths of *Ophicephalus* are not found either in the other two Early Chalcolithic sites or in the succeeding Harappan assemblage. This may indicate some significant change in the habitat of the above fish, in the margins of the Little Rann of Kutch in the beginning of the 3rd Millennium BC. Was the change due to the increase of salinity beyond the tolerance level of the above fish in this period? But Nal Sarovar data indicates an increase in precipitation in this period implying better input of fresh water in the region, which would have been helpful in maintaining the habitat rather than its deterioration. Hence, we may have to look at changes that took place in other ecological factors for the absence of *Ophicephalus* fish remains in the Harappan deposits of north Gujarat. Palaeoenvironmental studies of the sedimentary samples from the Ranns of Kutch are therefore very much called for, to understand the local changes in the environment and associated changes in the human settlements in north Gujarat.

Summary

The following broad points can be summarized from the above discussion:

The southeast monsoon is the most important factor that influences the climate in western India, particularly Gujarat. The middle and late Holocene palaeoclimatic records from Rajasthan and Gujarat show significant variations in the monsoonal precipitation. Among these the records from Nal Sarovar in the southern margin of north Gujarat is more important for understanding the palaeoenvironment of the region.

Palaeoclimatic interpretations of the Nal Sarovar data indicate a very dry phase with regular short wet spells in the middle Holocene from 6.6 to 4.8 ka (~4600 to ~2800 BC). This relatively dry phase corresponds with the development of Mesolithic culture and the beginning of the Early Chalcolithic life in the region.

Although no concrete evidence of events leading to independent domestication is found either in the Mesolithic or in the Early Chalcolithic assemblage at Loteshwar, the fluorine analysis of bone samples indicated little time gap between the end of the Mesolithic and the beginning of the Early Chalcolithic period. The cultural developments during the Mesolithic and the beginning of the Early Chalcolithic period in the region could be viewed as an adaptive mechanism in articulation with the environment whose resources were constantly challenged in a predominantly dry climate.

The wet phase that followed for about 4.8 to 3 ka (~2800 to ~1000 BC) corresponds to the spread of Early Chalcolithic as well as Harappan culture in north Gujarat. This period shows adaptation of diverse food producing communities to a slightly wetter environment and use of better technologies for exploitation of locally available environmental resources.

Since the deterioration of the climate towards a more dry phase started only around 3 ka (~1000 BC), the decline of Harappan culture in Gujarat by around 1900 BC onwards may not have been necessarily influenced by the climatic changes.

Ophicephalus otoliths from the Mesolithic sites close to the Little Rann of Kutch suggest a muddy estuarine environment with enough fresh water from 6.6 ka (~4600 BC) onwards. A change in the environment around the Ranns of Kutch in the first half of the 3rd Millennium BC is indicated by the absence of Ophicephalus otoliths in the Harappan assemblage.

Acknowledgements

Sincere thanks are due to Ajita K. Patel of the Department of Archaeology and Ancient History, The M.S. University of Baroda, William R. Belcher of the University of Wisconsin, Madison, USA, Richard H. Meadow of Harvard University, Dr. Anupama Kshirasagar, Deccan College, Pune and P.C. Mankodi of the Department of Zoology, The M.S. University of Baroda, for their help in preparing the paper.

Notes

1. The faunal collection from Loteshwar is studied by Ajita K. Patel of the Department of Archaeology and Ancient History, The M.S. University of Baroda.

2. Fluorine analysis of bone samples was carried out by A. Kshirasagar of the Department of Archaeology, Deccan College, Pune.

3. The unduly high value, 0.7, for the sample from Pit 2 may be due to the mixing up of bones from the Mesolithic deposit into which the pit was dug by the Chalcolithic occupants.

4. The genus level identification of the fish from the otolith samples was done by William R. Belcher of the University of Wisconsin, Madison, USA.

References

Ajithprasad, P. (2000): The Pre-Harappan Culture of Gujarat. In S. Settar and Ravi Korisettar (eds.), Indian Archaeology in Retrospect, Vol. II. Protohistory, *Archaeology of the Harappan Civilization.* ICHR, Manohar, New Delhi, pp. 129-172.

Ajithprasad, P. and V. H. Sonawane (1993): The Harappa Culture in North Gujarat: A Regional Paradigm. Paper presented at the Conference on *The Harappans in Gujarat: Problems and Prospects* Pune, Deccan College.

Barnard, A. (1979): Kalahari Bushman settlement Pattern. In Burnham, P. and Ellen (eds.) *Social and Ecological Systems,* London, pp. 131–144.

Bhan, K.K. (1994): Cultural Development of the Prehistoric Period in North Gujarat with Special Reference to Western India. *South Asian Studies,* 10: 71–90.

Brayson, R.A. and A, M. Swain (1981): Holocene Variations of Monsoon Rainfall in Rajasthan, *Quaternary Research,* 16: 135–145.

Casteel, R.W. (1976): *Fish Remains in Archaeology.* Academic Press, London, pp. 17–40.

Dhavalikar, M.K. and G.L. Possehl (1992): The Pre-Harappan Period at Prabhas Patan and the Pre-Harappan Phase in Gujarat. *Man and Environment,* XVII(1): 72–78.

Goudie, A.S., B. Allchin, and K.T.M. Hegde, (1973): The former extension of the great Indian sand desert. *Geographical Journal,* 139 (2): 243–257.

Hegde, K.T.M, V.H. Sonawane, D.R. Shah, K.K. Bhan, P. Ajithprasad, K. Krishnan and S. Pratapachandran (1988): Excavation at Nagwada 1986 and 87: A Preliminary Report. *Man and Environment,* XII: 55–65.

Irie, T., T. Yokohama and T. Yamata, (1967): Calcification of Fish Otolith Caused by Food and Water. *Bull. Jap. Soc. Sci. Fish,* 33(1): 24–26.

Majumdar (1960): *Historical and Cultural Chronology of Gujarat.* M.S. University Oriental Institute Publication, Baroda, Preface XVII

Majumdar, A. and V.H. Sonawane (1996–97): Pre-Harappan Burial Pottery from Moti-Pipli: A New Dimension in the Cultural Assemblage of North Gujarat, *Pragdhara,* 7: 11–17.

Pandarinath, K., Sushma Prasad and S.K. Gupta (1999): A 75 ka. Record of Plaeoclimatic Changes Inferred from Crystallinity of Illite from Nal Sarovar, Western India. *Journal of the Geological Society of India.* 54: 515–522.

Patel, G. (1977): *Gujarat's Agriculture.* Overseas Book Traders, Ahmedabad: 47.

Possehl, G. L. (1993): *Radiometric Dates from South East Asia* (Manuscript, Compiled by G.L. Possehl).

Rao, S.R. (1963): Excavations at Rangpur and other excavations in Gujarat. *Ancient India,* 18–19: 13–27.

Rao, S.R. (1985): Lothal: A Port Town (1955–62). *Memoirs of the Archaeological Survey of India.* No. 78 V. II. A.S.I., New Delhi.

Sankalia, H.D. (1965): *Excavations at Langhnaj: 1944–63, Part I Archaeology* Deccan College, Pune, pp. 41–64.

Sharma, C. and M.S. Chauhan, (1991): Plaeovegetation and environmental inferences from the Quaternary palynostratigraphy of Western Indian Plains. *Man and Environment,* XVI: 65–71.

Singh, G. (1971): The Indus Valley Culture seen in the context of post glacial climatic and ecological studies in northwest India. *Archaeology and Physical Anthropology of Oceania,* 6: 177–189.

Singh, G., Joshi, R.D. and Singh, A.B. (1972): Stratigraphic and Radiocarbon evidence for the age and development of three salt lake deposits in Rajasthan, India. *Quaternary Research,* 2: 496–505.

Singh, G., R.J. Wasson, and D.P. Agrawal, (1990): Vegetational and Seasonal Climatic Change Since the Last Full Glacial in the Thar Desert, Northwest India. *Rev. Palaeobot. Palyno,* 64: 351–358.

Singh, N., G.B. Pant, and S.S. Mulye (1991): Distribution and long term features of spatial variations of the moisture regions over India. *International Journal of Climatology,* II: 413–427.

Sonawane, V.H. (2000): Early Farming Communities of Gujarat, India. In Bellwood, P., D. Bowdery, D. Bulbeck, D. Bear, V. Shinde, R. Shutter, G. Summerhayes, (eds.) *Bulletin of the Indo-Pacific Prehistoric Association,* 19 Indo-Pacific Prehistory: Melaka Papers, Vol. 3: 137–146.

Sonawane, V.H. and P. Ajithprasad (1994): Harappa Culture and Gujarat. *Man and Environment,* XX (1–2): 37–49.

Subbaro, B. (1958): *The Personality of India .* The M. S. University Archaeology Series 3, M. S. University Press, Baroda, 132 pp.

Sushma Prasad, Sheila Kusumgar and S.K. Gupta (1997): A mid to late Holocene Record of palaeoclimatic changes from Nal Sarovar : a palaeodesert margin lake in western India. *Journal of Quaternary Science,* 12 (2): 153–159.

Swain, A.M., J.E. Kutzbach, and S. Hastenrath, (1983): Estimates of Holocene Precipitation for Rajasthan, India, based on Pollen and Lake Level Data. *Quaternary Research,* 19: 1–17.

Wasson, R.J., J.I. Smith, and D.P. Agrawal (1984): Late Quaternary Sediments, minerals and inferred geochemical history of Didwana Lake, Thar Desert, India. *Palaeogeography, Palaeoclimatology, Palaeoecology,* 40: 345–372.

Chapter 8

Monsoon Environments and the Indian Ocean Interaction Sphere in Antiquity: 3000 BC— AD 300

SUNIL GUPTA

Scope of the Study

While writings on early exchange networks and ancient sea trade in the Indian Ocean emphasize archaeo-historical evidence of contact/interchange, environmental perspectives are lacking in the conceptualization of the early Indian Ocean as a sphere of interaction. The Indian Ocean stretches, in an extended sense, from the Red Sea to the South China Sea, incorporating the 'core' littoral regions of the Arabian Sea and Bay of Bengal (Figs. 1, 3). The monsoons define the environment of this maritime space. I have attempted to integrate monsoonal determinants into archaeo-historical viewpoints on contact and exchange in the ancient Indian Ocean world. The chronological span of 3000 BC–AD 300 adopted here underscores the need for a deep time perspective. The study does not claim comprehensive understanding of the Indian Ocean in Antiquity. I intend to focus upon crucial littoral regions and maritime episodes to elaborate on the theme of this study.

An Overview: Monsoon Environments in the Indian Ocean

The monsoons can be broadly defined as wind systems that seasonally reverse direction. A strong monsoon system prevails over the northern Indian Ocean, manifested in steady, rain-bearing winds which blow south-north from May to September (Southwest Monsoon/Southerlies) and north-south from October to April (Northeast Monsoon/Northerlies) (Fig. 1). While monsoon activity is strongest over Indian Ocean lands, adjoining littoral regions come within its orbit. In Egypt, the intensity of floods in the Nile river has been determined for millennia by monsoon rains over Ethiopia and Central Africa (Hassan, 1998). To the east, monsoon-like conditions prevail as far as Japan, where seasonal switching of wind direction is as persistent as India (*Encyclopaedia Britannica:* Climate). Initially, the Southwest Monsoon front develops off the East African coast. In his book *The Monsoons,* Das (1998: 50) says, 'The major part of the low level jet penetrates into east Africa during May and, subsequently, traverses the northern parts of the Arabian Sea before reaching India in June. Observations suggest that the strongest cross-equatorial flow from the southern to the northern hemisphere during the Asian summer monsoon is in the region of the low level jet. This has intrigued meteorologists because it is not clear why the major flow of air from the southern to the northern hemisphere should take place along a narrow preferred zone off the east African coast.' Besides the effect upon ocean currents, another consequence of the Indian Ocean monsoon is the precipitation it brings to the Afro-Asian landmass, particularly Ethiopia, the Yemen highlands, the Indian subcontinent and parts of mainland Southeast Asia. Monsoon rains in the subcontinent annually recharge the rivers in the Indo-Gangetic plain and the Indian Peninsula. A large quantity of sediment brought down by the rivers, especially the Indus and the Ganga is discharged into the Arabian Sea and the Bay of Bengal. Convergences between the monsoonal regime and maritime initiatives are a major focus of this study. For instance, the discovery of the *hippalus* (the Southwest Monsoon wind) by the Greeks in the 3rd century BC was a breakthrough for western commercial interests in the Indian Ocean. Interactive littoral regions of the Indian Ocean can be situated within monsoonal contexts: the dominant Southwest and Northeast monsoon fronts and related localized wind flows, currents, tides, fluvial discharges, etc. The greater monsoonal zone, stretching from the Red Sea region to the South China Sea, emerged as a vast, interconnected maritime market by the beginning of the Christian Era.

4th–3rd Millennium BC: Beginnings of Coastal Exchange

The adaptation of prehistoric communities to coastal environments involved, initially, the understanding of the inshore arena, the lagoons, creeks and closed bays. Finding edible shells on inter-tidal flats and fish in near waters were crucial experiences for mobile communities inhabiting coastal tracts to get into a settled way of life. Longer ventures for maritime food resources must have gradually built up deep knowledge of coastal geomorphology, tides, shoals and currents. The genesis of the first watercraft—rafts and dugouts—is intimately associated with near shore ventures. Sedentism on the coast implies regular generation of marine and agricultural food resources, demographic intensity to the point where the littoral population

demands better subsistence strategies and emergence of nucleated settlements (Bailey and Parkington, 1988: 1–8). The long period of adaptation and exploitation of maritime environments created 'evolved' seafaring communities. At some point in time a number of these coastal communities came to be 'appropriated' by inland agricultural societies expanding their concerns and ambitions to the coastlands. There has been little argument against this explanation. A powerful anti-thesis focuses upon the rise of the Andean Civilization in Peru on the basis of marine resources. The theory of Maritime Foundations of Andean Civilizations (MFAC) hypothesizes the growth of nucleated Neolithic settlements on the Peruvian coast sustained by the plentiful catch of the schooling fish—anchoveta. Moseley and Feldman (1988: 129–130) explain the transition from inshore to deep-water capabilities of the ancient Peruvians: 'Valdiria and Machalilla phase occupational midden deposits on La Plata island (Ecuador) establish a minimal 20 km offshore voyage capability prior to 1000 BC in the area north of the anchoveta belt. Within the anchoveta belt, voyages of 9, 14 and 16 km are established for the 1st millennium AD by artefacts recovered in guano deposits...By the 1st century AD, two varieties of sea-going craft are represented in Moche iconography of desert coast. These are the totora reed boats and balsa-log sailing rafts. Both vessels are described more than a millennium later in various ethnohistoric sources...' The MFAC model may not be as widely applicable as the conventional model of civilization growth based upon agriculture. However, it does point to the generative potential of coastal material cultures. We can envisage a strongly autonomous role for early seafaring communities in the formation of extended coastal networks, and making productive linkages with inland populations.

How does this conception hold for the monsoon lands? Prospections in different parts of the Indian Ocean rim have provided a conspicuous record of early settlement on coastlands. The shell middens on the Tihama coast of western Yemen and the Batinah coast of northern Oman show evidence of foraging communities taking to subsistence on marine foods (shell and fish) in the 5th–4th millennium BC (for Yemen see Al-Ansary 1996: 238–45 and location 2 in Fig. 1; for Oman see Biagi *et al.*, 1989: 1–8; Cleuziou and Tosi, 1989: 15–47 and location 4 in Fig. 1). Explorations on the southern Tihama (west Yemen coast) reveal shell midden sites situated close to old riverbeds and creeks which were once lush with grasses and mangroves (Tosi in de Maigret, 1986: 363–369). The Tihama

is the gateway to the highlands of interior Yemen and the long Arabian coastline upto Oman. The Yemen highlands were occupied by Neolithic folk between 6000–3000 BC (Fedele in de Maigret *et al.*, 1985: 431–37). The highlands of Yemen experience the Southwest Monsoon, receiving an excess of 300 mm rainfall every year. Agriculture developed on the intermontane plateaus (overlooking the Tihama) in the early 4th millennium BC and by the close of the 3rd millennium BC, the Bronze Age of Yemen was well underway, manifested in nucleated settlements and terraced fields (Edens *et al.*, 2000: 854). Contextual evidence suggests that the 'marine Neolithic' people of Tihama and the highlanders

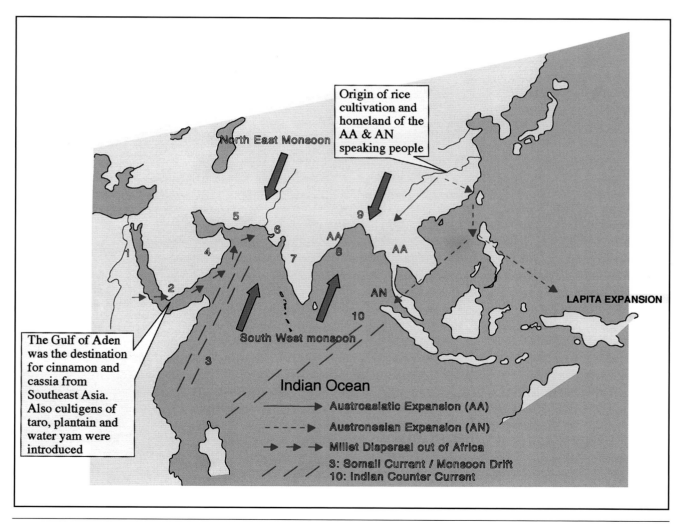

Fig. 1. Composite map showing the Monsoon regime, dispersal routes of plant domesticates and human expansions in the Indian Ocean world. See Table 1 for references to numbers in map.

Number	Place name; notes
1	Egyptian Red Sea Coast, Pharaonic port-site of Wadi Gawasis.
2	Yemen; Shell Midden sites on Tihama Coast; Neolithic settlements in the highlands.
3	Rufiji Delta on the Tanzanian coast; confluence of the Somali and Indian Ocean Counter Current.
4	Oman; Shell Midden sites on the Batinah Coast.
5	Makran-Sind Coast.
6	Gujarat Coast; Saurashtra; Earliest Millet Cultivation Area in the Indian Subcontinent; Location of port of Lothal.
7	Konkan–Kanara Coast.
8	Orissa Coast; Location of the Neolithic site of Golbai Sasan.
9	Estuary of the River Ganga.
10	Edge of the Bay of Bengal; Mingling of Austronesian and Austroasiatic Techno-Cultural Traditions.

of Yemen were in contact with Bronze Age Egypt. The Pharoahs of Egypt sent out trading expeditions to a land called Punt. The earliest hieroglyphic records in Egypt mentioning Punt are dated around middle 3rd millennium BC (Kitchen, 1993: 587–608). However, an earlier period of contact must have preceded a regular interchange. Artefactual indicators suggest that the Egypt-Punt contact had been established in the 4th–3rd millennium BC transition (Zarins, 1996: 89–104). The Puntic expeditions sought, among other things, for the incense called myrrh. Various hypotheses have been advanced on the location of Punt. The 'God's Land', as ancient Egyptians called Punt, is believed to have been located, variously in the Sudan littoral (Sayed 1977: 41–72), the Eritrean highlands (Kitchen, 1993: 603–606) and in Somalia (Chittick, 1976: 117–34). In my view, there are strong reasons to consider western Yemen as the core area of Punt (location 2 in Fig. 1). Western Yemen is synonymous with ancient Punt because historically speaking, this region supplied the best myrrh to the west.[1] Secondly, the Neolithic highlanders of Yemen who are found practising agriculture in the 4th millennium BC, must have acquired knowledge of tapping myrrh quite early.

In the 4th–3rd millennium BC, the Neolithic of Yemen was much more advanced than contemporaneous Neolithic cultures in Somalia or Eritrea.[2] Thirdly, obsidian found in predynastic levels in Egypt (Nagada I–III) has been related through trace metal analyses to sources in Eritrea-Ethiopia and Yemen, suggesting early maritime connections with the latter region (Zarins, 1996: 89–104). The reasons cited, with their attendant explanations, point to western Yemen being congruent with the land of Punt. The journey from Egypt to the myrrh-producing area of Punt must have involved crossing a stretch of the Red Sea, either directly from the Egyptian Red Sea coast or from southerly embarkation points. Plant domesticates dispersed from Africa also crossed the waters to Arabia, most likely through rudimentary maritime networks forming in the southern Red Sea in 4th–3rd millennium BC transition. Harlan (1993: 53–60) imputes the date 5th millennium BC for dispersal of sorghum cultigen out of Africa. The route of transmission, as constructed by Harlan (in Possehl, 1997: 87–100) crossed the southern Red Sea in the vicinity of the Bab el-Mandeb (straits which connect the Red Sea to the Gulf of Aden) and touched the Tihama coast, thence diffusing into southern Arabia (Fig. 1). The role of the coastal communities of southern Tihama as creators of early maritime and inland networks is strongly suggested by the above episodes centred on the Red Sea region. Unfortunately, there is nothing in the archaeological record about watercrafts in use in the southern Red Sea in such early times. Analogies can be drawn from later evidence. Raft-like sail boats depicted on a Middle Kingdom frieze (middle 2nd millennium BC) have been interpreted as Puntite vessels beaching on the Egyptian Red Sea coast with trade goods (Kitchen, 1993: Fig. 35.7; Fig. 2 b). Still later, in the 1st century AD, the author of the *Periplus Maris Erythraci (henceforth PME)* observed rafts crossing the Bab el-Mandeb straits (*PME* 7). Rafts are basic watercraft forms and the earliest watercraft used by the sedentary communities of Tihama must have been not unlike the ones seen by the *Periplus'* author. The raft-like sail boats on the Middle Kingdom frieze must have evolved from simpler types in the southern Red Sea.

A consensus holds on the idea of a coastal route of transmission of the sorghum cultigen from Yemen to Oman and beyond, into the Indian subcontinent (Cleuziou and Tosi, 1989: 15–47; Possehl, 1997: 87–100; Weber, 1998: 267–74). Early evidence of sorghum cultivation in Yemen is associated with Neolithic-Bronze Age transition around 2000 BC (de Maigret *et al.*, 1985: 423–454; 1986). A pottery vessel with impressions of sorghum dated 2500 BC has been found in highland Yemen (Harlan, 1993: 53–60). In Oman—the crucial link between south Arabia and India—sorghum has been recovered from contexts dated 2330–2250 BC at Hili 8 (Cleuziou and Tosi, 1989: 15–47). Italian excavators have proposed earlier dates (early 4th millennium BC) on the basis of sorghum seeds recovered from shell middens (the Qumrun complex) on the Batinah coast of Oman (Biagi *et al.,* 1989: 1–8). According to the Italians, the penetration of millet cultigens of east African origin into interior Oman (Hili 8) was effected through the midden sites on the Oman coast. While the Italian dates for sorghum in the Omani shell middens are not unanimously accepted (see Possehl, 1997: 94–95), most scholars agree on the idea of a marine conduit for penetration of sorghum. The dispersal route of sorghum along the coast of Arabia overlaps considerably with the Monsoon Drift (Fig. 1). The Gulf of Aden is subjected to both the southwest and northeast trade winds. The coast off East Africa also experiences the full intensity of the monsoons. As pointed out, a strong monsoon front forms off the Swahili coast in June. The developing Southwest Monsoon influences the Somali Current which is permanently present off the Somali coast. The flow of the current is determined by the Indian Monsoon. To quote Sahni (1997: 12): 'The Somali Current is one of the most prominent current systems of the Indian Ocean, and is part of the ocean's dynamic response to the bi-annual pattern of wind stress over a large part of it. This is noticed mainly along the Somali coast, where during the Southwest Monsoon it flows northeastwards at the rate of 0.5 to 2 knots in the upper 200 m or so of the ocean. In response to the return of the northeasterly trade winds the Somali current reverses to flow southwestwards at the rate of 0.5 knots.' According to Das (1998: 52) : 'The Somali Current may be considered to be a western boundary current of the Indian Ocean. But, its peculiar feature is a reversal in direction with the onset of the summer monsoon.' The extension of the Somali Current into the Arabian Sea is called the Monsoon Drift. The Somali Current may have facilitated functioning of early coastal exchange networks and transmission of plant domesticates along the Arabian seaboard.

Early dates for sorghum cultivation in Yemen, Oman and India range between 2500–2000 BC. It is important to distinguish between processes of dispersal and the beginning of sorghum cultivation in the Arabian Sea rim. The chronological range suggests that sorghum cultivation was already established in Arabia-

India by middle-late 3rd millennium BC. We must allow time for the processes of dispersal, which preceded the full-fledged cultivation of sorghum in Arabia. Harlan (1993: 53–60) has suggested the 5th millennium BC as the period of dispersal of sorghum out of Africa. Therefore, there is a lag of more than a millennium between the first dispersal of sorghum and evidence for its cultivation in the Arabian Peninsula. To my mind, the sorghum seeds recovered from the Qumrun midden complex (Oman) from contexts dated to the 4th millennium BC signify the dispersal of millets along the Arabian coast. The Qumrun finds are perhaps the crucial link between Harlan's date of original dissemination and the beginning of sorghum cultivation in the Arabian Peninsula. Beyond the Batinah, the African millet dispersal route leads to two major coastal regions in the western part of the Indian subcontinent. These are the Makran-Sind coast of Pakistan and the coast of Gujarat state in western India (locations 5 & 6 in Fig. 1). On the basis of archaeobotanical evidence, Weber (1998) proposes that millets of African origin entered the western part of the Indian subcontinent in two phases. The first phase of dispersal brought the finger millet to western India around middle 3rd millennium BC. The second phase in early 2nd millennium BC saw the infusion of the large millet (*sorghum bicolor*) and the pearl millet (Weber, 1998: 267–274). Weber's dates signify cultivation

and therefore, in the context of dispersals, we need to focus upon contact between the Arabo-Persian and Indic littoral regions in the preceding centuries: the 4th–3rd millennium BC transition.

However, the artefactual indicators are much too meagre for us to know whether the Oman littoral, Makran-Sind and Gujarat were linked by a regular exchange network in these early times. The excavators of the shell midden RH5 in Oman report intrusive wood remains of probable Baluchistan source in the 4th millennium contexts. However, they go no further than calling these driftwood (Biagi *et al.,* 1989: 1–8). The early occupation of the Makran coast of Baluchistan is indicated by basal levels at Miri Qalat tentatively dated to late 5th millennium BC (Besenval, 1997: 199–216). Excavations at the site of Balakot on the Sind coast reveal that occupation started in early 4th millennium BC (Frank-Vogt, 1997: 217–235). Miri Qalat and Balakot seem to have been settled in by pastoral groups from interior Baluchistan. Further east, the Gujarat coast was occupied by Chalcolithic communities, very likely from north Gujarat, sometime in the 4th millennium BC (Possehl, 1999: 611–14). Possehl also suggests that some of the pottery from the earliest levels of Prabhas Patan on the Saurashtra coast of Gujarat may be of Arabian provenance.[3] Shinde (1998: 173–82) finds some of the pottery in the basal levels at Padri on the Gulf of Khambhat littoral similar to

Table 2: References to Fig. 2

A	Queen Hatsheput's Merchant Ships for Punt. 2nd millennium BC.
B	Raft-like sailing vessels of Puntites (?) on Egyptian Red Sea Coast. 2nd millennium BC.
C	Clay Model of Boat from Ubaid Levels. Eridu, Mesopotamia. 4th millennium BC.
D	Graffiti of dhow-like sailing vessel. Harappan Civilization. 3rd millennium BC.
E	Depiction of a watercraft made of reeds on an amulet from Mohenjodaro. 3rd millennium BC.
F	Clay model of a sailing vessel from Harappan port of Lothal. 3rd millennium BC.
G	Seal from Mohenjodaro showing a river (?) boat. 3rd millennium BC.
H	Sketch of hull of a sewn plank boat.

Fig. 2. Early watercrafts in the Indian Ocean. See Table 2 for references to letters.

the Baluchistan-Sind Chalcolithic horizon of early 3rd millennium BC. In the absence of valid archaeological indicators for contact, evidence of early watercrafts along the Arabian-Persian-Indian coastlands is important for understanding the maritime context for dispersal of millet. A retrospective scenario for the southern Red Sea (Yemen coast) discussed above focuses upon rafts, possibly propped with leather bags, as facilitators. In the Oman and the Persian Gulf region, row boats without sails were probably the first watercraft to be used by early sedentary communities. Bas-relief representations of row boats have been found on the rock outcroppings at Jebel Jusasiyah on the island of Qatar (Nayeem, 2000: 372–75). The carvings are presumed to be of Protohistoric date. Ethnographically, the boats depicted at Jebel Jusasiyah are similar to the *sasha,* a row boat made out of palm leaf ribs and in use on the Oman coast till recent times (Heyerdahl, 1982). Interestingly, sails have been added at a later time to some of the Qatar row boat carvings, indicating a shift to exploitation of wind power (in Protohistoric times?). Certain (wood planked?) watercrafts seem to have been caulked with bitumen in Oman-Persian Gulf since early 4th millennium BC. Sumerian literary records of early 2nd millennium BC refer to ceremonial boats caulked with bitumen plying in water channels between Eridu and Uruk. Zarins (1992: 60–65), discussing the evidence, contends that a freshwater lake (on the banks of which the settlement of Eridu was located), became gradually dessicated and was abandoned following the Eridu occupation in the Early/Middle Uruk period, c. 3700–3400 BC. Going by this evidence, we can infer that the 2nd millennium Sumerian records allude to a time in the 4th millennium BC when the boats caulked with bitumen sailed the waters off Eridu. A clay model of a sailing vessel was excavated from Ubaid levels at Eridu (Barnett in Ratnagar, 1981; Fig. 2c). Evidence from archaeology shows a prolific use of bitumen as insets for implements and for making decorative objects in the Ubaid-Uruk contexts at Eridu (Zarins, 1992: 65). On the Oman coast, bitumen has been found at the RH5 midden on the Batinah coast, in context dated late 4th millennium BC (Cleuziou and Tosi, 1989: 28–29). The evidence of bitumen from RH5, in the form of tar at the base of a pot, has been interpreted by archaeologists in the field as binding material for watercraft. To quote Cleuziou and Tosi (1989: 30): 'We may safely conclude that some kind of a caulking material based on bitumen was cooked in the vessel. The easiest assumption we can draw is that this pot came to be used by a population that combined this new class of object with something quite different from susbsistence and probably even more revolutionary for a fisherman's way of life: the possibility of using bitumen for the buoyancy of watercrafts, allowing the assemblage of plankwood boats for the first time.' The bitumen found at the RH5 midden was brought to the site, presumably from asphalt seepages within the Oman Peninsula (Cleuziou and Tosi, 1989: p. 44, n.11). However, the main bitumen concentrations occur in the upper Gulf, in the Mesopotamian estuarine zone, in Kuwait and Failaka. The upper Gulf is the likely area for the initial development of bitumen caulking technology for watercraft, a contention drawing credibility from the varied use of bitumen in the Ubaid material culture. That boat-making technology may have spread from the Ubaid-Uruk centres to Oman remains a possibility, especially when faint outlines of exchange mechanisms in the Persian Gulf are visible in the 4th millennium BC (Al-Ansary, 1996: 238–245). Post-Ubaid settlements dated to middle 4th millennium BC have been discovered in eastern Arabia, the transition area between the upper and lower Gulf (Zarins, 1989: 74–103). While bitumen caulking of boats became established early in the Gulf-Oman littorals, it is a moot point when such vessels were used for crossing the Sea of Oman to India. Excavations at Ras al Junayz, a fishing and trading station of the Wadi Suq period (the 3rd–2nd millennium BC transition) on the Oman coast, have yielded some hundred pieces of bitumen from a room floor. The bitumen pieces carry impressions of rope and matting and are encrusted with barnacles, indicating sailing in the high seas (Newsletter of the Joint Hadd Project 1987).[4] Articles of trade of Harappan provenance were recovered from Ras al Junayz, suggesting that mariners from Ras al Junayz made the crossings to the Indian subcontinent. However, Ras al Junayz flourished during the 3rd–2nd millennium BC transition, much later than the period under review and at a time coinciding with the second phase of millet diffusion into the Indian subcontinent (Weber, 1998: 267–274). There is virtually no evidence (or indicators) for watercrafts in use on the Makran-Sind-Gujarat coasts in the 4th–3rd millennium BC transition. At the moment we can only assume the use of simple near shore watercraft on the basis of Harappan and ethnographic parallels (Ratnagar, 1981; Sahni, 1997: 9–18). In the period under review, regular interchange between the Arabian and Indian sea boards cannot be assumed, though rudimentary coastal networks seem to be operative on both sides of the Sea of Oman. The millet dispersals from Oman to the Indian subcontinent probably happened under circumstances of 'drift' : the occasional bitumen caulked vessel finding its way to the Makran or a wood-planked sail boat from Sind washing up on Arabian shores, like the driftwood from Baluchistan at the settlements of Qumrun.

The brief review of early settlements, maritime activity and dispersal of plant domesticates indicates that *conditions* for extended coastal networks to function had emerged in the 4th–3rd millennium BC transition along the western Indian Ocean rim. The knowledge of the monsoon regime, the seasonal switching of wind and current directions, must have been exploited for long movement of watercraft in coastal waters. It is under circumstances of growing interactivity along the Arabian Sea rim that the first infusion of millet cultigens into western India happened in middle 3rd millennium BC. A fuller understanding of early interactive processes along the western Ocean rim is precluded by lack of data from important coastal stretches (Tosi, 1990). The protohistory of south Arabian coastlands east of Aden, especially the dry Dhofar region of southern Oman, is not as well understood as the Tihama and Batinah. The Somali coast, which makes the southern coast of the Gulf of Aden, has been explored in parts. The vast Indian coastline south of Gujarat, namely the Konkan-Kanara-Malabar tracts along the Arabian Sea, have been little investigated from the point of view of early exchange. At the moment we know of Mesolithic communities which engaged in fishing in the creeks of the Konkan-Kanara tracts and the backwaters of Kerala. Dugouts and canoes still ply these waters (personal observation) and modern-day shell middens are survivals of a Protohistoric lifestyle (Deshpande-Mukherjee, 2000: 79–92).

Eastern Indian Ocean

In the eastern Indian Ocean the main littoral regions are those of the Bay of Bengal, the Gulf of Thailand and the Sulawesi Sea (Fig. 1). As discussed, the Southwest Monsoon front which originates off East Africa splits as it approaches the Indian Peninsula and a branch curves into the Bay of Bengal. The Southwest Monsoon front forms here later than in the Arabian Sea. However, its onset is equally intense with heavy precipitaion on the eastern Indian seaboard. Like the Indus sediments in the Arabian Sea, alluvial mud discharged by the river Ganga accumulates at a high rate at the head of the Bay of Bengal. The Bay has a steep continental shelf and a high tidal range (average 10.7 m). The high alluvium discharge from the major rivers reworks into coastal landforms through fluvio-marine processes. To quote Thima Reddy (1994: 44): 'The east coast is predominantly a plain with typical coastal depositional landforms. There are four major deltas in this region, namely those of the Mahanadi, the Godavari, the Krishna and the Kaveri. A number of elongated, narrow sand ridges exist parallel or subparallel to, and inland from the present shore almost all along the east coast.' East of the Bay of Bengal, a separate monsoon system combines Southeast Asia and northern Australia. This monsoon is considerably weaker than the Indian monsoon and is manifested in varying degrees in different parts of the region. Over Indonesia, the wind reversal is discernible but not as strong as over parts of mainland Southeast Asia like Thailand and Kampuchea. In the South China Sea, the southwest winds displace the cold dry northerlies from continental Asia in the summer, while the northerlies prevail in the winter months. In the Far East, the replacement of cold winter winds from Asia by warm and turbulent counter flows from the Pacific Ocean, especially over Japan, creates a monsoonal situation. In the eastern Indian Ocean, adaptations to maritime way of life were discovered at the shell midden site of Nong Nor (the 5th–4th millennium BC?) and subsequently at the Neolithic settlement of Kok Phanom Di, both situated at the head of the Gulf of Thailand (Higham and Thosarat, 1998: 40–66). The 'marine Neolithic' people of Thailand were to become part of the maritime tradition of Austronesian speakers, a Neolithic people who expanded across island Southeast Asia from original homelands in southern China. The community at Kok Phanom Di traded with rice cultivators in the interior (Higham and Maloney, 1989: 650–66; Higham and Thosarat, 1998: 61–63). These rice cultivators may have been Austroasiatic speakers, Neolithic people who also spread out of southern China and expanded into mainland Southeast Asia along land routes. Later in the discussion I shall focus upon the role of Austronesians and Austroasiatics as carriers of plant cultigens and spice products across the Indian Ocean (Fig. 1).

3rd–2nd Millennium BC: Extended Coastal Networks

The Egypt-Punt trade undergoes a critical shift in the 3rd–2nd millennium BC transition. The land-sea route to Punt is sought to be replaced (or augmented?) by a direct maritime connection from the Red Sea coast. Probably the first attempt to seek a direct route to Punt was by the Old Kingdom Pharoah Pepi II (2270 BC) who despatched men and material to build a port on the Egyptian Red Sea coast. The venture failed because of hostile desert tribes (Kitchen, 1993: 588–89). A 12th Dynasty port (1943–1898 BC) was discovered on the Egyptian Red Sea coast at Wadi Gawasis (location 1 in Fig. 1). An inscription on a Stelae found at Wadi Gawasis informs of a commercial voyage to the land of 'Bia-*Punt*' carried out from this Pharaonic port (Sayed, 1977: 41–72). The establishment of the port of Wadi Gawasis and building of a ship for Punt by the Pharoah's officers indicates a basic shift in strategy, showing the involvement of the state in creating infrastructure and bringing in experienced mariners. Egyptian sea voyages to Punt are recorded through the 2nd millennium BC (the Middle and New Kingdom periods), including the famous expeditions of Queen Hatsheput (1472 BC) and Ramses III (1184–1153 BC) (Fig. 2a). The Puntites themselves sailed to the Egyptian coast with commodities of trade. Theban tombs dated to middle 2nd millennium BC depict the Puntite missions, one of the friezes showing their arrival in raft-like sailing vessels (for review of evidence on Punt see Kitchen, 1993: 587–608; Fig. 2b). The evidence suggests that the Egypt-Punt sea trade had regularized in the 2nd millennium BC with maritime traffic originating from both ends. The effect of the Southwest Monsoon is discernible in the Red Sea. The Red Sea is a 'long, narrow central trench with a V-shape profile, more than 2000 m deep, bordered by extensive coastal shelves more than

50 m deep. Extending from Bab el-Mandeb for 2000 km, it terminates in the north in two diverticula, the Gulf of Suez and the Gulf of Aqaba' (Sidebotham *et al.*, 1989: 161). The monsoons strongly impact the southern part of the Red Sea littoral, bringing rainfall over the Yemeni and Ethiopian highlands. However, the southerlies are countered by dry northerlies in the northern part of the Red Sea. The direct sea voyage to Punt/southern Red Sea region would doubtless have been made using the incessant northerlies and the Indian Ocean monsoons which impact western Yemen and Ethiopia with regularity. The contrast between the large sailing ships of the Egyptians (Queen Hatsheput's mission) and the simple raft-like sailing vessels of the Puntites is conspicuous (Figs. 2a, 2b). The simple sailing boats of the Puntites must have evolved from rudimentary watercraft of the southern Red Sea. The Egyptian craft signify infusion of boat-making technologies into the Red Sea region. The Egyptians were sailing the Mediterranean since early 3rd millennium BC with the use of large sailing ships in the Nile as is evident from the boat of the Pharoah Cheops dated to 2530 BC (Gowlett, 1984: 184). The infusion of superior maritime technology by polities interested in furthering long-distance sea trade is a theme that was to be repeated in the Red Sea by the Israelis, the Ptolemies and the Romans.

The 3rd–2nd millennium BC transition is also the high point of sea trade in the northern Arabian Sea between Persian Gulf polities and the Harappan Civilization zone. This interchange is known mainly through deposition of exotic artefacts (the Harappan presence more conspicuous in the Gulf). The rich artefactual evidence of Harappan maritime contact comprises ceramics, weights, seals, carnelian beads, etc. The evidence has been documented and discussed thoroughly (Ratnagar, 1981; Chakrabarti, 1990; Potts, 1990: 323–33). The trade seems to have survived through much of the 2nd millennium BC. Barbar and Kassite levels at Bahrain and Failaka have yielded Harappan circular seals, weights and ceramics (Hojlund, 1989: 49–53). In the lower Gulf, Harappan elements survive in the material culture of Tell Abraq through the 2nd millennium BC (Potts, 1990: 323–33). Explorations on the Makran show little evidence of exchange activity, with the 2nd millennium BC Shahi Tump ceramic depositions found away from the coastal areas (Besenval, 1992: 25–35). On the Indian littoral, Pd. III levels at Prabhas Patan, dated to late 2nd millennium BC, correlate to the Kassite presence in the Persian Gulf. A steatite amulet with graffiti (recalling Persian Gulf seals modelled on Harappan examples), an obsidian flake (from Yemen?) and a gold object recovered from Prabhas Pd. III are indicative of external contacts. Also, the Prabhas III people continued to use cubical chert weights of Harappan inspiration (Dhavalikar, 1989: 348–50). Prabhas Patan is situated near the confluence of the Hiran river and the Arabian Sea, presenting itself as an outlet from the fertile hinterland of Saurashtra. Agricultural settlements in central Saurashtra in the 3rd–2nd millennium BC cultivated a variety of millets, including sorghum (Possehl, 1997: 91–94). Prabhas may have mediated the transmission of millet cultigens from Arabia into interior Saurashtra. A proliferation of Harappan-Late Harappan storage jars in coastal and interior Oman in the 3rd–2nd millennium BC transition indicates export of essential commodities, most likely food products, from fertile areas of Saurashtra and Sind (Cleuziou and Tosi, 1989: 15–47; for Saurashtran origin of Oman pottery see Sahni, 1997: 9–18). A structure in Prabhas Pd. III, identified as a warehouse may have been a cattle pen as a large number of cattle bones have been found in the vicinity. Historical and ethnographic analogies point to Saurashtra being a regular supplier of agricultural and farm produce to west Asia since early times.[5] Possehl (1999) speaks of the collapse of high craft Harappan centres of Lothal,

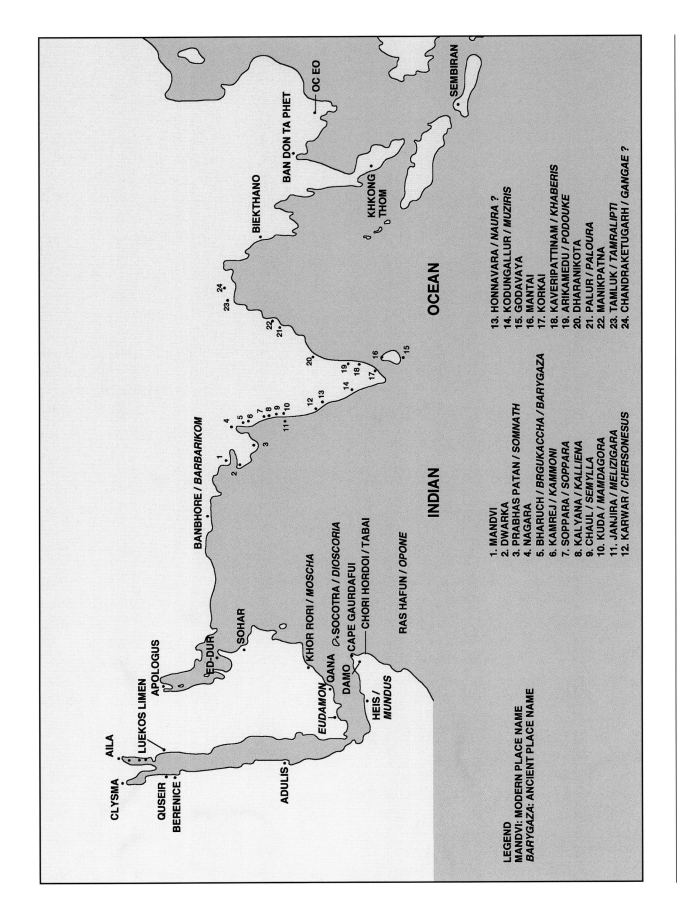

144

Fig. 3. Early historic ports and market-towns across the Red Sea–Indian Ocean.

Kuntasi and Dholavira in Gujarat in the early 2nd millennium BC while maintaining that the Chalcolithic agricultural communities of Saurashtra carried on as usual. The decline of high craft centres reflects the collapse of elite trade with Mesopotamian polities while Saurashtran trade in essentials carried on with arid regions of Oman. Prabhas seems more likely to be a 'Chalcolithic' port rather than an urbanized Harappan establishment, dedicated to the more resilient networks across the Sea of Oman. In a way, the vigour of the Late Harappan contact with Oman reflects the basic, enduring nature of exchanges between western India and the Arabian Peninsula. Extended coastal networks, created by Neolithic communities responding to imperatives of need (food, boat-building material), came to be used by resurgent Harappan and Mesopotamian Civilizations for fulfilling high aesthetic needs (beads, ivory). When the trade in elite goods declined, the fundamental modes of exchange in aspect of necessities remained in place.

Bronze Age sailors in the northern Arabian Sea must have experienced unpredictable wind flows and shallow seas. Both monsoons (Southwest and Northeast) are modified by regional wind flows about the Persian Gulf (Ratnagar, 1981: 162). The explorer Thor Heyerdahl constructed a reed ship modelled on a vessel depicted on an Harappan seal (Heyerdahl 1982; Fig. 2e). Emulating ancient seafarers of the Sea of Oman, Heyerdahl undertook a voyage from the Tigris-Euphrates delta to the port of Karachi in Sind and then sailed with the monsoons across the Arabian Sea upto Djibouti in the southern Red Sea region. The uncertainty of sailing in the northern Arabian Sea was experienced by Heyerdahl during the voyage of his reed ship *Tigris* as it approached the Makran coast of Pakistan when the Northeast Monsoon was expected. To quote: 'The weather forecast for the Gulf of Oman and the Arabian Sea had been Northeast wind at ten to twenty knots, but it was

dead calm. But strangest of all, the famous monsoon had not blown on our sails for a single whole day. The most usual wind came from Iran (*NW*), but we also had winds that varied between Southeast and Southwest. No matter from which side the wind blew, we struggled to keep a fairly steady course...' (Heyerdahl, 1982: 263). The northern Arabian Sea as a whole is shallow due to the deposition of silt from the Indus river, with shoal formation especially endemic off the coast of Kutch and in the Gulf of Khambhat (Fig. 4). The *Periplus Maris Erythraei,* a sea guide compiled in the 1st century AD for sailors voyaging from Egypt to India informs that the Indus river brings down 'an enormous volume of water; so that a long way out at sea...the water of the ocean is fresh from it' (*PME* 38). That this discharge goes 'a long way out at sea' is indicated by suspended particles from the Indus found along the Kutch coast and at the mouth of the Gulf of Kutch (Hashimi and Nair, 1988: 56–57) (Fig 4). It has been estimated that the Indus releases over 200 cubic kilometres of water with 450 million tonnes of suspended load annually (Hashimi and Nair, 1988: 56–57; Flam, 1999: 35–69). In fact, the Indus silt gradually moves further out to sea, eventually depositing in a great alluvial fan in the Arabian Sea basin.[6] Archaeological and ethnographic indicators point to operation of vessels with moderate displacement. Ratnagar (1981: 160–70) asserts that flat-bottomed boats operated off the western coast of India as these must have been suitable for berthing in shallow creeks. On the basis of ethnographic evidence, Sahni (1997: 14–18) proposes that timber boats with sails, not unlike the dhows of today, crossed the Sea of Oman in the high noon of Harappan-Gulf sea trade (Fig. 2 d, g, f).

So the western Indian Ocean had two contemporaneous networks in operation in the 3rd–2nd millennium BC: the Egyptian-Punt trade linking the northern and southern Red Sea and the Gulf—Harappan/Chalcolithic sea trade across

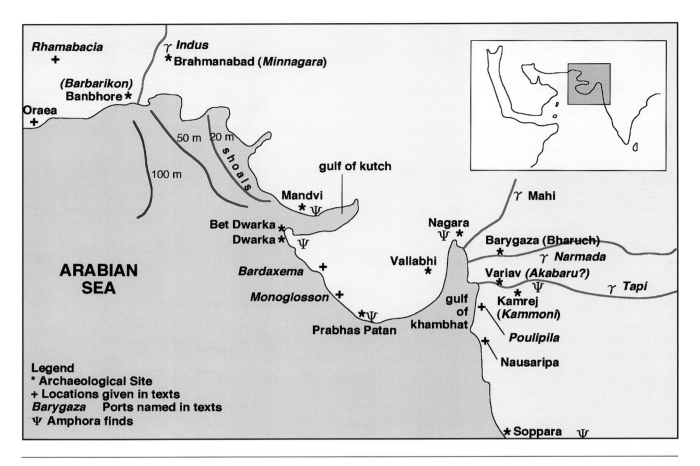

Fig. 4. Early historic ports and market-towns in Lower Indus/Coastal region.

the Sea of Oman. The potential linkages between the two interactive regions must have been through the Gulf of Aden. We have discussed this maritime region as a conduit for diffusion of millet domesticates from Africa to the Indian subcontinent. Possehl has urged for revival of historical focus upon the path of the Monsoon Drift, associated with seasonal wind and current flows from East Africa to the northern Arabian Sea (see above). To quote Possehl (1997: 94): 'The millets document an additional leg to the maritime life of the times, extending the sailing down the Arabian coast at least as far as the mouth of the Red Sea. Even broader geographical horizons for this inter-regional interaction come to us from a report of copal from Madagascar or Africa in a burial at Tell Asmar in Mesopotamia.' On the basis of available indicators (millet, obsidian, copal) and the evidence for occupation

of the Arabian coast by Neolithic communities, we can envisage simple coastal networks depending upon near shore watercrafts exploiting the Somali Current off East Africa and its extension along the Yemen-Oman coast (see discussion on sorghum above).

Opposite the Arabian Peninsula, the coast lands of the Indian Peninsula were still settled by communities following a Mesolithic-Neolithic lifestyle. In the eastern Indian Ocean the Austroasiatic and Austronesian expansions reached the edge of the Bay of Bengal, near the Malaysian peninsula and Sumatra. Neolithic cultures flourished on both sides of the Bay of Bengal through the 2nd millennium BC. The last of the Austroasiatic expansions—represented by the Ban Kao Neolithic culture occured along the Malaysian Peninsula. On Indian shores, we find Neolithic concentrations in the Krishna valley

146

(Andhra Pradesh) and in coastal Orissa. Golbai Sasan, a Neolithic site on the Orissa coast has yielded cord-marked pottery with impressions of rice in its basal levels (location 8 in Fig. 1). Unfortunately, there are no carbon determined dates for Golbai Sasan. On the basis of contextual data, Chakrabarti (2000) has proposed that basal levels of Golbai Sasan may go back to early 2nd millennium BC. Cord-marked pottery in the context of domesticated rice is the primary archaeological expression of Austroasiatic movements in mainland Southeast Asia. This pottery and evidence of rice at Golbai Sasan has been viewed by a number of archaeologists as part of an Austroasiatic expansion through land bridges between southern China and northeast India (Higham, 1996: 294–308). However, the land connection with the Neolithic civilization of northeastern India is far-fetched for there is no cord-marked pottery tradition in Bengal through which such expansion could have taken place. Golbai Sasan is located merely 20 km from the Bay of Bengal and is close to the lagoonal Chilka lake, the latter being an important sanctuary for ships in historical times (Kar *et. al.,* 1998: 107–14). As I shall argue below, the maritime area on the southeast of the Bay of Bengal emerged as a possible staging point for long-distance ventures of the Austroasiatic-Austronesian people in Indian Ocean lands. The lack of positive evidence makes deep sea crossings by rice cultivators appear unlikely. A plausible maritime route for migration of Austroasiatic people to Orissa would be along the coast, rounding the head of the Bay of Bengal from departure points in Burma. The Golbai Sasan evidence has to be closely interpreted and investigated for its maritime dimension.

The interest of expanding inland polities is manifested in the two main trade networks of the 3rd–2nd millennium BC, the Egypt-Punt trade and the Gulf-Harappan trade. The monsoons are exploited for sailing. However, the maritime exchanges in the Red Sea and the northern Arabian Sea still do not involve crossings of the high seas. We can describe these as extended coastal networks. Long distance coastal networks may have been operative in the Bay of Bengal rim in the 2nd millennium BC, as indicated by the deposition of cord-marked pottery of Southeast Asian affinity on the Orissa coast.

2nd–1st Millennium BC: Deep Sea Trading Networks

The connection between Egypt and Punt survives in the 1st millennium BC though the records are now erratic. A stela ascribed to the XXVI Dynasty (7th century BC) refers to 'rainfall upon the mountain of Punt', an allusion either to the Ethiopian or Yemen uplands which are impacted by the Southwest Monsoon. The land of Punt becomes an abstract geographical nomenclature for the Ptolemaic Greeks who rule Egypt in the latter part of the 1st millennium BC (Kitchen, 1993: 602). Biblical sources (dated early 1st millennium BC) allude to a major commercial initiative from the northern Red Sea. According to Old Testament records, the merchant ships of King Solomon set out from the port of Ezion-Geber in southern Israel to bring back exotic commodities such as gold, ivory, peacock, tamarind, sandalwood and cotton from the famed port of Ophir (1 Kings 9: 26–28; 2 Chronicles 8: 17–18). This event is placed in early 1st millennium BC. Ophir is sought to be located, variously, on the Red Sea littoral, in south Arabia and in India. Wherever Ophir was, the goods acquired from there by Solomon's ships included commodities of definite south Indian origin. For goods from Ophir (parrot, tamarind), the Old Testament uses terms etymologically derived from Old Tamil.[7] The historical context is strongly suggestive of crossings of the Arabian Sea with the aid of monsoons from deep in the south of India to some point in the Gulf of Aden. Solomon's initiation of trade with Ophir is achieved by the induction of Phoenician sailors and ships built from the timbers of Lebanon.[8] Like the sailing of Egyptians to Punt, Solomon's mission to Ophir suggests that induction of superior maritime skills into the Red Sea region required political will.[9]

Another historical episode which hints at an interchange between eastern and western Indian Ocean networks revolves around the demand for two eastern spices—cinnamon and cassia—in the northern Red Sea region from early 1st millennium BC. Cassia is mentioned in Biblical records dated early 1st millennium BC and in ancient Greek sources throughout 1st millennium BC (Casson, 1984: 225–246). How were these spices, grown on trees native to certain parts of mainland Southeast Asia, acquired by Hebraic and Greek societies in such early times? Various explanations have been offered for the source of cinnamon and cassia by ancient Greek historians (Herodotus, Dioscorides, Pliny) and modern commentators. While most locate the area of acquisition in the Gulf of Aden region there is no agreement about the source. The arguments are, variously, that cinnamon-cassia was brought to the Gulf of Aden region from either India or Southeast Asia; that the spices were actually grown in East Africa; that cinnamon-cassia was actually some other botanical product and not the spices as we know them today. Casson (1984: 225–246; 1989: 122–125), offers a detailed review of evidence and concludes that cinnamon and cassia reached the west, in all likelihood, from its native habitat in Southeast Asia; the best coming from southern China. Neolithic communities of mainland Southeast Asia, while developing sophisticated rice cultivation technologies, had also become proficient in plant cultivation and foraging. This is apparent at the early rice-growing settlement of Hemudu in lower Yangtze basin dated to 5000 BC. Remains of leaves of the tree *cinnamomum chingii* have been found at Hemudu (Chang, 1981: 177–186). We also find Neolithic trade networks, for jade

artefacts, across southern China as early as the 3rd millennium BC (Huang, 1992: 75–83). Contextual evidence suggests that southern Chinese Neolithic communities were amply conditioned for growing and trading botanical products, including cinnamon, by early 1st millennium BC.

The equally pertinent question is, how could cinnamon-cassia be transported to the Gulf of Aden? I draw attention to one of the greatest maritime initiatives in motion from midddle 2nd millennium BC. Austronesians based in the Sulawesi Sea region began to cross vast stretches of sea to settle in South Pacific islands. The Austronesian seafarers are identified in the archaeological record as Lapita pottery people.[10] Their material culture reveals them to be traders, involved in long-distance transfer of commodities in the Pacific and Sahul regions between 1000 BC–AD 500. Obsidian sourced in the South Pacific island of New Britain is found from Borneo to Fiji, a span of 6500 km! (Bellwood, 1992: 128). The obsidian deposition indicates Austronesian maritime prowess and capacity for long-distance trade. Austronesian seafarers could also set their sights west to the Indian Ocean. A study of the flora of sub-Saharan Africa shows that three food plants (plaintain, taro and water yam) were introduced from Southeast Asia at some undetermined date (Blench, 1996: 417–436). These cultigens represent early imports from Southeast Asia. The context is strong enough to hypothesize Austronesian involvement in consigning cinnamon and cassia from Southeast Asia to the Gulf of Aden region in the 1st millennium BC. The likely staging area for Austronesian maritime ventures must have been the southeastern littoral of the Bay of Bengal. This area, as pointed out, saw a coming together of Austroasiatic and Austronesian-speaking people (location 10 in Fig. 1). Linguists have related the archaic Austronesian substratum in Nancowry to 'proto-Austronesian' language groups. Nicobar, where the language is spoken, has been viewed as a relic area and Nancowry a survival from the time when Austroasiatic and Austronesian people still preserved the linguistic affinity which bound them to Austric.[11] Evidence of archaic Austronesian languages has also been found on the western coast of Sumatra (Bellwood, 1992: 114). Is this a glimmer of evidence for early Austrasiatic-Austronesian maritime bases in the Bay of Bengal sphere, bases used for staging ventures to the Indian Peninsula and the Gulf of Aden?

The literary sources, and the botanical indicators, suggest that the Gulf of Aden region became a focal area integrating lines of maritime exchange between the eastern and western parts of the Indian Ocean. The Bronze Age of south Arabia is dated to the 2nd millennium BC and towards the close of this millennium urban settlements emerge on the Yemen coast (Edens *et al.*, 2000: 854–62). The Iron Age of south Arabia was well underway by early 1st millennium BC and towards the middle of the millennium we see the rise of competing littoral states like Ausan and Qataban (van Beek, 1958: 141–52). The early states of south Arabia expanded their interests to East Africa. The Arabian outreach to the Swahili coast must have to do with increasing Arab knowledge of the monsoon phenomena. Chami (1999: 205–14) has proposed that the coastal tracts of Tanzania, particularly the Rufiji river delta, acted as a bridge between Iron Age communities of coastal Africa and the Arabian-Mediterranean World (location 3 in Fig. 1). Chami also supports the theory (first proposed by Miller, 1969s) that the East African coast—the Ausanitic/Tanzanian tracts—received cinnamon and cassia brought across the Indian Ocean from Southeast Asia (Chami and Msemwa, 1997: 673–77; Chami, 1999: 205–14). The Miller/Chami conception defines an alternate route to the traditional idea of the 'Sabean Lane': the maritime route along Southeast Asia—south India/Sri Lanka-south Arabia axis (Blench, 1996: 417–36; for Indo-south Arabia contact see Kirk, 1975: 20–34). The Miller/Chami theory of a

trans-oceanic route for transfer of cinnamon-cassia is credible for the following reasons (a) the Indian Counter Current links Southeast Asia with the Somali Current off the Tanzanian coast and (b) the Somali Current/Monsoon Drift forms a maritime corridor from just north of Madagascar to the Arabian Sea (*Bartholomew World Atlas*: xxv). This route must have coexisted with the classic Sabean Lane. In either case, trans-oceanic movements would have been only possible by sailing the high seas with the knowledge of monsoonal wind and current flows. Chami (1999: 205–14) identifies the Rufiji delta with the market town of Rhapta, which the Greek sea guide dated to 1st century AD, *Periplus* mentions as the last staging area of trade on the East African coast. Chami's field operations in the Rufiji delta have yielded strong artefactual evidence of East African contact with the Mediterranean and Indian worlds. Rufiji/Rhapta is located at a unique point of convergence of the Somali Current and the trans-oceanic Indian Counter Current and therefore may have become a focal area of maritime exchange linking southern Arabia, India, the Red Sea regions and Southeast Asia. Chami focuses upon the cinnamon-cassia trade in the context of the long-distance trade networks of early centuries AD. The discussions in this paper stress a much earlier period, placing the opening of sea lanes from Southeast Asia to East Africa sometime in the first half of the 1st millennium BC. The Biblical references to acquisition of spices from the Gulf of Aden indicate such a possibility and anticipate regular land falls made by Austronesians much before they decided to occupy the island of Madagascar sometime in the 1st millennium AD (for Austronesian colonization of Madagascar, see Reade, 1996; for review of African archaeology, see Sinclair, 1990: 1–40).

The close of the 1st millennium BC saw increasing integration of the littoral regions of the extended Indian Ocean, primarily due to the growing knowledge of the monsoons and development of maritime technologies (navigation aids, ships) designed to exploit the monsoonal phenomena for deep sea voyaging. The *Periplus Maris Erythraei* refers to the port of Eudamon Arabia (modern Aden), stating that the port was an entrepot for exchange of goods from Mediterranean lands and India (*PME* 26). The existence of Eudamon, which flourished in 3rd–2nd millennium BC, is suggestive of high sea trade across the Arabian Sea. The Greek knowledge of monsoonal sailing in the western Indian Ocean was acquired either from the Arabs of the Gulf of Aden or Indian mariners who frequented the gulf. The story of an Indian mariner, shipwrecked in the Red Sea and taken to the court of the Ptolemies, is often associated with the obtaining of the secrets of sailing in the Arabian Sea by Red Sea Greeks (Karttunen, 1989: 19–22; Salles, 1993: 494). The *Periplus* uses the term *hippalus* for knowledge of monsoon sailing. The knowledge of sailing across the Arabian Sea was known before the hippalus 'rediscovery'. Who discovered the secrets of sailing across the ocean with trade winds? The question remains open to interpretation. The conditions surrounding the 'pre-hippalus' discovery of the monsoons have yet to be answered.

Early 1st Millennium AD: the Indian Ocean Interaction Sphere

Various maritime traditions (Western, Arabian, Persian, Indian and Southeast Asian), with a long period of development, engage each other in the Indian Ocean in the BC–AD transition. The knowledge of monsoon voyaging is now 'universal' and large watercraft are introduced. Older coastal networks remain functional. Exchange activity heightens as emergent polities create infrastructure and make investments in long-distance maritime trade. The gathering momentum for long-distance trade in the various littoral regions of the Indian Ocean led, finally, to full-fledged trade wind-supported voyages in the BC–AD transition. Alongside, the older coastal networks continued to be active. At some point in time, yet undetermined but before 1st millennium AD, a maritime technology spread along the Indian Ocean rim: that of sewn planked boats (Fig. 2h). The *Periplus* speaks of sewn boats plying between Oman, a port at the mouth of the Persian Gulf, and Qana, the frankincense port on the Yemen coast. The sea guide also refers to sewn boats on the East African coast (*PME* 15). Ethnographic data on coastal watercraft from eastern Africa and Oman reveals the survival of sewn boats in modern times. Chittick (1980: 297–309) recorded sewn boats on the Somali coast during his reconnaisance of the Horn in the seventies. Vosmer (1994) has studied these boats on the Batinah coast of Oman. Sewn boats are quintessential coastal craft, designed to operate directly from sea shores, the structural flexibility imparted by the stitching allowing the vessels to absorb the shock of the surf as they beach. Kentley (1996: 247–60) notices similar stitching methods in sewn boats of the eastern Indian Ocean, the boats on the Indian side of the Bay of Bengal being double-stitched due to the stronger surf (Fig. 5). We cannot be sure where and when sewn boats were introduced, except to say that their stitching technique diffused through most of the Indian Ocean (Kentley, 1996: 247–60; Manguin, 1996: 181–98).

The imperatives of sea trade led to the founding of ports in differing geomorphological and environmental settings (Table 1). Adverse marine or coastal environments were not a factor in the instituting of early port-sites. For instance, we find two major ports in antiquity, the Harappan harbour at Lothal and the great Early Historic entrepot of Brghukaccha (the Barygaza of Greeks) located on the littoral of one of the most violent seas in the world: the Gulf of Khambhat (Fig. 4). High tides and endemic shoal formation were the major dangers for sailors in

Fig. 5. Picture showing a modern sewn boat in Eastern India. Boats mentioned in the *Periplus* may be similar to this.

the Gulf of Khambhat. In section 46 of the *Periplus Maris Erythraei* we are warned of the dangers of sailing in the Narmada estuary which formed the entrance to the great harbour of Brgukaccha/Barygaza: 'For this reason entrance and departure of vessels is very dangerous to those who are inexperienced or who come to the market-town for the first time. For the rush of waters at the incoming tide is so irresistible, and the anchors cannot hold against it; so that large ships are caught by the force of it, turned broadside on through the speed of the current, and so driven on the shoals and wrecked...' The *Periplus*' reference to high tides in the Gulf is supported by the fact that this maritime zone experiences one of the highest tidal ranges in the world (Nayak and Sahai, 1983: 152). On the eastern Indian coast, the estuarine zone of

the river Ganga is also fraught with high tides, unpredictable currents and hidden shoals (location 9 in Fig. 1). A seaman aboard the *Charles Cooper*, a steamship sailing for Calcutta in 1860 describes the passage through the Hughli estuary as the 'most difficult and dangerous passage in the world' where 'the current runs with great rapidity over treacherous shoals and quicksands'. (Bound and McLeod, 1992: 464). Some of the largest Early Historic ports on the eastern Indian sea board were located in the lower Ganga region (Table 1; also Chakravarti, 1992: 155–60).

The two examples from the western and eastern Indian coastlands illustrate the strength of commercial imperatives of sea trade over harsh monsoonal microenvironments. The opening of deep sea routes saw the introduction of large watercraft and establishment of ports with wharfs, warehouses and crafting centres (Fig. 3). The setting of Bronze Age-Iron Age harbours in the Indian Ocean was determined by political considerations, resource areas in proximity and need for dominance over strategic sea lanes.[12] Ptolemaic period ports on the Red Sea coast of Egypt were refurbished by Roman authority and land routes from the Nile to the coast were upgraded by watch towers and watering stations (Sidebotham, 1986a; Zitterkopf and Sidebotham, 1989: 155–89). The Hadhramaut region in eastern Yemen became rich with the shift of Roman sea trade to this region. The Hadhramauti ports of Qana and Khor Rori were founded in the beginning of the Christian Era to take advantage of new opportunties offered by Roman mercantilism. These ports, under the direct control of Hadhramauti rulers, traded in frankincense and offered berthing facilities for ships crossing the Arabian Sea back and forth from India (Sedov, 1996: 11–35). Explorations on the Somali coast have brought to light coastal trading stations flourishing at the turn of the Christian Era. Deposition of Roman, Indian and Persian ceramics indicates the external contacts of the African Horn (Chittick, 1976: 117–34; 1979: 273–77; 1980: 297–309; Smith and Wright, 1990: 106–14). Suhar, a Late Iron Age port on the Oman coast, was founded in the beginning of the Christian Era. Excavations have yielded the Indian Red Polished Ware (ceramics inspired by the Mediterranean sigillata tradition) in the basal levels (Kervran, 1996: 37–58). The Persian Gulf was experiencing a resurgence in maritime activity in the early centuries AD, with trade ships from India and south Arabia reaching ports near the Hormuz and at the head of the Gulf (Potts, 1990; Salles, 1993: 493–523). On the Indian subcontinent, coastal settlements and ports came up or were expanded in the BC–AD changeover, reflecting the consolidation of monsoonal trading activity (Gupta, 1998a). The fairly detailed list of ports and coastal settlements provided in the

Periplus Maris Erythraei (1st century AD) and the *Geographia* of Ptolemy (2nd century AD) reflect more than western geographical curiosity about the subcontinent. The Greek records are a testimony to the rapid opening of the Indian coastlands in the early centuries AD (Fig. 3). Archaeological fieldwork undertaken to locate the '*Periplus*' ports reveals that Early Historic trading stations and harbours flourished in diverse microenvironments. There were the large riverine ports like Bharuch (*Periplus'* Barygaza-on-Narmada), ancient harbours at Chaul (*Periplus'* Semylla) and at Kuda (*Periplus'* Mandagora) situated near deep creeks of the Konkan and lagoonal ports such as Arikamedu (*Periplus'* Poudouke) near the Bay of Bengal and Muziris on the Malabar Coast (Figs. 6, 7, 8).[13] Coast lands are places of opportunity, places for interactivity between interior and overseas and places where syncretic traditions emerge. These processes are evident in the Indian context. The opportunties opened by Roman sea trade raised prosperity levels in western India and stimulated Indo-Roman syncretic expressions in material culture (inspiring the Indian Red Polished Ware) and ideas (adoption of Alexandrian horoscopic precepts) (Gupta, 1998a; Gupta, 1998b: 87–102). The sea route from Egypt to Peninsular India is marked by 'sudden' deposition of Mediterranean artefacts, especially remains of amphorae (Gupta, 1994b). The opening of the subcontinent to sea lanes is reflective of a strong Indian response to the regularization of deep sea trade using monsoon winds.

Similar responses are noticed in Southeast Asia, where a number of coastal trading stations emerged in the early centuries AD (Glover, 1996: 129–58). The Indo-Southeast Asian interchange is expressed in the wide deposition of Indian pottery (the Rouletted Ware specifically) and beads of stone and glass similar to Indian types. The proliferation of Roman artefacts on the east coast of India, especially at the coastal site of Arikamedu, is indicative of direct western

Fig. 6. A view of the Early Historic port at Kuda, Maharashtra, Western India.

Fig. 7. A view of the ancient Indo-Roman trading station at Arikamedu, South India.

Fig. 8. A view of the site of Kamrej near the river Tapi, Gujarat, Western India.

involvement in maritime networks of the eastern Indian Ocean.[14] Crafting technologies, especially of glass and stone beads, spread from South India to Southeast Asia in the early centuries AD. A particular glass bead type, the Indo-Pacific beads, were manufactured about the Bay of Bengal littoral and exported in large numbers to the Far East: southern China, littoral Korea and western Japan. Deposition of Indo-Pacific beads in Japan shows sudden rise in the Yayoi Period (the 1st–2nd century AD). We have argued that the spurt in bead imports into Yayoi Japan reflects activation of exchange networks in the eastern Indian Ocean in the early centuries AD (Gupta, 2000: 73–88). All these developments were integral to what Glover (1990: 10) describes as 'the first appearance of a world system' manifested in 'a vast network of trade stretching from the Mediterranean basin and the Red Sea to South China.'

Conclusion

Historical stages in the integration of the Indian Ocean are reflected in the notional chronological divisions in our discussion. However, we must be careful to distinguish deep time perspectives from a purely 'developmental' view of the making of the Indian Ocean interaction sphere. There were constants. Even when a fundamental switch to deep sea voyaging had taken place by close of the 1st millennium BC, rudimentary coastal networks continued to function; for example, the rafts propped by leather bags which the *Periplus'* anonymous author saw in the vicinity of Qana. The survivals into present times of palm rib row boats on the Oman coast and dugouts along the western Indian seaboard testify to the endurance of coastal Neolithic traditions. The expansions opened conduits of trade. A deep time perspective allows us to view the transitions from expansion

154

to interchange (Hodder, 1987: 1–8). Broad contours of growth of long-distance maritime networks in the Indian Ocean are proposed below:

Settlement of coasts by hunter-gatherer groups and adaptation to staple marine resources like edible molluscs and fish. Construction of rudimentary watercraft (rafts, dugouts, canoes) for fishing in near waters. Attempt to establish offshore capabilities to search for marine resources (Midden sites in Tihama, Batinah and Gulf of Thailand littoral).

Increased range of coastal movement of watercraft as nucleated littoral settlements emerge. Marine resource exploitation is combined with exchange activity. Improvements in boat technologies (dispersal of millet cultigens from Africa to Arabia. Austronesian expansion across island Southeast Asia).

Use of monsoon winds and currents for long distance sailing (Egypt-Punt trade in the Red Sea; Harappan sea trade with Persian Gulf cultures). Formation of extended coastal exchange networks.

Switch to deep sea voyaging with the help of the Indian Ocean monsoons (Roman sea trade with Indian Ocean lands). Older coastal networks remain functional. Exchange activity heightens as emergent polities create infrastructure and make investments in long distance maritime trade.

The Indian Ocean emerges as an interactive arena by early 1st millennium AD.

Notes

1 Myrrh from Yemen was being supplied along land routes to Nabataea (modern Jordan) and Palestine since early 1st millennium BC. Yemen, the famed land of myrrh was sought to be conquered by the Romans. A military expedition by the Roman Prefect to Egypt, Aelius Gallus, was mounted in 26 BC. Gallus reached close to Marib (northern Yemen) before he turned back for reasons unknown. For discussion on Gallus' invasion see Sidebotham (1986 b: 590–602). For wholesale trade in myrrh from Yemeni ports in the early centuries AD see van Beek (1958: 141–52).

2 In the 4th-3rd millennium BC transition, Neolithic communities practising agriculture flourished in the uplands of Yemen (Al-Ansary, 1996: 238–239; Edens et al., 2000: 854–62). The discovery of a cultivated field in the Yemen highlands dated to early 4th millennium BC indicates the early sedentism of foraging communities. The process of sedentism and beginning of farming in neighbouring Eritrea, Sudan and Somalia do not seem to be as advanced as in Yemen. For discussion on the Neolithic of Ethiopia and Sudan see Phillipson (1996: 307–308). For field explorations of prehistoric and Neolithic (?) cave sites on the northern Somali coast, see Brandt and Brook (1984: 119–121).

3 It shares nothing with Burj basket-marked, Togau or Kechi Beg ceramics. It is certainly not of the Amri-Nal assemblage. In the end we do not know very much about the people who made and used the Pre-Prabhas ceramics. It seems clear that they are part of the early 3rd millennium BC and that some of their pottery lies outside the Amri-Nal tradition. It may be that their origins lie in the Arabian Gulf or Saudi Arabian coast…' (Possehl, 1999: 613–14).

4 For discussion on use of bitumen in the Persian Gulf in Bronze Age, see Zarins (1992: 65–66).

5 The increased deposition of Harappan pottery in Oman in the 3rd–2nd millennium transition most likely reflects transport of essential products (cereals, oil) from agriculturally rich western India to the semi-arid Oman region. Harappan wares found at Ras al Junayz have been traced to the Indus port of Lothal (Gujarat) through comparative XRD analysis by Dr. V. Gogte of Deccan College, Pune (personal communication). Sahni (1997: 9–18) has observed that Harappan black slipped jars and jars with incised rims found in the lower Gulf are similar to ceramics from contemporaneous sites in Saurashtra/Gujarat. The Sorath Harappan sites of Gujarat were strongly oriented to a farming economy, as exemplified by excavations at Rojdi in central Saurashtra (Possehl, 1999). Farm and agricultural products from Harappan and Late Harappan sites in Saurashtra were the likely fillings in the transport ceramics found in Oman. Evidence of Iron Age storage pottery from western India in Oman indicates that the trade in essentials carried on (for discussion see Gupta, 1998a). Sahni's (1997: 9-18) ethnographic survey of the current dhow trade between western India and the UAE (Oman Peninsula)

reveals that essentials (wheat, rice, meats) are still supplied from the former region.

6 *Encyclopaedia Britannica* 1985. Vol.1 p. 509. '...submarine basin of the southern Arabian Sea, rising to meet the submerged Carlsberg Ridge to the south, the Maldive island chain to the southeast, India and Pakistan to the northeast and the Arabian Peninsula to the west...The floor of the basin is covered by sediment deposited by the Indus river in the form of a great alluvial fan whose thickness diminishes to the south.'

7 For various locations suggested for Ophir see Schoff (1912/95: 97, 151, 260). Archaeological corroboration for the Biblical reference to Ophir and its goods is provided in an inscribed potsherd (dated to 8th century BC) found at the old port-site of Tell Qasile, Israel. The ostracon informs: 'Gold of Ophir for Beth Horon, shekels III' (Puskas, 1988: 21). See *Biblical Archaeologist,* September 1984 for the ostracon.

8 The transport of ship-building timber from Lebanon into the Red Sea seems to have occurred more than once in history. A palynological study on the uplands of Lebanon carried out by Yasuda *et al.,* (2000: 127–36) points to the human hand in the denudation of cedars in the 3rd millennium BC. Yasuda *et al.,* relate the denudation to the large requirement of timber by the Mesopotamians, following the story in the Gilgamesh Epic. It is not far-fetched to assume that Lebanese timber also fulfilled the demand of Egptian ship-builders.
As late as early 2nd millennium AD, the Crusaders built ships on the Sharm el Sheikh from timber brought from Lebanon (Perry, 1997).

9 An ethnographic record in the form of a traditional history of Jews of western India—the Bene Israelis—suggests another early crossing of the Arabian Sea. The Bene Israelis of western India trace their origin to members of a Hebrew tribe that fleeing persecution in their native land, arrived by sea at the Konkan coast of western India where their ship was wrecked. Most Bene Israeli historians date this event to early 1st millennium BC. The Bene Israelis (this community name is a recent adoption) believe themselves to be the descendants of the survivors of the shipwreck. A twin tumuli at the village of Navgaon near Alibag township on the Konkan is revered by the Bene Israelis as the burial-site of the dead from the shipwreck (Samson, 1917/reprint 1984: 51–54). This tumuli still exists and has never been excavated. The Bene Israelis consider themselves to be members of one of the ten tribes of Israel which formed the Kingdom of Israel after the death of King Solomon in 950 BC. Bene Israeli historians have put forth a number of reasons why they are essentially of the nation of Israel and the causes that made them flee by sea to India. In this regard, an authoritative journal of the community '*Vellimadthem Otham*' (Vol.1, No.7, Sept-Oct 1959: 5) informs: 'According to one theory which has been to some extent substantiated by many unique rituals and customs observed by us even this day, our forefathers belonged to the period of King Ahab (875–854 BC) and were staunch supporters of Prophet Elijah. It is probable, they left the country as a result of Queen Jezebel's (Ahab's wife) persecution of Prophet Elijah and his disciples...' Another body of opinion among the Bene Israelis holds that the Israeli diaspora to western India took place in the immediate aftermath of the disastrous invasion of Shalmanesar, the King of Assyria, upon Israel and his taking into captivity all the tribes of Israel. This event is put at 740 BC or 721 BC in the Biblical chronology (Samson, 1917/reprint 1984: 51–54). The '*Vellimadthem Otham*' (ref. above) posits a third probable date for the Israeli migration around 586 BC, when the First Temple at Jerusalem was destroyed by the Babylonians. However, some Bene Israelis believe their arrival in India to be as late as the 2nd century AD (Ezekiel, 1981: 29–30).

10 The Early Historic trade ports also symbolize political competition and rivalry; as between Saba and Hadhramaut on south Arabian coast and between the Kshatrapas and Satavahanas on the Konkan coast. For theoretical approaches to peer polity interaction, see Renfrew and Cherry, 1986. For role of ethnic communities in supply of resources to the Early Historic ports in India, see Stiles (1993: 153–67) and Stiles (1994).

11 For review of archaeological evidence of '*Periplus*' port-sites in India see Gupta, 1994a: 217–225; also Gupta, 1993: 119–127; Gupta, 1996: 52–58; Gupta, 1998a. A number of scientific studies have reconstructed the geomorphology of the Indian coastlands. For geophysical reconstruction of Muziris on the Malabar, see Mathai and Nair (1988: 106–119). For geomorphology of the Gujarat coast see Nayak *et al.,* (1989) and Shaikh *et al.,* (1989: 41–48). For remote sensing of the entire Indian coast see Sahai (1992).

12 For excavations at Arikamedu see Wheeler (1954) and Begley *et al.* (1996). Gupta (forthcoming) discusses the ancient Indo-Southeast Asian interchange in context of the

Indian Ocean; for Roman trade impact on Sri Lanka see Bopearachchi (1997: 377-91).

13 Bellwood (1992: 128) describes the material marking out Lapita presence: 'The most remarkable features of Lapita assemblages include striking dentate-stamped or incised pottery, a range of tools, body ornaments and fish-hooks made of shell and a far-flung exchange network involving obsidian from sources in the Admiralty Islands and New Britain.'

14 Reid (1988: 31– 32) explains the linguistic syncretism: 'If Nancowry really is an Austroasiatic language its morphology clearly shows it has Austronesian connections. These could not be the result of language contact with perhaps Malay to the south, since it is highly unlikely that a language would borrow morphology without borrowing the forms to which they are attached. Neither could Nancowry be an Austroasiatic language with a Malay substratum – the affixes are too archaic to be considered to be Malay. The only alternative is to consider Austronesian and Austroasiatic to be genetically related...' For discussion on Reid see Blust (1996: 122-123).

References

Al Ansary, A.R. (1996): The Arabian Peninsula, in *History of Humanity, From the Third Millennium to the Seventh Century BC* Vol. II, A.H. Dani and J-P. Mohen (Eds.), pp. 238–45. Paris: UNESCO.

Bailey, G. and J. Parkington (1988): *The Archaeology of Prehistoric Coastlines.* Cambridge University Press, Cambridge, 154 pp.

Begley, V., P. Francis Jr., I. Mahadevan, K.V. Raman, S.E. Sidebotham, K.W. Slane and E.L. Will (1996) *The Ancient Port of Arikamedu, New Excavations and Researches, 1989–92.* Ecole Francaise D'Extreme-Orient, Pondicherry.

Bellwood, P. (1992): Southeast Asia before History. In Arling, N. (ed.): *Cambridge History of Southeast Asia Vol. I.* Cambridge University Press, Cambridge, pp.55–136.

Besenval, R. (1992): Recent Archaeological Surveys in Pakistani Makran. In Jarrige, C. (ed.): *South Asian Archaeology 1989.* Prehistory Press, Madison, pp. 25–35.

Besenval, R. (1997): The Chronology of Ancient Occupation at Makran: Results of the 1994 Season at Miri Qalat, Pakistan Makran. In Allchin, R. and B. Allchin (eds.): *South Asian Archaeology 1995 Vol. I.* The Ancient India and Iran Trust, Cambridge, pp. 199–216.

Biagi, P., R. Maggi and R. Nisbet (1989): Excavations at the Aceramic Coastal Settlement of RH5 (Muscat, Sultanate of Oman) 1983–85. In Frifelt, K. and P. Sorensen (eds.): *South Asian Archaeology 1985).* Scandinavian Institute of Asian Studies, Occasional Papers No. 4., Copenhagen, pp. 1–8.

Blench, R. (1996): The Ethnographic Evidence for Long Distance Contacts between Oceania and East Africa. In Reade, J. (ed.): *Indian Ocean in Antiquity.* Keagan Paul International, London, pp. 417–436.

Bopearachchi, O. (1997): Sea-borne and Inland Trade of Ancient Sri Lanka: First Results of the Sri Lanka-French Exploratory Programme. In Allchin, R. & B. Allchin (eds.): *South Asian Archaeology 1995.* Oxford & IBH Publishing Co, New Delhi, pp. 377–391.

Bound, M. and D. McLeod. (1992): The Charles Cooper in Calcutta: A Maritime link between the United States, India and the Falklands Island. In Nayak, B.U. and N.C. Ghosh (eds.). *New Trends in Indian Art and Archaeology, S.R. Rao's 70th birthday felicitation Volume II.* Aditya Prakashan, New Delhi, pp. 461–468.

Brandt, S.A. and G.A. Brook. (1984): Archaeological and Palaeoenvironmental Research in Northern Somalia. *Current Anthropology,* 25 (1): 119–121.

Casson, L. (1984): *Ancient Trade and Society.* Wayne State University Press, Detroit, 284 pp.

Casson, L. (tr) (1989): *Periplus Maris Erythraei.* Princeton University Press, Princeton, 320 pp.

Chakrabarti, D.K. (1990): *The External Trade of the Harappans.* Munshiram Manoharlal, Delhi.

Chakrabarti, D.K. (2000): *An Archaeological History of India.* Oxford University Press, Delhi.

Chakravarti, R. (1992): Maritime Trade in Horses in Early Historical Bengal: A Seal from Chandraketugarh. *Pratna Samiksha. Journal of the Directorate of Archaeology, West Bengal,* 1: 155–160.

Chami, F. (1999): Graeco-Roman Trade Link and the Bantu Migration Theory. *Anthropos* 94, 1999/1–3: 205–215.

Chami, F. and P.J. Msemwa. (1997): A New Look at Culture and Trade on the Azanian Coast. *Current Anthropology,* 38 (4): 673–677.

Chang, K.C. (1981): The Affluent Foragers in the Coastal Area of China: Extrapolation from evidence on the Transition to Agriculture. *Senri Ethnological Studies 9,* 1981: 177–86.

Chittick, N. (1976): An Archaeological Reconnaissance in the Horn: The British-Somali Expedition, 1975. *Azania* XI: 117–134.

Chittick, N. (1979): Early Ports in the Horn of Africa. *International Journal of Nautical Archaeology,* 8.4: 273–277.

Chittick, N. (1980): Sewn boats in the western Indian Ocean, and a survival in Somalia. *The International Journal of Nautical Archaeology and Underwater Exploration,* 9.4: 297–309.

Cleuziou, S and M. Tosi. (1989): The South-Eastern Frontier of the Ancient Near East. In Frifelt, K. and P. Sorensen (eds.). *South*

Asian Archaeology 1985. Scandinavian Institute of Asian Studies Occasional Papers No. 4, Copenhagen, pp. 15–47.

Das, P.K. (1998): *The Monsoons*. National Book Trust, Delhi.

de Maigret, A., G.M. Bulgarelli, F.G. Fedele, B. Marcolongo, U. Scerrato and G. Ventrone. (1985): Archaeological Activities in the Yemen Arab Republic 1984. *East and West*, 1985: 423–454.

de Maigret, A. (1986): Archaeological Activities in the Yemen Arab Republic 1985. *East and West*, pp. 337–395.

Deshpande-Mukherjee, A. (2000): An Ethnographic Account of Contemporary Shellfish Gathering on the Konkan Coast, Maharashtra. *Man and Environment* XXV (2): 79–92.

Dhavalikar, M.K. (1986): Note on Prabhas Patan in *Encyclopaedia of Indian Archaeology*, Vol. II. Indian Council of Historical Research, Delhi, pp. 348–350.

Edens, C., T.J. Wilkinson and G. Barratt. (2000): Hammay el-Qa and the Roots of Urbanism in Southwest Arabia. *Antiquity* 74: 854–862.

Ezekiel, B. (1981): *The Bene Israelis of India, Images and Reality*. University of America Press, Washington D.C.

Flam, L. (1999): The Prehistoric Indus River System and the Indus Civilization in Sindh, *Man and Environment*, XXIV (2): 35–69.

Frank-Vogt, U. (1997): Reopening Research on Balakot: A Summary of Perspectives and First Results. In Allchin, R. and B. Allchin (eds.): *South Asian Archaeology 1995 Vol. I*. The Ancient India and Iran Trust, Cambridge, pp. 217–225.

Glover, I.C. (1990): *Early Trade between India and Southeast Asia*. Hull: Centre for Southeast Asian Studies.

Glover, I.C. (1996): Recent Archaeological Evidence for Early Maritime Contacts between India and Southeast Asia, in *Tradition and Archaeology, Early Maritime Contacts in the Indian Ocean*, H.P. Ray and J-F. Salles (eds.), pp. 129–158. Delhi: Manohar.

Gowlett, J. (1984): Australia—colonization of the island continent. In *Ascent to Civilization—Archaeology of Early Man*. Galley Press, UK.

Gupta, S. (1993): The Location of Kammoni (*Periplus* 43). *Man and Environment*, XVIII (2): 119–127.

Gupta, S. (1994a): Archaeology of Indian Maritime Traditions. *Man and Environment*, XIX (1–2): 217–225.

Gupta, S. (1994b): The Dressel 2-4 Amphora in Indo-Mediterranean Sea Trade, in *Archaeology as an Indicator of Trade and Contact, Theme Papers of the World Archaeological Congress—3*. December 4–11, 1994. New Delhi.

Gupta, S. (1996): Kuda—Mandagora of the Greeks, *India Magazine*, Aug-Sept 1996: pp. 52–58.

Gupta, S. (1998a): *Roman Egypt to Peninsular India: Archaeological Patterns of Trade, 1st century BC—3rd century AD*. Unpublished Ph.D. thesis. Deccan College, University of Pune.

Gupta, S. (1998b): Nevasa: A Type-site for the Study of Indo-Roman Trade in Western India, *Journal of South Asian Studies*, 14: pp. 87–102.

Gupta, S. (Forthcoming). Early Indian Ocean in the context of the Indian Relationship with Southeast Asia. Paper submitted to the *Centre for Study of Civilizations*, New Delhi.

Harlan, J.R. (1993): The tropical African Cereals. In Shaw, T., P. Sinclair, B. Andah and A. Okpolo (eds.), *The Archaeology of Africa*, Routledge, London, pp. 53–60.

Hashimi, M. and R. Nair (1988): Dynamic Tidal Barrier of Gulf of Kutch and its relevance to the Submergence of Ancient Dwarka. In Rao, S.R. (ed.): *Marine Archaeology of Indian Ocean Countries, Proceedings of the First Conference on Marine Archaeology of Indian Ocean Countries*. National Institute of Oceanography, Goa, pp. 56–57.

Hassan, F. (1998): Holocene Nile Floods, Birket Qarun, Faiyum, Egypt. *Newsletter of the YRCP-ECGP Program, Japan Ministry of Education*, March 1998.

Heyerdahl, T. (1982): *The Tigris Expedition*. Unwin Paperbacks, London.

Higham, C. (1996): *The Bronze Age of Southeast Asia*. Cambridge University Press, Cambridge, 381 pp.

Higham, C. and B. Maloney, (1989): Coastal Adaptation, sedentism and domestication: a model for socio-economic intensification in prehistoric Southeast Asia. In Harris, D. R. and G.C. Hillman (eds.): *Foraging and Farming, the Evolution of Plant Exploitation*, One World Archaeology Series—13. Unwin Hyman, London, pp. 650–666.

Higham, C and R. Thosarat. (1998): *Prehistoric Thailand*. River Books, London.

Hodder, I. (1987): The Contribution of the Long Term. In Hodder, I. (ed.): *Archaeology as Long Term History*. Cambridge University Press, Cambridge, pp. 1–8.

Hojlund, F. (1989): Some New Evidence of Harappan Influence in the Arabian Gulf, in *South Asian Archaeology 1985*, K. Frifelt and P. Sorensen (eds.), Copenhagen: Nordic Institute of Indian Studies, pp. 49–54.

Huang, T.N. (1992): Liangzhu—a late Neolithic jade yielding culture in Southeastern coastal China. *Antiquity* 66: pp. 75–83.

Kar, K.K., K.K. Basa and P.P. Joglekar. (1998): Explorations at Gopalpur, District Nayagarh, Coastal Orissa. *Man and Environment*, XXIII (1): 107–14.

Karttunen, K. (1989): *India in Early Greek Literature*. Finnish Oriental Society, Helsinki, 293 pp.

Katsuhiko, O. and S. Gupta. (2000): Far East, Southeast and South Asia: Indo-Pacific Beads from Yayoi Tombs as Indicators of Early Maritime Exchange. *Journal of South Asian Studies*, 16: 73–88.

Kentley, E. (1996): The Sewn Boats of India's East Coast. In Ray, H.P. and J-F. Salles (eds.): *Tradition and Archaeology, Early Maritime Contacts in the Indian Ocean*. Manohar, Delhi, pp. 247–260.

Kervran, M. (1996): Indian Ceramics in Southern Iran and Eastern Arabia, Repertory, Classification and Chronology, in *Tradition and Archaeology, Early Maritime Contacts in the*

158

Indian Ocean, H.P. Ray and J-F. Salles (eds.), Delhi, Manohar, pp. 37–58.

Kitchen, K.A. (1993): The Land of Punt. In Shaw, T., P. Sinclair and A. Okpoko (eds.): *The Archaeology of Africa.* Routledge, London and New York, pp. 587–608.

Kirk, W. (1975): The Role of India in the Diffusion of Early Cultures. *Geographical Journal* 141: 20–34.

Manguin, P-Y. (1996): Southeast Asian Shipping in the Indian Ocean during the First Millennium AD. In Ray, H.P. and J-F. Salles (eds.): *Tradition and Archaeology, Early Maritime Contacts in the Indian Ocean.* Manohar, Delhi, pp. 181–198.

Mathai, T. and S.B. Nair. (1988): Stages in the Emergence of the Cochin-Kodungallur Coast-Evidence for the Interaction of Marine and Fluvial Processes. In Jain, D.V.S. (ed.): *Palaeoclimatic and Palaeoenvironmental Changes in Asia.* Indian National Science Academy, New Delhi, pp. 106–119.

Miller, J. (1969): *The Spice Trade of the Roman Empire.* Clarendon Press, Oxford, 294 pp.

Moseley, M.E. and R.A. Feldman (1988): Fishing, Farming and the foundation of Andean Civilization. In Bailey, G. and J. Parkington (eds.): *The Archaeology of Prehistoric Coastlines.* Cambridge University Press, Cambridge, pp. 125–134.

Nayak, S.R. and B. Sahai (1983): Morphological Changes in the Mahi Estuary, in *Proceedings of the National Conference on Application of Remote Sensing to National Resources, Environment, Land Use,* pp. 152–154. Bombay: Indian Institute of Technology.

Nayak, S., M.C. Gupta, A. Pandeya, M.G. Shaikh, H.B. Chauhan and S.R. Rao (1989): Application of Remote Sensing for Monitoring of Coastal Environment. Paper read in seminar on *Remote Sensing of Environment.* Space Application Centre, Ahmedabad.

Nayeem, M.A. (2000): *The Rock Art of Arabia.* Hyderabad Publishers, Hyderabad.

Perry, I. (1997): Migration of Ship-building from the Mediterranean to the Red Sea: Shipwreck of Sharem el-Sheikh. In Mathew, K.S. (ed.): *Shipbuilding and Navigation in the Indian Ocean Region AD 1400–1800.* Munshiram Manoharlal, Delhi.

Phillipson, D. (1996): Africa, excluding the Nile Valley. In A.H. Dani and J-P. Mohen (eds.): *History of Humanity Vol. II.* Routledge and UNESCO, London & Paris, pp. 299–319.

Possehl, G.L. (1997): Seafaring Merchants of Meluhha. In Bridget and Raymond Allchin (eds.): *South Asian Archaeology 1995.* Delhi, pp. 87–100.

Possehl, G.L. (1999): *The Indus Age—the Beginnings.* Oxford and IBH Publishing, New Delhi, 1063 pp.

Potts, D.T. (1990): *The Arabian Gulf in Antiquity* Vol. II. Clarendon Press, Cambridge.

Puskas, I. (1988): Indo-Mediterranica I, *Acta Classica* XXIV: 15–22.

Ratnagar, S. (1981): *Encounters, the Westerly Trade of the Harappan Civilization.* Oxford University Press, Delhi, 294 pp.

Reade, J. (1996): *The Indian Ocean in Antiquity.* Keagan Paul International, London.

Renfrew, C. and J.F. Cherry (eds.) (1986): *Peer Polity Interaction and Socio-Political Change.* Cambridge University Press, Cambridge, 179 pp.

Sahai, B. (1992): Remote Sensing of Coastal Environment—An Overview. Paper read in *Workshop on Utilisation of Coral Reef Maps,* Space Applications Centre, Ahmedabad, January 7, 1992.

Sahni, L. (1997): Ethnoarchaeology of Harappan Sea Trade, *Man and Environment,* XXII (1): 9–18 pp.

Salles, J-F. (1993): The Periplus of the Erythraean Sea and the Arab-Persian Gulf. *Topoi,* 3/2: 493–523.

Samson, D.J. 1917/ reprint (1984): The Bene Israel: Why? Where? And When?, *Souvenir of the Shaar Harahamim Synagogue,* October 1984.

Sayed, Abdel Monem A.H. (1977): Discovery of the Site of the 12th Dynasty Port at Wadi Gawasis on the Red Sea Shore, *Revue d'Egyptologie,* Paris. Tome 29: 140–178.

Schoff, W.H. (1912/95): *The Periplus of the Erythraean Sea.* Munshiram Manoharlal, Delhi.

Sedov, A.V. (1996): Qana' (Yemen) and the Indian Ocean: the Archaeological Evidence. In Ray, H.P. and J-F. Salles (eds.): *Tradition and Archaeology, Early Maritime Contacts in the Indian Ocean.* Manohar Publishers, Delhi, pp. 11–35.

Shaikh, M.G., S. Nayak, P.N. Shah and B.B. Jambusaria (1989): Coastal Landform Mapping around the Gulf of Khambhat using Landsat TM Data. *Photonirvachak, Journal of the Indian Society of Remote Sensing,* 17 (1), 41–48.

Shinde, V.S. (1998): Pre-Harappan Padri Culture in Saurashtra: the Recent Discovery. *Journal of South Asian Studies,* 14: 173–182.

Sidebotham, S.E. (1986a): *Roman Economic Policy in the Erythra Thalasa.* E.J. Brill, Leiden, 226 pp.

Sidebotham, S.E., J.A. Riley, H.A. Hamroush and H. Barakat (1989): Fieldwork on the Red Sea Coast, The 1987 Season, *Journal of the American Research Centre in Egypt,* Vol. XXVI: pp. 127–166.

Sidebotham, S.E. (1986b): Aelius Gallus and Arabia, *Latomus,* Tome XLV, fasc. 3: pp. 590–602.

Sinclair, P.J.J. (1990): Archaeology of Eastern Africa: An Overview of Current Chronological Issues. In Sinclair, P.J.J. and G. Pwiti (eds.): *Urban Origins in Eastern Africa.* The Central Board of National Antiquities, Stockholm, pp. 1–40.

Smith, M.C. and H.T. Wright (1990): Notes on a Classical Maritime Site: the Ceramics from Ras Hafun, Somalia. In Sinclair, P.J.J. and J.M. Rakotoarisoa (eds.): *Urban Origins in Eastern Africa.* Central Board of National Antiquities, Stockholm, pp. 106–114.

Stiles, D. (1993): Hunter-Gatherer Trade in Wild Forest Products in the Early Centuries A.D. with the Port of Broach, India. *Asian Perspectives,* 32 (2): 153–167.

Stiles, D. (1994): Hunter-Gatherer Trading Systems of India: Past and Present, in *Archaeology as an Indicator of Trade and Contact, Theme Papers of the Third World Archaeological Congress.* December 4–11, 1994, New Delhi.

Thimma Reddy, R. (1994): Coastal Ecology and Archaeology: Evidence from the East Coast of India. *Man and Environment,* XIX (1–2): 43–55.

Tosi, M. (1987). *Newsletter of the Joint Hadd Project.* Rome: IsMEO.

Tosi, M. (1990): The Indian Ocean Perspective of the Indus Civilization. Lecture delivered at the Deccan College, Pune. April 27, 1990.

van Beek, G.W. (1958): Frankincense and Myrrh in Ancient South Arabia, *Journal of American Oriental Society,* 78 (3): 141–52.

Vosmer, T. (1994): Traditional Boats of Oman: links past and present. Paper read at *International Seminar on Techno-Archaeological Perspectives of Seafaring in the Indian Ocean.* March 1994, NISTADS, New Delhi.

Wheeler, R.E.M. (1954): *Rome Beyond the Imperial Frontiers.* G. Bell and Co, London, 192 pp.

Weber, S.A. (1998): Out of Africa: The Initial Impact of Millets in South Asia, *Current Anthropology,* 39 (2): 267–274.

Yasuda, Y., H. Kitagawa and T. Nakagawa (2000): The earliest record of major anthropogenic deforestation in the Ghab Valley, northwest Syria: a palynological study. *Quaternary International* 73/74 (2000): 127–136.

Zarins, J. (1989): Eastern Saudi Arabia and External Relations: Selected Ceramic, Steatite and Textual Evidence 3500–1900 B.C., in *South Asian Archaeology 1985,* K. Frifelt and P. Sorensen (eds.), Copenhagen: Scandinavian Institute of Asian Studies, pp. 87–103.

Zarins, J. (1992): The Early Settlement of Southern Mesopotamia: A Review of Recent Historical, Geological, and Archaeological Research. *Journal of the American Oriental Society* 112.1, 1992: 55–77.

Zarins, J. (1996): Obsidian in the Larger Context of Predynastic/Archaic Egyptian Red Sea Trade, in J. Reade (ed.): *The Indian Ocean in Antiquity.* Keagan Paul International, London, pp. 89–106.

Zitterkopf, R. and S.E. Sidebotham (1989): Stations and Towers on the Quseir-Nile Road. *Journal of Egyptian Archaeology,* 75: pp. 155–89.

Chapter 9

Influence of Drainage Patterns on Historic Cultures of Rajasthan

LALIT PANDEY

Availability of water has always played a major role in the development or the decline of civilizations, particularly in regions with a monsoon climate. The regions of monsoon climate are distinct from other climatic regions in as much as the monsoon regions have a well marked 'rainy season' unlike other climatic regions. As such regions receive heavy rainfall during the monsoon season, a number of natural water reservoirs are generally formed in the depressions away from the main rivers. These reservoirs are annually replenished by the monsoon rains. Hence, in monsoon regions the earliest settlements came up along these depressions in addition to the river banks which are the usual places for the growth of settlements. This is established by the fact that a large number of historic sites are found far away from river beds in the vicinity of such depressions.

Very high fluctuation in annual rainfall in a region as well as considerable local variation even in neighbouring regions are common characteristics of monsoon rainfall. As a result, depressions in some regions might not be replenished by annual rainfall for several years, while another depression that may have been dry for many years might be recharged. If such changes persist, they might result in a settlement shift. This is more likely to be the case when the civilization has not developed effective water storage techniques. In addition to natural reservoirs, artificial reservoirs are also constructed by building dykes or embankments.

So the earliest civilizations that developed along the depressions—whether natural or artificial—developed techniques of water management to ensure a regular supply of water throughout the year. The survival and development of such civilizations depended upon the effectiveness of water management techniques developed by them. The fact that many such sites have been found which remained in existence for a considerable time is proof that they had evolved effective water management techniques.

Several Vedic hymns vividly describe the phenomenon of monsoon rain. The *Atharva Veda* describes the thunder of clouds and the flash of lightning which accompanies the thunder, and their effect on vegetation, plant, animal and human life. The *Chhandogya Upanishad* (Limaye and Wadekara, 1959) describes the rain as a sacrificial hymn. Each stage of the rain forms one of the five stages in the sacrificial hymn or the *Pancha Bhakti Sam*: the gathering of the white clouds is the first stage; the forming of rain-bearing clouds is the second stage; the rainfall is the third stage; thunder and lightning is the fourth stage; finally the rain ceases and that is the fifth stage (2.15.1). This same Upanishad refers to the soul as, 'the "Vidhriti", i.e., the dyke or the embankment for the safety of these worlds.' (8.4.1). The imagery here suggests the technique of constructing embankments for artificial reservoirs. The early Brahamana and Buddhistic texts declare that it was the duty of the state to develop necessary methods of irrigation to protect the people from the fluctuations of the monsoon.

These fluctuations of the monsoon were believed to be due to the vagaries of the god of rain. Panini's *Ashtadhyayi,* a Sanskrit text of the 6th century BC, also narrates that rivers and wells were considered a major source of irrigation. Panini also describes the good qualities of alluvial soil (Agrawala, 1953). Writing in the 4th century BC, Kautilya, the Prime Minister of Emperor Chandragupta Maurya, says in his *Arthashastra* that 'Land which is not dependent on the god of rain is the best for agriculture' (Sastri, 1988).

Water storage management and irrigation, were the means by which the land could be freed from dependence on the god of rain; and it was the prime duty of the king to arrange for these. Two types of embankments were constructed. The first type appeared to be tanks which were fed by natural springs of water, while the other seemed to imply storing water in reservoirs by means of embankments. The Sudarshana lake in Saurashtra which was formed by constructing an embankment, provides the best archaeological evidence of embankments during the Mauryan period which also suggests that irrigation facilities were provided by the state in those times.

Rajasthan provides an interesting case for the study of the relationship between the shift in drainage patterns due to meterological and geological changes and the resulting changes in civilization and settlement patterns. In Rajasthan, there have always been semi-arid, semi-desert, and even desert regions lying contiguous to well-watered regions. These areas have never remained continuously favourable to settlement over an extended period. The reasons for this have been: a constant shift in the desert, the shift in the Ghaggar drainage system, and changes in the rainfall which is a normal occurrence in the regions of monsoon climate.

On the basis of archaeological research, it can be said that agriculture-based permanent settlements began to appear only after 3000 BC.

The stimulus for the colonization of the desert was provided by the pressure of the population in the more favourable areas on the margins of the desert. In northeastern Rajasthan the agricultural way of life was established before 1000 BC. Yet studies in the palaeoenvironment of the northeastern and western part of Rajasthan show no evidence of the existence of even a rudimentary village settlement anywhere in the desert outside the Ghaggar valley. The origin and development of Pre-Harappan, Harappan, PWG, and Rang Mahal culture in the Ghaggar valley was the result of the fluctuations in the hydrology of the river which were controlled by geomorphic factors operating in the Himalayas and the Indus-Ganga divide as suggested by Raikes (1968) and Ghosh (1989).

Outside the area of the Indus Civilization, the southeastern region of Rajasthan which is generally known as Mewar, is one of the major regions which provides the evidence of settled community life. This region also provides very useful information about the historic culture of Rajasthan which developed here in the periods (i) from 300 BC to the 3rd century AD and (ii) from the 3rd century AD to the present times.

The hill ranges and the river system of southeast Rajasthan played an important role in the development of historic cultures. The Aravalli hills proper lie in the present district of Udaipur. They derive their name from the Rajasthani word 'Ada-Vala' meaning 'a beam that is lying cross-wise'. In the Udaipur, Banswara and Dungarpur districts, the Aravallis have a thick vegetation cover which has always supported human and animal life. Southeast Rajasthan has two important drainage systems. The first system consists of the Banas river and its tributaries which forms a part of the Ganga water catchment system. Their water flows into the Bay of Bengal. The second system consists of the Mahi river and its tributaries Som, Jakan and Vakal. Their water flows into the Gulf of Cambay (Fig. 1).

The terrain of southeast Rajasthan is mostly hills with undulating plains which form a number

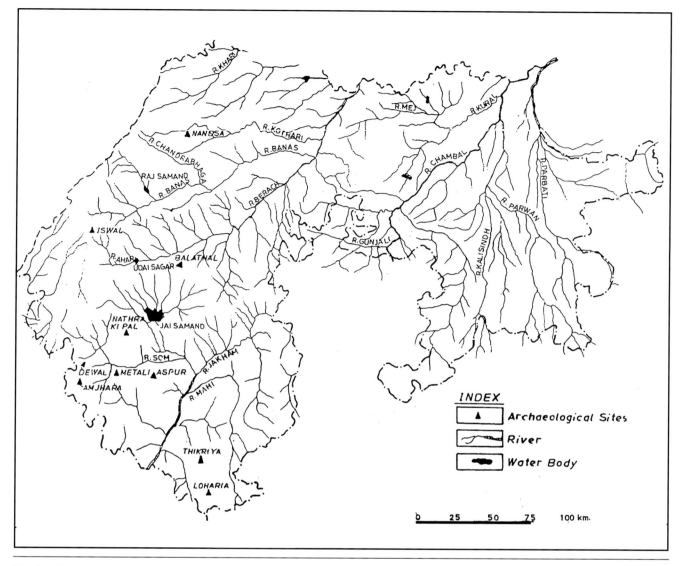

Fig. I. Drainage patterns, southeastern Rajasthan.

of micro water sheds. The rivers form flat basins. The watersheds and basins contain a number of natural depressions which are regularly charged by monsoon rains (Fig. 2). This creates ideal conditions which favour development of agricultural cultures away from the river beds. So a good number of sites are also found in the hinterlands as distant as ten kilometres from the main river (Ghosh, 1989; Misra, 1967) (Fig. 3).

The findings of the Ahar, Gilund and Balathal excavations provide information that Tan Ware and Reserve Slip Ware closely resemble the pottery of the Harrapan culture of Gujarat. Our

present knowledge of the archaeology of Dungarpur and Banswara is limited to only eight sites. Of these two are Black-and-Red Ware Chalcolithic sites and six are historical. A more detailed archaeological study of the region will establish that these two districts were the gateway for Gujarat and Kutch Harappan influence (IAR, 1979–80). The drainage pattern of Dungarpur and Banswara districts will also provide clues which support this hypothesis. The undated Nasik cave inscription (which epigraphists place in the 2nd century AD) of a Saka king Nahapan and the Junagadh Inscription of Rudradaman suggest that

163

Fig. 2. Natural depression at Iswal District, Udaipur.

Dungarpur, Banswara, Chittor and also Ajmer were under the Saka dominion. The Nasik cave inscription also provides information that the brother-in-law of the king, after inflicting a crushing defeat on the Malavas, went to the Pushkar lake for a ceremonial construction (*Epigraphica Indica*, Vol. 8). The region around Pushkar was under the political control of the Uttambhadras who were rivals of the Malavas. The Malavas were originally from Punjab and had migrated to Rajasthan to maintain their freedom (Sharma, 1966).

According to Allchin: 'Prior to the introduction of well-digging and other methods of water storage and irrigation in early historic times, perennial sources of fresh water must have been rare indeed—the majority of lakes in Rajasthan being saline and the rivers were seasonal torrents. Both men and animals must have been drawn to Pushkar during dry season.' Besides the inscriptional evidences, the discovery of Saka coins at Pushkar proves that the region around Ajmer was a centre of socio-political activities during the early centuries of the Christian era. The following inscriptions of the 3rd century AD provide information that the region around Chittor and Bhilwara was a major centre of the rulers of Malavas and other regions who had faith in Vedic sacrifices:

Nandsa Inscription of Sri (?) Som Sogi of AD 225 was found in a lake.

Bhattisom Sogi's Inscription of Nandsa was located half a kilometre away from Sri Som Sogi's Inscription (*Epigraphica Indica* Vol. 27).

Barnala Inscription of AD 227 (*Epigraphica Indica* Vol. 26).

Barnala Inscription of AD 228 (*Epigraphica Indica* Vol. 26).

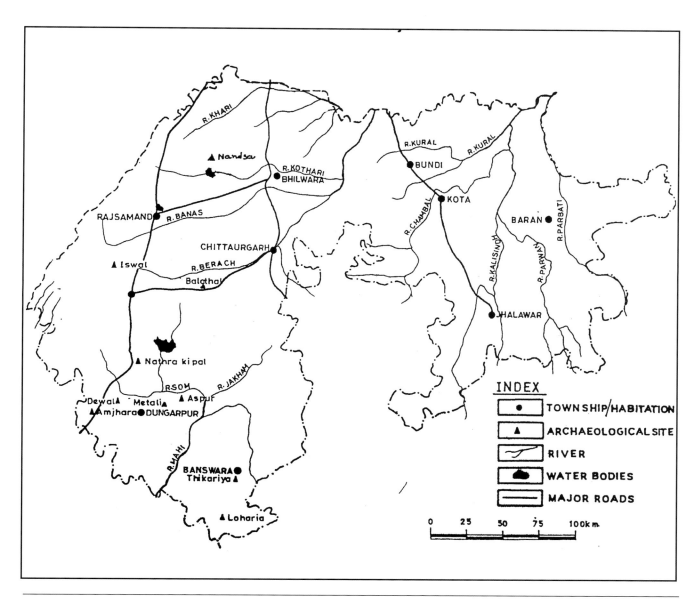

Fig. 3. Archaeological sites, southeastern Rajasthan.

The inscriptions 1 and 2 were also found in a lake. The Nandsa Inscription narrates that the leader of the Malavas Sri (?) Som Sogi established the Yupa inscription in a lake. According to the inscriptions, this lake is no less holy than Pushkar. The inscriptions do not make it clear whether the lake is a natural depression or an artificial one. But what does clearly emerge from these inscriptions is that by that time human settlements remain confined to river banks alone. The remains of five ancient embanked reservoirs surrounding the site at Nagar prove that there were proper facilities of water supply. This ancient site also provides the remains of an ancient square-shaped stepwell (Jain, 1972).

Together with this inscriptional evidence there are also some archaeological sites which are located on natural depressions. Balathal, Iswal (Udaipur District) and Natharapal (Rajsamand District) also corroborate this (Fig. 3).

D.D. Koshambi says that during the early centuries of the Christian era, the baron might have added value to the land by undertaking works beyond the means of a single village like building dams, canals, etc. (Kosambi, 1965). During the early medieval period *Vraksa Ayurved, Krisiparasar* and *Brihad Samhita* were also written. These texts provide useful information regarding plant diseases, fertilizers and meterology (Bose, 1971). The law books which were written during the early Christian era, laid down rules regarding irrigation. They provide useful information regarding the various types of water storage techniques like tanks, wells, pounds embankments, etc. Digging of wells had become a major activity in the medieval period of Rajasthan. The archaeological evidence of the later period also corroborates this. It is interesting to note that the term *vapi* is derived from the Sanskrit root *vap* which means to sow. Therefore, it is clear that the stepwells were made for irrigating the fields but they would be equally useful for supplying drinking water and also for irrigating gardens (Sharma, 1987).

Besides the stepwells, the use of the Persian wheel and leather bucket had become widespread during the 9th and 10th century AD in Rajasthan. The Pratapgarh (near Chittorgarh) inscription of AD 946 mentions two terms: Arahatina and Kosavaha which are related to irrigational activities. Arahatina is the instrumental form of Arhata, a Persian wheel, the Sanskrit form being Araghatta. The other term Kosavaha is applied to as much land as can be irrigated by one kosa, or a leather bucket (*Epigraphica Indica* Vol. 14, p. 176).

Therefore, on the basis of the above archaeological and literary evidence, it can be said that during the Historic period the people of Rajasthan maintained the growth of civilization by using water management techniques in agriculture.

Conclusive evidence of the deliberate digging of ponds or tanks, in a Harappan context is forthcoming from the recent excavations at Dholavira (Kutch, Gujarat), (Mohan, 1997). During the course of the present exploration in southeastern Rajasthan, it was found that the utilization of the natural depressions and artificial structures for the storage of rainwater is still in practice. Thus there is a direct link between the development of artificial devices for storing water and the growth of civilizations away from the river banks. This highlights the necessity of a fresh approach to the study of the civilizations of the historic period of Rajasthan.

References

Agrawala, V.S. (1953): *India as known to Panini*, University of Lucknow.

Allchin, F.R. (1979): *A Source Book of Indian Archaeology*, Vol.I, Mushiram, Manoharlal Publishers, Delhi.

Atharva Ved: Roth, R. and W. Whitney, edited by Rama Chandra Sarma (with Sayana's commentary) (English translation and Notes), Cambridge, 1955.

Bose, D.M., S.N. Sen and B.V. Subbarayappa (eds.) (1971): *A Concise History of Science in India*, Indian National Science Academy New Delhi.

Chakrabarti, Dilip, K. (1989): *A History of Indian Archaeology from the Beginning to 1947*, Munshiram Manoharlal Publishers, Delhi.

Chhandogya Upanishad: Limaye and Wadekara (eds. and trans.), V.S.M. Poona, 1959.

Epigraphica Indica: Vols.-VIII, XIV, XXVI, XXVII, Govt. of India, New Delhi.

Ghosh, A. (ed.) (1989): *An Encyclopaedia of Indian Archaeology*, Vols.I & II, Munshiram Manoharlal Publishers, Delhi.

Indian Archaeology: A Review (1979-80): Archaeological Survey of India, New Delhi.

Jain, K.C. (1972): *Cities and Towns of Ancient India*, Motilal Banarsidas, Delhi.

Kautilya Arthashastra: Shama Sastri, R. (ed. and trans.), Mysore, 1929, Kangle, R.P. (ed. and trans.), University of Bombay, 1988.

Kosambi, D.D. (1965): *The Culture and Civilization of Ancient India in Historical Outline*, Routledge and Kegan Paul, London.

Misra, V.N. (1967): *Pre- and Proto-history of the Berach Basin, Southern Rajasthan*, Deccan College, Pune.

Mohan, Vijneshu, (1998) : *India, 50 years of Independence*, Delhi.

Raikes, R.L. (19680): Kalibanga: Death from Natural Causes, *Antiquity* 4: 286-91.

Sharma D. (1966): *Rajashthan Through the Ages*, Govt. of Rajasthan, Jaipur.

Sharma, Gopinath (1975): Rajasthan Studies, Educational Publishers, Agra.

Sharma, R.S. (1987): *Urban Decay in India*, MLBD, Delhi.

Varahmihira Brihad: Bhatt, Ramkrishna, M. (ed. and trans.) (1981), Samhit (Part-1), Delhi.

Chapter **10**

Adaptive Responses to a Deficient Monsoon in Gujarat, Western India: Changing Pattern in Organization of Labour and Settlement Rearrangement

SWAYAM PANDA

Introduction

Most literature on 'pastoralism' operates on a premise of the negation of the agricultural mode of production. This premise arises primarily from a distinct ecological, cultural and political system associated with the pastoral people. The overt recognition of the distinctiveness of the pastoral communities is often encountered in literature with an extreme polarization—desert and sown, nomads and sedentary—that ultimately creates pristine categories of modes of production. The structural matrix of the practical economic activity of an ecological population within a given ecological niche functions within a flexible framework of the combination of two or more subsistence modes to meet different economic and cultural needs. The risk aversion is fundamental to subsistence economies, and within this constraint, the subsistence producers adopt an extremely flexible approach towards the organization of production. The organization of production in turn depends on the ecological constraints or endorsements particular to the eco niche. This realization ordains a regional sensitivity in formulating the perceptions about a particular organization of production.

The organization of production in our region deals with two relevant components, i.e., agriculture and pastoralism. The study focuses on the organization of production in three different sets of population, which imbibe these two components either directly or indirectly. The common component in these three sets of adaptation strategies is livestock rearing. The household is the primary unit of production and consumption. Hence, the economic success of the respective production regimes is evaluated here by measuring the ability to cope with a characteristically uncertain environment through a designed institutional support system to use the resources of land and labour effectively.

The Ecological Matrix

The geographical boundary of Gujarat can, for purposes of analysis in this paper, be perceived in terms of two distinct 'regions': the peninsular part of Gujarat and Mainland Gujarat. These two regions not only manifest clear ecological differences but adaptation strategies followed by the people in each region differed from one another in a historical perspective. The alluvial plains, most suitable for agriculture are mostly found in Mainland Gujarat. Mainland Gujarat is crisscrossed with mighty rivers such as Saraswati, Rupen, Banas, Sabarmati, Mahi, Narmada and Tapi. In contrast, Peninsular Gujarat, except for Bhadar and to some degree Satrunji, does not have a mighty river system to support agriculture through irrigation. Good nutrient soil is mostly confined to Mainland Gujarat and is found along riverine piedmonts. Good cultivable soil in Kutch is confined to a few patches, insignificant for supporting any large-scale agrarian economy. Good nutrient soil is available in Kathiawar Peninsula though the depth of the soil varies from place to place. The coastal alluvium is limited to only the southwestern corner of the peninsula. The rest of the coastal strip is covered either by saltpan or by swampy mangroves. The central part of the Kathiawar Peninsula is a hill tract; which at some places has good nutrient soil, but the depth of the soil is shallow.

The amount of rainfall in Mainland Gujarat is more than in Peninsular Gujarat. The whole of Kutch is classified as an arid ecozone based on low and uncertain rainfall. Similarly, except for the southeastern corner of the Peninsula, the larger part of Kathiawar again is classified as a semi-arid zone. It will not be out of place here to evoke a similar opinion by Whyte: '...western Gujarat (Saurashtra and Kutch) is more pastoral than agricultural. The rainfall is low and often uncertain, there are periods of scarcity of grazing and fodder cycles of five or six years.' (Whyte, 1975: 227). More than seventy species of grasses and fifty species of legumes are identified from Peninsular Gujarat that forms the main source of fodder in this semi-arid zone (Ganguli et al., 1964). The grassland in Peninsular Gujarat is found in two types of habitats, i.e., (a) plains formed by an old and young alluvium including coastal sands and sandy habitats with deep soils (b) rugged hilly projection and general formation with shallow soils (Rao, 1970). As a contrast to the available cultivable soil in Peninsular Gujarat, the predominance of grasslands in the region shifts the focus to pastoral possibility.

On the one hand, low rainfall jeopardizes agricultural prospects and on the other, the porous soil and the high atmospheric temperature fail to retain rainwater in the open-air reservoirs, small pools, and lakes. Depending upon the climatic conditions and location of the reservoirs, as much as eight to ten feet of water is lost through evaporation alone annually; and in case of shallow reservoirs, this can be a large faction of the total quantity of stored water (Shukla et al., 1962). Compounded with the fluctuation of rainfall, these areas become inhospitable for a purely agricultural population. With the temporal increase in humidity in rainy seasons and with reservoirs and ponds to store the runoff rainwater, the regions where good cultivable soil is available become temporarily suitable for agriculture. However, due to the precarious character of the rainfall, agriculture is not considered as a dependable means of livelihood. This explains the marginal character of the major portion of Peninsular Gujarat. The highland zones and mountain valleys of Kutch and Kathiawar in the most liberal estimation would be suitable for dry cropping within a mixed economy. Hence, the pastoral component plays an important role in the life of this region.

The most important factor in defining the marginality in this region is the uncertainty of the rainfall which often results in a drought situation. The frequent droughts in this region have a decisive impact on the economy. Drought and aridity are some of the inherent characteristic features of monsoonal climates where the rainy season is restricted to a few months. Drought and aridity are related concepts for it is the dryness that in its extreme intense form and long duration contributes to the aridity (Meher-Homji, 1997). Hence, the drought and aridity is defined after due consideration of the nature and duration of dryness. Venkateswaralu (1993: 1) identifies three categories of drought: meteorological, a situation when there is a significant decrease in the normal precipitation over an area; agricultural, when the soil moisture and rainfall are less adequate during the growing season to support healthy crop growth; and hydrological, when the prolonged drought leads to marked depletion of surface water and consequent drying up of the reservoirs, lakes, streams, rivers and fall in the ground water level. Meteorological and hydrological droughts have a direct effect on the pastoralists. Nonetheless, the pastoralists through their extended symbiotic relationship with the agricultural population in the region also feel the impact of agricultural droughts. Hydrological droughts force the pastoralists to abandon the drought-affected region for years together. As vegetational growth and germination depend on the moisture content of a particular soil, the spatial movement of pastoralists needs synchronization with the monsoon rainfall pattern.

The pastoralists generally return to their home province immediately after the commencement of the monsoon. Mallik and Govindaswamy (1962–63) divide the drought into early, middle-season and late drought depending on the time of its occurrence in relation to the kharif cropping season. Their analysis indicates that Gujarat in general is prone to the middle-season droughts, as the late monsoon is much less dependable than the early monsoon showers. High-assured rainfall pockets are found around Junagarh, Rajkot, and Palitana, whereas areas of low rainfall comprise Ranpur, Chuda, Dhanduka, and Dholera (Biswas and Basarkar, 1982). In areas such as Junagarh and Rajkot where the rainfall is relatively assured, the cultivable land available is limited. Agriculture is possible in a limited way in the tablelands and mountain plateau where the depth of the soil is shallow. In geographical terms, these areas are suitable for mixed economies, as some of the land is appropriate for cultivation and the surrounding land is suitable for grazing. Though these regions are discounted from severe seasonal droughts, the occurrence of moderate drought cannot be ruled out (Sahu and Sastry, 1992). Consequently, the level of agricultural production remains unreliable.

Such famines occur either due to the failure of the monsoon, or due to heavy untimely rainfall that destroys the crops. A large number of wells and tanks were constructed in the past famine years as the rulers of the land patronized these endeavours to provide work for the rural folk. Hence, many of the wells and tanks now used for agricultural purposes are the by-products of the recurring famine years in the last few centuries. Agricultural expansion in the recent past has increased the population considerably. Several acreages of pasture and wastelands have been reclaimed to satisfy the demands of this growing population. This phenomenon of agricultural expansion on the one hand has created an illusory agricultural past, while simultaneously reducing traditional pastures. The investment in artificial irrigation over the centuries has successfully brought a considerable landscape under cultivation. Yet, this has not resulted in total abandonment of the pastoral component in their economy.

The Land-Use Reallocation: A Historical Perspective

The environmental condition of Peninsular Gujarat heavily skews the opinion towards pastoralism. If we believe that in the last five thousand years, the environment of the region has not undergone any substantial change, the cultural history then should reflect a pastoral hegemony; and it does. Kathiawar and Kutch were under the pastoral chiefs at least up to the 12th century AD (Swayam, 2000: 64–89). The agricultural expansion in these two regions is comparatively a recent phenomenon. The first large-scale agricultural colonization of Kathiawar can be assigned to the Solanki period. Agricultural expansion in Kathiawar was gradual under Mughal and Maratha rule. Even today, most of the agricultural landholding in Kathiawar is not irrigated. By and large, agriculture here is rain fed. In Kathiawar Peninsula most of the villages depend on runoff rainwater stored in tanks. Colonial sources reveal that most of these tanks were built up as a restoration measure to fight the famine situation.

Yadavas, Gujjars, and Ahirs are the most prominent names among the past pastoral communities of Western India. A review of historical events shows a gradual expansion of agriculture into Peninsular Gujarat. This expansion has made a decisive impact on the social structure and cultural framework of the region. At present Rabaris, Bharwads, Sindhi Muslims, and Charans are some of the easily identifiable pastoral communities. The ethnic identity of many of the past pastoral communities has been redefined in the context of the cultural process which spans the last two millennia. As an essential component of this historical process, the occupational dynamics have drawn a section of pastoralists into other artisan trades. They have acquired specialized knowledge in different crafts to cater to different requirements of the settled population. For example, Gujjar Sutars have taken up carpentry; Gujjar Sonis have mastered goldsmithy; Gujjar Salats practise masonry; Kamalias, a sub-group of Bharwads has taken to weaving; Banjaras and Kathis have involved themselves in trading. There are many more examples of such occupational shifts from one sphere to another. Hence, occupational diversification here forms an integral part of the social and cultural dynamism involving both pastoral as well as agricultural populations.

The cultural process involving the shifts in occupation has remained contiguous throughout. Though the process has included different people and has shifted its venue, the influence has remained valid in different spatial and temporal milieu. The most recent part of this contiguous process of subsistence change can be traced back to the colonial period where invincible historical record exists. British sources are an undeniable testimony to the sedentarization process of a few pastoral communities towards the end of the 19th

century. The British Gazetteer mentions that a section among the Kathis, who practise agriculture in different parts of Gujarat at present, were involved in trade with their large herds of pack bullocks before a road transport system was developed by the British administration. Charans at present have taken to agriculture, and have settled down in different parts of Gujarat with the aid of land grants sanctioned by the state government. Ahirs and Gujjars have settled down and assimilated themselves in the mainstream agricultural population to a point beyond any cultural distinction.

Since Independence, the government has been making sincere efforts towards the settlement of the nomadic pastoral communities in Gujarat. The 'Rabari-Bharwad settlement Yojana' (1955–56) was the first political effort in independent India to catalyze the settling process that was already set forth by the colonial administration. This policy facilitated the transferring of the government-owned fallow lands to Rabaris and Bharwads, who form the majority of the pastoral population in Gujarat, by creating 'Cooperatives for Cultivation and Pastoralism' (*Ksheti-Gopalana Mandal*). At the end of this five-year plan, 224 such cooperatives were formulated, and the government of Gujarat transferred 24,902 acres of fallow land to them. In subsequent years, *'Saurashtra Maldhari Yojana'*, *'Gujarat State Gopalak Cooperative Societies'* and the *'Gujarat State Dairy Development Council'* encouraged the settling down of the pastoralists through various policies sponsored by them. The paucity of more direct evidence to reconstruct the historical process of change along with the lack of any serious academic attempt to record and understand these dynamics adds bitterness to our realization that now we have little scope for studying the pastoral production system that was in vogue before the monetary economy grew in Gujarat.

Making the Most Out of a Semi-Arid Eco-System: The Changing Configuration of Labour

The preceding paragraphs make a critical assessment of the historical position of two important components of the organization of production in Peninsular Gujarat. The rest of this analysis focuses on the pastoral component and examines the adaptational challenges it has faced because of agricultural expansion in the region. Three variants of pastoralism were identified. The most rare among these is semi-nomadic pastoralism—a classic form of their past way of life which has become obsolete in the age of modernization. This is a lifestyle where pastoralists depend solely upon livestock rearing and move between semi-permanent villages known as *wands* and transit camps known as *tandas*. The second category refers to those who are engaged mainly in pastoral activities, but are attached to agricultural settlements. They can be considered as the pastoral specialists, like other specialized groups in an agrarian village. The third variant is a mixture of pastoralism and agriculture, where the pastoral segment of the economy is considered crucial both in economic as well as in cultural terms. I have elaborately discussed the organization of labour elsewhere (Swayam, 2001) in each of these adaptation strategies. I present here a brief summary to connect this adaptive response to the changing historical and ecological domains outlined in the preceding paragraphs.

The Maldhari Lifestyle: The Archetypal Model

This particular lifestyle of the Maldharis is considered as classic because most of them believe that 'a long time ago' all of them followed such a lifestyle. The sedentarization process in the last few centuries has resulted in some deviations in this traditional lifestyle. Only a few of them still lead a lifestyle that can be considered close to this classic form. Such Maldhari populations are mostly found in parts of Kutch, north Kathiawar, and around Dwarka in Jamnagar district. Tuna wand was taken as a representative settlement of this category. Wand is a term particularly used in the context of pastoral and semi-permanent settlement. Its residents occupy the wand for a few specific months of the year. For the rest of the year, they migrate to other regions with their domestic herd, leaving behind nothing but the basic residential structure. They return to the wand in the congenial months to bring life back for a few months again. While on their way to pastureland in the summer months, they camp at regular intervals. These mobile units of pastoral groups along with their belongings are known as tanda. Thus, the life cycle of the pastoralists of Tuna wand oscillates between their wands in Kutch and their tandas in Mehsana and Nalakantha.

Tending the herd is considered a male domain while the household is entrusted to the women folk. Men take their herds to the pasture in two shifts. The first is early in the morning, after the first meal of the day. The milking is primarily the responsibility of men, but women and children help the male members of the family. When pastures become scarce within the daily commutable distance around the wand, the whole family relocates itself along with the herd.

The life of the pastoralists shifts to the tanda mode. The wand is divided into small herding groups, which stay together for the rest of the year. The periodicity of their transhumance to and out of the wand is regulated by the commencement of the rain. They migrate in a group of three to five families together. Tandas move towards the summer province slowly and steadily. The general direction of movement of the tandas remains the same every year, but the routes followed by a particular tanda may differ. The division of the

Women busily gathering crop in Western Rajasthan, India. (Photo by Takeshi Takeda)

labour within the family remains the same. Women in the tandas cook the food early in the morning. After male members walk out of the camp with their herds, they finish the household chores, and pack the tent along with other belongings. They load it on their pack animals and then move to the next destination. Though the decision to relocate the camp is taken by the leader (*Mukhi*) of the group, the responsibility to shift the camp rests primarily on the women; and children help their mothers in this.

The wands in the Kutch generally move to Mehsana and Panchmahal hills through north Gujarat. Some of them even move further to Malwa. The pastoralists move out of the agricultural tract of the north Gujarat plain before the rabi crops come up and choose a tract of pastureland where water is available in plenty, to spend the driest months of the approaching summer. The tandas either move into the forests of Panchmahal or to the banks of the Nal Sarovar. Here they put up a camp for a longer duration. Towards the month of March, the areas that are used by the pastoralists as their summer province, receive an influx of several tandas. They settle there together and graze their cattle on the adjacent grasslands for the rest of the summer. These camps last until the arrival of the monsoon, and almost resemble a small village without any permanent structure. In these camps, again the daily schedule of the wand is followed. In a nutshell, the herd is the domain of male members and the household is the domain of women.

The division of labour within the family thus is relatively simple. We need to delve into the labour management principles followed by these pastoralists in the larger context of the subsistence strategy of the community. The pastoralist communities in Gujarat are also a cohesive unit with their own religio-political structure to administer community matters within themselves. Mostly, the pastoralist communities here are organized under a few religious leaders who enjoy considerable power in civic and economic domains. They act in close alliance with the headmen of a number of wands who form their clientele for magico-religious services. The power structure of the community is further broken up into smaller units during the transit days when the leadership of the tanda is vested in *Mukhi*. The labour management principles within the community are enforced through a set of norms that are strongly embedded within the cultural practices of the group. Some of the cultural norms that have direct or indirect bearing on the labour management are discussed below.

The division of labour across the sex and age-sets is outlined above. The major subsistence responsibilities rest on the shoulders of the elder male members of the family—the basic production unit in the overall economy. The socialization and training of children and adolescent members is done in accordance to this norm. Male adolescent members accompany the elders with the herds and learn the requisite skills as they grow into adulthood. Similarly, girls are trained by the elder women in the family and in the group, in matters related to household responsibilities. On the economic front, thus, the elders and adolescent members constitute the labour force. Primary labour is thus sourced from within the family. Individual families manage the herds. The growth and loss of the herd directly affects the household economy. The viability of the household economy depends on the experience and calibre of the male members of the family. The families are mostly nuclear families headed by the eldest member of the household; other adolescent and matured members form an integral part of the household economy. When adolescent members mature to attain adulthood, they part with a few animals from the patriarch's herd to establish an independent unit. The formation of independent herding units becomes complete after marriage. Though in most cases, the independent herding unit is formed only after the marriage of the adult male member; there are a few exceptional cases where unmarried adult members are seen raising their independent herding unit, functioning within the joint family.

The household economy thus depends on the number of male members within the family and the growth and prosperity of the herd depends on the adequate care of the animals. When, a sufficient work force is available to look after the herd, the growth is assured. When there is a dearth of adolescent male members within the family, labour may need to be hired from outside the family. On such occasions, poorer members of the community within the wand or from other wands are enlisted to share the responsibility of tending a growing herd. The headman of the wand plays an active role in such a bid to satisfy the need for more labour by a household. The terms and conditions of service are worked out with the intervention of the headman of the wand who takes the needs of both the sides into account. If there is a violation of the agreement, it becomes easier for the headman to pass the verdict. This kind of a mutual understanding helps the redistribution of the herd and the labour within the community. The members who have lost their animals due to some calamity, thus get the chance to rebuild their herd within a few years. On the other hand, the man whose herd has grown out of the proportion to the available working members within the family gets a chance to manage a larger herd and maximizes the return in such a situation.

The relationship between the employee and the employed in the above arrangement is contractual. The employee in such a situation is free to write off the contract without extending the contract once the contract period is over. The employee is rewarded with a fixed number of animals each year during this period of service. Once he feels that the number of animals he owns is sufficient for an independent herd of his own, he walks out with his share of animals. As contractual arrangements work within a period, rearrangement needs to be made when the situation demands. Another way to arrange the labour for a household is through marital relationship. The practice of bride price is found among these pastoralist groups. Generally, the bride price is paid either in livestock or in cash. The payment of bride price in cash is a recent phenomenon, while payment by livestock is an age-old practice. If a patriarch owns a proportionately larger herd that he along with his family members can manage, has an unmarried girl, he looks for a prospective groom who is not in a position to pay the bride price. Then he enters into a contract with the parents of the prospective groom and accepts the fixed price through service. The groom lives with the bride's family and handles the herding responsibility. In such an arrangement the groom not only pays the bride price through service, he also accumulates a sufficient number of animals of his own to start an independent unit after the marriage. This kind of arrangement works more or less on the same lines as a contractual arrangement, but an agreement on a marital relationship in the future forms an additional conditionality.

The Contemporary Maldhari Lifestyles: Contextual Adaptations

Owing to agricultural expansion and alterations in the ecological resources, the lifestyle of most of the pastoral population has undergone some obligatory changes. In spite of the disparities in the lifestyle adopted by them, domestic animals form the hub of their life, a necessary parameter for establishing the integrity of *Maldhari* identity. Let us now discuss the framework of labour management in these contemporary forms of pastoralism in Gujarat.

■ **The Pastoralists Within Agricultural Settlements**

It is difficult to say whether it is due to compulsion or compatibility that pastoral populations have been attracted towards the agricultural population. It seems to me that a combination of confrontation and convergence of interests work incessantly to define the premise of symbiosis between the agricultural and pastoral population. While the mode of land use becomes the point of confrontation, the mutuality in the economic advantage brings both the groups to the bargaining platform that decides the nature of interaction between the agricultural and the pastoral population. The existence of such an ideal symbiotic relationship is hardly noticed anywhere. Wherever the coexistence of pastoralists and agriculturists is found the balancing of confrontation and convergence of interests is a rolling process. The principles of social and economic interaction are often broken and realigned in accordance to the group dynamics. In the context of marginal ecologies, such a situation is the most natural outcome wherever both pastoralists and agriculturists share the eco-niche. In Gujarat, the Bharwads and Charans are the most numerous pastoral communities who have accepted this as their lifestyle. Goalas, in the east and central India may be cited here as a known parallel. The detail of a case study of a typical village, Takhalgarh in south Kathiawar, is discussed below to throw some more light.

Rajputs and Kathis together with other caste groups form the majority of the population in this village. The Rajputs own most of the land around the village, followed closely by the Kathis, a known cultivator caste in the region. We find six Bharwad families in this village occupying one end of the settlement. None of these Bharwad families own land except the plot on which their residential structure stands. They earn their livelihood primarily from raising herds of domestic animals. Though they are mostly sheep/goat keepers, almost all families have some cattle to fulfil the milk requirement of the household.

The responsibility of managing the herd is distributed among the members of the community. The total livestock of Bharwads is divided into eight herding units. These herding units consist of single species; hence, the fulfilment of the specific requirements of each species becomes easier. The total number of sheep, goats, and cattle belonging to the Bharwad families are 710, 310 and 99 respectively. This total livestock is divided into two herding units of cattle, two of goats and four of sheep. The size of the cattle herd is 50 cattle per unit. The herd size of the goatherd varies in the range of 150–200 while that of the sheep varies between 120 and 250. In this arrangement, the benefits are distributed among the owners in proportion to the number of animals they own. Thus, this mutuality provides flexibility to manage the herds to get the advantage of multi-species pastoralism.

The responsibilities of managing the herd are divided among the family members. The children in the Bharwad community grow up with animals and at the age of ten, they are capable of extending valuable assistance to the elder members of the family. Further training is acquired from the elder members of the community. The smaller animals like sheep and goats are kept in the courtyard. However, cattle are penned outside the settlement. Only the calves are kept under the supervision of the female members of the family in the courtyard when their mothers are being taken out for grazing. The cattle penning area is simply a temporarily fenced premise outside the settlement where the cattle are penned at night and one of the family members is made to sleep there to keep guard against any mishap.

The entire responsibility of the household rests with the women. Young girls help their mothers in the household chores. The elder child, be it a boy or a girl, attends to the children in the family. While the Bharwad men are away in pasture herding the cattle, women churn the milk and prepare butter out of it. Part of the butter is used in the cooking of vegetables and the rest of it is clarified to prepare ghee. The calves and rams are attended to by the women and children in the sheds attached to the house. Boys and girls, along with the elder members or women, collect twigs and tender grass to feed the calves and rams.

In the rainy season, only the adult men migrate to distant pastures. They generally commute to the hilly terrain of Amreli district. They migrate with the minimum possible baggage and utensils, just enough to satisfy their basic needs. They migrate in groups of three to five and choose a place to settle down. As women do not migrate with men, they have to prepare their own meals apart from grazing the herd. In the morning, they cook some crude bread and eat it with pickles. A bowl of milk supplements the meagre diet in these hard days of camping life. They take the herd to the pasture after the meal. In the evening, all of them come back and share a community life in the camp. When they cook their evening meals in the hut, they talk to each other and often spend the evenings singing songs. The milk yield of the cattle increases in the rainy season, which enables the women to prepare ghee and store it for future use.

The economic life of the Bharwads of Takhalgarh thus oscillates between their permanent residence in the village and the monsoon camps. When they are at the village, the archetypical division of labour is strictly followed. However, when they are in the monsoon camps, without the women folk, the norms of traditional division of labour between sexes is ignored. The effective management of the available workforce is achieved through the internal arrangement of herding units. It is important for each family to raise a multi-species herd to provide flexibility to their economy. In the given ecological context, a multi-species herd cannot be grazed together as cattle depend primarily on grazing while sheep/goats depend on top feeds. Another difficulty in raising a multi-species herd individually by a particular family is lack of enough labour within the household. While the ecological condition puts constraints against herding together large and small animals, none of

the families has enough members to graze them separately. In such a conflicting situation, distribution of the large and small animals possessed by different households to form single species herding units and then the allocation of available labour is a brilliant solution. The Bharwad families here function as a cooperative, where the domestic animals and the available labour are managed effectively through mutual understanding.

As these pastoral communities are found close to the agricultural population, the exchange of labour between them is also noticed. The Bharwad women work in the fields of village cultivators at the time of sowing, weeding and harvest, as daily labour. Bharwad men take on the herding responsibility of non-lactating cows and immature bullocks of the agricultural population. Local cultivators do not prefer to feed these animals because of their economic redundancy. Thus, the agricultural population hires skilled labour from local Bharwads to support these animals until they become economically useful. Such an arrangement brings additional payment for pastoralists. It also helps them to assume a space in the overall scheme of the mutual dependency of different sections of professionally specialized groups in the village.

■ **Pastoralists Involved in Limited Agriculture**

In the previous section, we have illustrated the scope of pastoralism as a specialized occupation within the agrarian economy. In a marginal ecology, animal husbandry forms an indispensable component of agricultural economy. In an agrarian economy, animal husbandry has a serious limitation of growth especially in the eco-niche where though agriculture is dependable to some degree, the supply of fodder and water to sustain a large number of animals is limited. The mobility of the pastoral component enables it to drive domestic animals to far off pastures in the lean months of the year. Hence, the presence of a pastoral population proves beneficial to the agricultural population. In such a situation, the pastoralists attached to an agricultural settlement are assimilated into the local caste structure.

An inherent problem in the social and cultural assimilation process is the political subjugation of the minority community. Any confrontation with the dominant community has serious consequences for the pastoral community whose growth is necessarily limited, as only a limited number of animals can be sustained in the given ecology. The threat of regular confrontations, and subjugation, and the ecological limitation of the growth of pastoralism as a component within the agrarian universe provides an imperative to establish exclusive pastoral settlements away from the agricultural settlements. There are several such settlements in Peninsular Gujarat. These settlements are found in the regions where agriculture is less dependable. The settlement density of these regions is low and plenty of pastureland is available around the settlement. Some of these settlements may be heterogeneous, but the pastoral population forms the majority. Pastoralism is the mainstay of the economy of the region, yet limited cultivation takes place in order to produce a part of the grain requirement. In such a situation, confrontation is rare and the pastoral community has greater political autonomy. Let us consider a case study of such a village from central Kathiawar to illustrate this.

The village of Resamia consists of Bharwads and Rabaris while Rajputs constitute only 5% of the total population. The Bharwads in Resamia rear sheep and goats. A few of them have recently taken loans to raise cattle. The cattle-rearers outnumber the sheep and goat-keepers among the Rabaris. The herd size for sheep and goats varies from 100 to 200 heads. The herd size of cows and buffaloes is comparatively smaller: the size of a buffalo herd lies between 15 and 25 and that of cows varies from 20 to 30. Most of the families keep a mixed herd of buffalo and cows chiefly because they want to mix buffalo's milk with cow's milk in order to attain the required fat level prescribed by the dairies.

Most of the families own land in the valley surrounding the settlement. Bajri and jowar are the main crops grown by them and the majority of the families are engaged in both farming as well as pastoralism. The annual productivity of the crops varies from year to year. Years of consecutive drought are very common in the area. Therefore, the practice of pastoralism is in no way less significant to them. Their life revolves around their herds rather than being rooted in their land holding. The division of labour in the community is the basic premise that reflects this cultural ethos in the crudest form.

In the month of June after the first rains of the monsoon, the land is prepared for growing kharif crops (*Bajri* or *Jowar*). The land is manured, ploughed and levelled and is sown in the middle of June. As soon as the crops come up in the fields, the men take the herds away from the settlement to the surrounding pasture. There they camp with their herd until the harvest. The men generally abstain from going to places far-off from the village for the rainy season camps. They shift to these camps leaving behind their children, women, and old people in the village. After the sowing is over, no hard work is required in the field. The women, children, and old people of the family do the weeding operation collectively. In the meantime, men take care of the herd in their rainy season camps. As the herd remains away from the cultivated fields, the chances of the destruction of crops by the domestic herds decrease.

While the crops grow in the field, men in the camp spend their time debating the success or failure of the crops in that particular year. When they gather in the evening the talk mostly revolves around their calculations and forecasts about the yield from their fields. The women come regularly to the camp bringing food and news about the village. As the crops ripen, in the fields, the time for harvesting and taking the yield to the granaries draws closer. The men return to the village for this task and their schedule of herding changes. The harvest season begins in the month of October when the rainfall is almost negligible. The sky is clear and the moon shines brightly on the pastures. The men are therefore able to take their herds to the pasture for a second shift at night. The herd is taken before midnight to graze in the adjacent hills until late at night. After a few hours of grazing, when the animals have had their fill they settle down for the night. The herdsmen too settle around a fire and sleep for a while. At dusk they bring the herd back to the camp. On the way back, they water the herd at the nearest watering hole and close the enclosure of the temporary cattle pen at the camp to prevent any animal from straying away in their absence. The responsibility of supervising the penned animals is given to one of the men, and the rest hurry to the village to harvest the crop. This is done in rotation to allow each of the men to attend to their requirements in the harvest.

Once the crops are cut and piles are made in the field, the rainy season camp is abandoned. The herds are brought back to the village and are penned in the field. Men then follow their usual schedule of herding and in their leisure time, the crops are harvested. The grains are carried home and the hay and other by-products are piled in the field. The cattle graze on the stables left out in the fields after the crops are harvested. The peripheries of the valley and the foothills of the surrounding mountain ranges are covered with thick patches of grass. The cows are given the hay to supplement their grazing. The whole family lives together happily for a couple of months. Several festivals are celebrated and marriages arranged in this season. After the grasslands and the agricultural by-products close to the village are depleted, they are ready once again to shift to the summer camps.

Migration to the summer camps is a major event in the community's annual life cycle. The whole village is divided into several groups of families ready to migrate to different areas. Once the group is finalized, each group approaches the owner of the 'Vid' (the expanse of pasture land in which private ownership subsists) to settle the amount for the lease for a period of four to five months. The group collects the contribution from all the members and pays the settled sum to the owner of the 'Vid'. Once these formalities are over, the whole family moves to the summer camp. Sometimes old people are left in the house with a few cows and enough grain for their consumption, and they look after the household in the absence of other family members.

The summer camps are enjoyable for the men who get good food cooked by the women folk. They give their sole attention to herding. The division of labour in the camp is now evenly distributed between both the sexes. Women cook and take care of the young calves. The children who have come of age accompany their father or the elder members of the family to the pasture. Here they get their practical lessons in the art of herding. The women use their leisure time for stitching and embroidering clothes. The old people engage themselves in rope making. Children play throughout the day in and around the camp. Sometimes the herds of calves are sent to the pasture under the supervision of the adolescent boys and girls. In their mobility pattern, the village serves as the anchor from which their whole subsistence activity is monitored. Their social and cultural lives are always linked with the village.

In the beginning of this section, the contextual specificities are explained. It essentially differs from both the above-mentioned pastoral ways of life. Agriculture as a supplementary component in the economy brings some of its inherent constraints into the pastoral way of life. One of the notable impacts is a relatively more permanent settlement with well-built house structures. In case of such a dual economy, we find a new equation in the management of labour. Their involvement in dry farming puts an

A woman removing chaff using a winnowing fan in
Western India. (Photo by Takeshi Takeda)

additional demand on the existing labour in the cultivating season. Hence, we notice a change in the division of labour and a convenient scheduling of herding and agricultural operation to meet this demand. Men handle the additional responsibility of cooking in the monsoon camps. In the same vein, women handle the additional responsibility of a part of the agricultural operation along with their traditional domain of housekeeping.

Such an adjustment of division of labour between the sexes is not sufficient. When crops mature in the field, it requires a large workforce to harvest them quickly before they fall to the ground. Labour from outside may be hired for meeting the demand, but this is too expensive. Hence, to employ the available workforce within the family is the only option. In case this is not sufficient, mutual arrangements are made to meet the demand. As the entire workforce in a viable pastoral economy is fully employed in raising livestock, a temporary withdrawal of labour from the pastoral sector is inevitable. At the same time, the herding requirements cannot be ignored entirely. Cattle are grazed usually in two sessions, once in the morning and again in the afternoon. The minimum herding requirement for cattle is a major grazing session with an adequate amount of drinking water. This minimum requirement needs to be attended to before the labour moves to harvesting. This is achieved by grazing the herd at night. Thus, the daytime is freed for harvesting. When the animals rest in the camp in the daytime, there is no work other than keeping a vigil over the herd. One or two can do this work, while the rest of the labour is available for harvesting. This rescheduling of herding and harvesting continues until all the members in the group have harvested their crops.

The Settlement Rearrangement

The effect of the frequent drought situation and gradual agricultural expansion has resulted in desperate settlement reorganization to accommodate the changing social structure of organization of production. We have already observed the cultural response to the deficient monsoon, now let us explore a little more the process of settlement reorganization in the last couple of centuries.

The Survey of India (undivided) had conducted a detailed geographical survey in this region towards the fag end of the 19th century. The 1 inch=1 mile topographical sheets of this survey records the exact location of the deserted settlements/ruins in this region. This also provides the original name of the deserted settlements/ruins wherever it is available. The names of these deserted settlements and their respective geographical coordinates are compiled from these topo-sheets into a catalogue. The locational analysis of these sites shows up some important aspects of settlement reorganization. The sites are confined to the Kutch and Kathiawar Peninsula only and are more or less evenly distributed in the Kutch region where they are located either in the hilly regions or deep in the widespread grasslands. A comparison with the present settlement pattern of Kutch shows that the abandonment of these sites has given way to small agricultural settlements nested within the available small pockets of cultivable land. The location of these deserted sites barely suggests the possibility of any agricultural activity. The sites are concentrated more in the western grassland zones of Kutch than in the eastern part. This may suggest a shift of emphasis from pastoralism to dry land agriculture.

The situation in the Kathiawar Peninsula is not significantly different from that of Kutch. In Kathiawar the deserted sites are found more in the northern part than in the southern part of the peninsula. In the northern part, they stretch the whole length of the hilly landscape. It is important to note here that the presence of major drainage lines like Bhogava, Bhadar, Keri, Ghello, Kalubar in the eastern part and the Satrunji and

Machundri in the south prove congenial to the agricultural operation in these areas with variable success. The landscape of the Kathiawar Peninsula constitutes two series of mountains, i.e., the northern series and the southern series. The northern series includes the Barada hills near Porbandar in the west coast, the Mandav range running east-west in the north along the coast of the Gulf of Kutch, and the Thanga chain to the south of the Mandav range. The southern series includes the greater and the lesser Gir. These two belts of hills that cross the breadth of the Kathiawar constitute two distinct water partings, and form a narrow stretch of tableland which occupies the centre. The major drainage lines in the eastern and southeastern part of the Kathiawar collect the water from these hill ranges. The tableland in the centre interspersed with hill ranges and river valleys supports the agricultural lifestyle depending on the intensity of the rainfall. It is interesting to note that the distribution of the deserted sites leaves out all these areas where agriculture is a distinct possibility. The distribution of these deserted sites more or less matches the distribution of the grasslands in the northern part. In the southern part, the deserted sites are located in the foothills on the outskirts of the forest. Another interesting feature of these deserted sites is their location in the upper reaches of their associated drainage lines. The northern concentration is found in the catchments of the rivers Und, Susa, Phuljhar, Beti, and their numerous tributaries. Similarly, the southern concentration can be associated with the Hadali, Machunndri, Dantavar rivers, and their tributaries.

A large majority of these sites is strategically located on a well-drained elevated landscape overlooking the surrounding grasslands. These features match the settlement location strategy of the pastoral communities (Swayam, 2000: 293–295). The size of these settlements varies from 0.5 to 1.5 hectares. The depth of the habitational deposit in these sites in general is shallow ranging between 20 to 80 cms. One can make out the plan of the settlement and each of the household that constitute it from the eroded boulder conglomerates. Along with these eroded building materials, a small amount of ceramics forms the archaeological assemblage of these sites. The ceramics thus become the only artefact to suggest the antiquity of the sites. Based on the ceramics, some of the sites can be dated to the early historic period, while others can be safely assigned to the medieval and more recent times. The exact temporal variation of the sites awaits a detailed study to reveal more information on the settlement history of this region. Whatever may be their temporal horizon, the most recent abandonment can be dated back to the last part of the 19th century.

Another important aspect of the recent changes in the settlement pattern is the multiplicity of some of the settlements either due to internal growth or to the merger and relocation of settlements. The inferences about the fission, merger, and relocation of the settlements are reconstructed by a detailed study of the settlement names in this region. The growth and multiplicity of the settlement was inferred from the suffix *Nana* or *Mota* to a particular settlement name. It is a general practice in the region to add *Nana* at the end of the original settlement name for a new breakaway settlement of the original one. The nomenclature in case of more than one breakaway settlement is a little different. The new settlements in such a case take a descriptive prefix to the original settlement. For example, four affiliated settlements of the original settlement Bara in western Kutch take a prefix according to the direction of their location: *Vachla, Athamna, Ugamna,* and *Dakshna.* There are several such sets of descriptive prefixes used for naming the new settlements. The relocation of settlements is also encountered. The relocation may be complete or partial, depending on the magnitude of the participation of the concerned population in such an event. In case of complete relocation, the old site is found in ruins, which is referred to as the old (*Juna*) settlement. The new settlement has the term *Nava (new)* as a prefix or suffix to the original name of the settlement. In case of partial relocation of the settlement, both the old and the new settlements are occupied, and they are addressed, just as the *Mota* and *Nana*, with the prefix *Nava* or *Juna*. A few cases of settlement relocation by merger are also noticed. In such a case, two or more small settlements were abandoned to form a fresh settlement in a new location. For example, the Chhala-Jodhpur in northern Kathiawar is a new settlement formed by the merger of the Chhala and Jodhpur, two erstwhile small settlements now found in ruins within a radius of a few kilometres of the present location of the settlement. Thus, the study of the nomenclature of the settlements suggests three kinds of changes in the settlement organization, i.e., split, relocation and merger.

The cases for settlement mergers are very few and mostly confined to the northern part of the Kathiawar. This is particularly noticed in the northern slopes of the Mandav hill range where new settlements have emerged due to merger of two or more small settlements. The old abandoned sites are in general located in the foothills of the upper reaches of the local drainage line. In contrast, the new site is now located down the stream in the plateau. The plateau unlike the surrounding hills has a better soil depth; hence it is used for agriculture. The surrounding hills are used as the grassland to sustain the domestic herd. As the access to the grasslands is important in their economy, the relocation of the settlement never exceeds a daily commutable range. Similarly, the phenomenon of complete relocation as is discussed above also reflects this strategic shift of emphasis from pastoralism to a mixed economy. The cases for complete relocation of the settlements are noticed in the north, northeastern, and central part of the Kathiawar Peninsula.

The split or the fission of a settlement is the most prominent and noticed, almost universally. This is an entirely different phenomenon. The difference lies in the choice of location by the breakaway group in establishing the new settlement. In general, the same parameters are applied. Hence, this phenomenon is linked with the internal growth of the population at a particular settlement. Once the population growth threatens the resources of a particular locale, a resizing of the population helps in restoring the ecological balance to sustain a certain level of production. This phenomenon is thus noticed in case of eco zones sustaining the agricultural mode of production as well as that of a mixed economy. A close look at this phenomenon suggests that the lower reaches of the river valleys in the eastern part of Kathiawar have successfully supported the agricultural mode of production. Similarly, more arid and hilly regions in Kutch, and in the north, and western part of the Kathiawar go well with a mixed economy.

As mentioned earlier, the gradual effort for agricultural expansion over a millennium with a sustained and successful political agenda to promote sedentarism in Kutch and Kathiawar has achieved some lasting results. Political support for the construction of an irrigational infrastructure, has transformed the ecological possibility of an agricultural mode of production into a socio-economic reality. Water conservation and water harvesting has supplemented the water requirement for agricultural operations in the river valleys and tablelands. Against this background of technological innovation and socio-political dynamics, let us juxtapose these two sets of information, i.e., the distribution of the abandoned settlements and the recent settlement growth and relocation.

The result of the efforts towards agricultural expansion is most visible in the eastern part of the Kutch. In the western part of the Kutch, the numerous abandoned sites and the relocation of the settlements to more congenial locales for agricultural operation indicates the shift of the economy from pastoralism to either a mixed economy, or, in a few cases, to agriculture. The existence of several pastoral settlements (*wands*) in this region further suggests that a section of the population still follows the pastoral mode of life. Hence, the agricultural expansion in Kutch has remained partial. The situation in north Kathiawar is similar to the Kutch. Though the agenda of agricultural expansion has realigned the settlements to favour a limited agricultural activity, the pastoral component is not fully abandoned due to ecological constraints of frequent droughts. This situation can as well be held true to the southeastern part of the Kathiawar Peninsula. The settlements in the foothills of the Girnar were abandoned and relocated in the plains available along the river valleys and the coast. The situation changes towards the eastern part of the Kathiawar Peninsula. Numerous major drainage lines in the east, and the tableland between the north and south series of hills provide enough cultivable land. Irrigation facilities such as tanks, reservoirs, wells, and bore-wells provides the crucial support in the dry months and in the case of monsoon failure. This area therefore is devoid of abandoned settlements. The only change in the settlement pattern in this eastern part of Kathiawar is due to the split or fission of some settlements due to natural internal growth.

Conclusion

The application of a suitable technological skill and effective management of the manpower are two important cultural aspects of any successful economy. Therefore, the adaptational response brings comprehensive modifications in these aspects. The settlement rearrangement in the region is thus a step towards creating fundamental provision for an agricultural economy. The pastoral settlements that were located in the elevated landscapes within the pastureland now required a composite eco-niche, which would provide them access to both the pasture and cultivable land. This necessitated a change in the location of the earlier settlements located ideally for pastoral activities only. Within the known technological know-how of agriculture, the land around the erstwhile pastoral settlements was impossible to cultivate. In the contemporary context, with the advanced knowledge of plant species and farming techniques, however, the agricultural activities can be performed without shifting the settlement location. Thus in the context of the pre-industrial society, the settlement rearrangement seems to be the only available rational decision.

The assimilation and acculturation can be perceived as two distinct cultural processes, which are responsible for the occupational transition from pastoralism to agriculture and other artisan groups. Assimilation is a more probable situation in the context of subjugation of a minority population by the majority population. The pastoral population that is drawn closer to the agricultural population for exploiting the possibilities of mutual advantages is the natural prey to the assimilation process. The constraints to the growth of the pastoral mode of production at a close proximity to the agrarian economy are initially countered by making adjustments in labour management. As we have seen above in case of the Bharwads in Takhalgarh, the labour management follows more of a cooperative policy. This can be held as a contrast to the individual orientation of the labour management policy in case of their archetypal lifestyle. The viability of such a cooperative effort to maximize the productivity starts falling apart when the number of pastoralists exceeds the threshold.

Beyond this threshold, the economy becomes unstable and non-sustainable. A homeostasis can be maintained only by ascertaining a critical number of pastoralists in a particular agricultural settlement. The rest of the pastoral population needs to migrate to other settlements or take to some different means of livelihood. Due to the political subjugation and increasing reliance on the dominant population, the pastoral specialists find it difficult to maintain a distinct cultural identity. Pastoralists gradually inculcate cultural values of the dominant population and ultimately give in to the pressure of culture change. As the cultural distinction diminishes, they are accepted as a component of the occupational structure of the village. Finally, they form one of the lower rungs of the existing social structure. Hence, the cooperative effort in the labour management is of critical importance in determining the direction of change. It acts as the inertia to the transition.

In the acculturation process, though the pastoral population maintains a distance from the agricultural population, it establishes a lethal contact with agriculture. The contact is considered lethal as it is seen as the first step in the acculturation process of pastoral populations. Their involvement in agriculture puts severe constraints on their migration circuit. This indirectly affects the herd size. The constraint in the mobility and growth of the herd defines the limitation of the pastoral economy. The gains from the agricultural sector to some extent neutralize the loss. In a semi-sedentary existence such as this, one cannot increase the herd size beyond a critical level, as the migratory circuit is smaller. Hence, involvement in agriculture becomes essential. The production in both the agricultural and pastoral sectors can be maximized only through harmonizing the schedule of operations in both sectors.

Another important challenge is to organize labour to take care of both the agricultural and pastoral sectors. As we have noticed in the case study of Resamia, the labour requirement can be met by temporary withdrawal of labour from one sector and deputing it to the other sector with a decisive shift in the priority of a selected

component in their mixed economy. Such a labour management is possible if pastoralists work in a cooperative spirit. It is important here to distinguish between the 'social cooperation' that exists in their archetypal lifestyle and the 'cooperative management of labour' in the present context. Tandas are small groups of pastoralists who, though they accompany each other in the migration circuit, hold individual responsibility for herding. There is hardly any mutual exchange of workforce within the group. In case of a shortage in the workforce, labour is usually hired on contract. The 'cooperative labour management' principle on the contrary works on a broader economic platform. Thus, the cooperative management of labour provides the cutting edge in both forms of the current adaptation strategy followed by pastoral communities in Gujarat.

Thus, the organization of labour is one of the important cultural apparatus to respond to ecological constraints. Many more innovations are possible in the organization of labour to defy the external impulses that disturb the economic viability of a particular mode of production. Apart from the economic dimension of the organization of labour in such situations, it has a social significance as well. Any form of labour management has a distinct cultural consequence. The lowest denomination of labour can be defined at the individual level. Every individual, in accordance to his skills and physical ability, is considered as a single unit of labour. Hence, the organization of labour in principle deals with the individuals at different levels of social and cultural complementarity. For example, the division of labour within different members of the family is the most fundamental social complementarity between members of a small group of individuals. As we move up in the ladder of social institutions, the social role of the organization of labour assumes more complexity. More complex social consequences of a particular form of organization of labour can be perceived in the clan, community, and regional levels. Often we find some form of cultural mechanism providing a moral support to this organization of labour. The institution of marriage reinforces the economic significance of the division of labour between its members with a cultural recognition. Similarly, different norms and mores can be identified which essentially match a particular organization of labour by spreading the social solidarity both horizontally and vertically. Thus, the symbiotic relationship between the pastoral and agricultural communities may be looked at from this perspective. Both, the pastoral and agricultural population of Gujarat, share a common cultural attitude towards rainfall and droughts. The common sentiment regarding the hope for a good monsoon, as well as the trepidation of a drought, is echoed in the narratives and poetries in the oral tradition. The frequent occurrence of drought helps in the perpetuation of this psychosis of hope both in agricultural and pastoral communities. As the hope for a better monsoon lingers in the social psychology of different communities in the region, the organization of labour takes numerous innovative forms in the negotiation of interests of different modes of production. In the name of negotiation, a disproportionate growth of one community at the cost of the other generates a social surplus to germinate a civilization. In the game of negotiation, it is equally probable to engage a civilization for a blatant consequence.

References

Biswas, B.C. and S.S. Basarkar (1982): Weekly rainfall probability over dry farming tract of Gujarat, *Journal of Arid zone Research Institute*, 21 (3): 187–194.

Ganguli, B.N., R.N. Kaul, K.T.N. Nambiar (1964): Preliminary Studies on a Few Top-Feed Species. *Annals Of Arid Zone*, 3(1–2): 33–37, CAZRI, Jodhpur.

Malik, A.K. & Govindaswamy (1962–63): The Draught Problem in India in Relation to Agriculture. *Journal of Arid Zone Research Institute*, 1: 106–113.

Meher-Homji, V.M. (1997): Droughts, aridity and desertification in the Indian subcontinent. *Journal of Arid Zone Research Institute*, 36: 1–18.

Rao, R.S. (1970): Studies on the flora of Kutch, Gujarat (India) and their utility in the economic development of semi-arid region, *Annals Of Arid Zone*, 9 (2): 125–142, CAZRI, Jodhpur.

Sahu, D.D. and S.N. Sastry (1992): Water availability pattern and water requirement of Kharif crops in Saurashtra region, Gujarat. *Journal of Arid Zone Research Institute*, 31, (2): 127–133.

Shukla, R.N., A.V. Deo, S.S. Katti, S.B. Kulkarni (1962): Water evaporation retardation by surface films. *Annals Of Arid Zone*, Vol 1, No (2): 127–131.

Swayam, S. (2000): *The role of cattle pastoralism in Gujarat (2500–2000 BC): an ethnoarchaeological perspective*. Unpublished Ph.D. Thesis, Deccan College Post-Graduate and Research Institute, Pune.

Swayam, S. (2001): Sedentarization and adaptation strategies of pastoral communities in Gujarat: some reflections on the change in labour management. *Journal of Indian Anthropological Society*, 20 (2).

Venkateswaralu, J. (1993): Effects of drought on Kharif food grains production: a retrospect and prospect, *Journal of Arid Zone Research Institute*, 32 (1): 1–12, Jodhpur.

Whyte, R.O. (1975): The nature and utilization of the grazing resources in India. In L. S. Leshnik and G.D. Sontheimer (eds.): *Pastoralists and nomads in south Asia*. Otto Harrassowitz, Wiesbaden, pp. 220–234.

A man with an impressive moustache and a turban in Rajasthan, India. (Photo by Takeshi Takeda)

Chapter **11**

Climatic Fluctuations: A Plea of Concerted Efforts in the Mid-Ganga Valley Through ALDP

PURUSHOTTAM SINGH

Introduction

It is generally agreed that climate is one of the most important factors in controlling the origin, development and decline of human cultures as it governs, to a considerable degree, the organic resources of a region and related human activities. Thus, the reconstruction of past climate is important not only to archaeologists but to students of various other disciplines as well. In this context, the study of lacustrine sediments is particularly important as these sediments are better preserved and stratified and, therefore, can produce better chronological controls on the evidence collected.

Study of Lacustrine Sediments in Western Asia

Perhaps one of the earliest attempts in this direction was that of van Zeist of the University of Groningan, Netherlands and H.E. Wright of the University of Minnesota who in 1960 tried to reconstruct the prehistoric environments of southwestern Asia by the study of pollen or sediment cores obtained from Zeribar lake situated at an elevation of 1300 metres, about 160 km northwest of Kermanshah near Iraq border in the Zagros mountains of western Iran (van Zeist and Wright, 1963; van Zeist, 1969). This lake measures about 3 – 5 km. and is only a

few metres deep. This investigation led them to conclude that between 30,000 and 14,000 years ago the climate of southwestern Asia was not only colder than it is at present but that it was also considerably drier (Singh, 1974). van Zeist further remarks that between 10,000 and 6000 cal. yrs. BP the precipitation in the autumn, winter and spring could hardly have been less than it is at present, but the summers were drier. Since summer dryness does not affect annuals such as wild cereals, one may assume that the areas which in early Post-glacial times provided suitable habitats for these species were to a large extent the same as the present-day ones (van Zeist, 1969: 45).

Study of Lacustrine Sediments in Rajasthan

In India the studies of Gurdeep Singh from the evidence coming from the salt lakes in western Rajasthan (Sambhar, Lunkaransar and Didwana) and one freshwater lake deposit at Pushkar in the Aravalli Hills in the early seventies indicated that although plant microfossils first appear in Phase II with the deposits of lacustrine sediments dated to around 10,000 cal. yrs. BP, the *Cerealia* type of pollen only in Zone B, Phase III, dated by C-14 method to c. 7500 BC and it continued thereafter throughout Phases III and IV. Phase V is immediately followed by aridity for which there is stratigraphic evidence that the salt lakes started drying. The climate did not ameliorate until about Phase V (early centuries AD to present). The pollen analysis of some archaeological soil samples from Kalibangan in conjunction with the data obtained from the lake deposits revealed that the climate was more humid in Rajasthan and from 3000 BC the climate shifted to the wetter condition and continued to be congenial upto 1800 BC when aridity begins. A short relatively drier climate prevailed between 1800–1500 BC. This climatic change resulted in the dessication and diversification of fertile area and adversely effected the once-thriving Harappan culture in this region (Singh *et.al.*, 1974).

The occurrence of *Cerealia* type of pollen in the context of 7500 BC together with the evidence found in the Vindhyas indicate that the Neolithic beginnings of India may go back to the 7th millennium BC.

This five-fold division of lacustrine sediments has been further divided into six phases by D.P. Agrawal who has suggested the following climatic sequence (Agrawal, n.d.):

Phase I   (pre-10,000 cal. yrs. BP)—intensely arid
Phase II (c. 10,000–9,500 cal. yrs. BP)—far more humid than today
Phase III (c. 9,500–5,000 cal. yrs. BP)—a little more humid than today
Phase IV (c. 5,000–3,000 cal. yrs. BP)—intensely more humid than today
Phase V   (c. 3,000–1800 cal. yrs. BP)—intensely arid
Phase VI (early centuries AD to today)—present conditions prevail

More recently, pollen analytical studies were undertaken by the scientists of the BSIP, Lucknow from samples obtained from Punlota lake of Nagaur district, and Bagundi and Pachpadra lakes in Barmer district—all in Rajasthan. Of these, Bagundi and Pachpadra are saline lakes, while Punlota is a freshwater lake. After a study of the pollen diagram of a 2.00 m deep sedimentary profile from Punlota lake, the diagram was divided into four pollen assemblage zones; I, II, III and IV, starting from bottom upwards. Details regarding each zone are as follows: (Chanchala: personal communication).

Pollen Zone I (200–175 cm): Around 9200 to 8050 cal. yrs. BP the region had predominantly non-arboreal vegetation under arid climatic conditions.

Pollen Zone II (175–85 cm): Around 8050 to 3810 cal. yrs. BP warm and moist conditions prevailed as reflected by the establishment of savannah type vegetation together with an increase in the fern as well as fungal spores.

Pollen Zone III (85–25 cm): Around 3810 to 1325 cal. yrs. BP decrease in warm and moist conditions is seen as the savannah phase is replaced by reduction in arboreal taxa as well as fern spores and grasses with simultaneous increase in Cyperaceae and cheno/Ann, etc.

Pollen Zone V (25–0 cm): Comparatively drier or more or less similar conditions are witnessed as in the preceding phase for the increase in Poaceae and Caryophyllaceae.

Pollen vegetation of three surface samples from Dagana region has also revealed the dominance of non-arboreal vegetation. The overall vegetation assemblage reveals the presence of marked open grassland-scrubby vegetation which more or less shows coherence with the extant vegetation in the region.

Karewas of Kashmir

Another area which has been studied in great detail is the Karewas of Kashmir. These fluvio-lacustrine sediments, topped by a mantle of loess giving rise to flat plateau-like surface in Kashmir valley, are known as Karewa in the local language. A decade of long, interdisciplinary research of these deposits has provided a lot of data on floral and faunal changes which makes it possible to

India is not a very rich country, but its people are calm and gentle. (Photo by Yoshinori Yasuda)

monitor past climate and environmental changes in a definite time frame (Agrawal, 1992). This study has indicated that the periods of Upper Palaeolithic, Neolithic, and Historical (1st millennium AD) are marked by large populations in the valley and also coincide with the periods of climatic amelioration. Thus the two studies noted above, indicate that enormous possibilities exist where useful results can be obtained by lake drilling with concerted efforts. However, it also needs to be emphasized that for a large country like India, the data obtained so far, is grossly inadequate and such lake drilling needs to be done in various other parts of the country in order to complete this picture.

Archaeological Potential of Sarayupar Plain

The Sarayupar plain forms a distinct geographical entity which is bounded on the south and the west by the Ghaghara river, on the east by the Gandak and extends upto the foothills of the Himalayas. It is drained by several rivers like the Ghaghara, the Kuwana, the Ami, the Rapti and the Chhoti Gandak. This region which is part of the vast Ganga plains, attained its present form during the post-Tertiary period when this deep trough was filled up by fine alluvium brought down from the Himalayas, with an average thickness of 1300-1400. The meandering of the rivers in Prehistoric

191

Indians have a great capacity to rise above all adversity. (Photo by Yoshinori Yasuda)

times formed innumerable ox-bow lakes, some of which are perennial. These lakes are rich in aquatic fauna and the lands around them were covered with wild grasses, many of which had edible grains. With the onset of the milder climate of the Holocene the marshy land gradually turned into good grassland which attracted small animals. Mesolithic hunters inhabited some of these areas, evidence of which have been found on several sites in Pratapgarh district. In course of time Neolithic man also inhabited these areas, evidence of which has been found in the lower levels of Sohagaura, Imlidih and Narhan.

As archaeologists, we have been responsible for locating and excavating ancient human settlements in the eastern part of Uttar Pradesh comprising the present-day districts of Basti, Siddharth Nagar, Gorakhpur, Maharajganj, Ballia, Mirzapur and Varanasi for well over two decades

(Singh, 1994 and 1996). The excavated sites are Narhan, Imlidih Khurd and Dhuriapar in District Gorakhpur and Bhunadih and Waina in Ballia district.

Located on the left bank of the Ghaghara, Narhan was subjected to archaeological excavations by us during 1984–89 (Singh, 1994). As a detailed report on these excavations has already appeared, the essential features may be briefly recounted as follows: Narhan has given a continuous sequence of cultures from about 1300 BC to the end of the Gupta period. The lowest deposits are dated between 1300–900 BC on the basis of radiocarbon dates. The earliest inhabitants lived in wattle and daub houses, used characteristic pottery called Painted Black and Red Ware and cultivated two crops a year (rice, wheat and barley alongwith several varieties of pulses and oilseeds). They domesticated cattle, sheep-goat and horse and supplemented their food by hunting deer and antelope, and by fishing. Similar data was subsequently obtained from Khairadih, located about 50 km downstream on the same river but on the opposite (right) bank.

Imlidih is located on the left bank of Kuwana, a minor tributary of the Ghaghara. The small mound has provided data regarding a Pre-Narhan settlement which is datable to c. 1700–1300 BC (Singh, 1991–92, 1992–93 and 1993–94). As in the case of Narhan, the inhabitants of the Pre-Narhan phase also lived in wattle and daub houses but used a different type of pottery termed as corded ware. As noted above, they also cultivated two crops a year comprising of the same cereals, pulses and oilseeds. They supplemented their food by hunting deer, and wolf and by catching fish and turtle for which river Kuwana is so famous even today. Dhuriyapar is again situated on the left bank of the same river (Kuwana) as Imlidih and was investigated by us during 1991 and the remains of Narhan culture were found at the lowest levels. Subsequent deposits on this site include those of Kusana and Gupta times. This site was deserted after the Gupta period but was reoccupied once again in the Medieval times (AD 900–1500).

The small settlement of Bhunadih is spread in an area of four acres on the right bank of Bahera nullah which eventually discharges itself in the Ghaghara. Test pits excavated by us during 1995–96 resulted in the documentation of Pre-Narhan and Narhan cultures.

Waina is located on the old bed of the Chhoti Sarayu, a tributary of the Ganga river in the Ballia district and it has given a similar cultural deposit as the above noted site. The occurrence of a few microliths in the lowest level of this site, a solitary polished stone axe (celt) from Narhan and vestiges of a Neolithic deposit at Sohagaura demonstrate that these sites were visited by Neolithic people, presumably around 2000 BC and since then permanent human settlements flourished in the valleys of Ghaghara, Kuwana and Chhoti Sarju.

The excavations of the large settlements of Agiabir located on the left bank of the Ganga, about 50 km upstream of Varanasi is currently under progress (Singh et al., 2000).

As stated above, the area under archaeological investigation is dotted with several ox-bow lakes (locally known as *Tals*) of varying dimensions, such as, Bakhira (Basti district), Ramgarh (Gorakhpur district) and Suraha (Ballia district). These *Tals* represent abandoned channels or deep natural depression wherein surface drainage collects in the absence of adequate outlets. Besides, there are several temporary swamps and *jhils*. The largest reservoir of Gorakhpur district is Chilua *Tal*, located about 5 km north of the district headquarters, covering an area of 25 km. Another natural reservoir known as Bhakhira *Tal* is located on the borders of the Gorakhpur and Basti districts. Ramgarh *Tal* measures 20 km and lies 5 km east of Gorakhpur city. Nadaur and Gaura *Tals* are ox-bow lakes. Some of these lakes are so large that they have given birth to small perennial rivers. For example, the Kuwana springs emerge from a lake that has its source in the eastern part of Bahraich and flows through Gonda and Basti. Similarly, the Ami has its source in the Bakhira *Tal*. A study of the lacustrine deposits of the middle Ganga plain will greatly help in understanding the man-plant relationship and in reconstructing past climatic conditions of Pre and Protohistoric north India.

References

Agrawal, D.P. (1992): *Man and Environment in India through Ages*, Books and Books, Delhi.

Agrawal, D.P. (n.d.): *Reconstructing the Past Climate and Environment*, A.S. Memorial Lecture delivered at Birbal Sahni Institute of Palaeobotany, Lucknow.

Singh, Gurdeep, R.D. Joshi, S.K. Chopra and A.B. Singh. (1974): Late Quaternary history of vegetation and climate of Rajasthan desert, India, *Philosophical Transactions of the Royal Society of London*, Biological Sciences, B 267/889: 467–501.

Singh, Purushottam (1974): *Neolithic Cultures of Western Asia*, Seminar Press, London and New York.

Singh, Purushottam *et. al.,* (1991–92): Trial Digging at Dhuriapar, *Pragdhara* No. 2, pp. 55–60.

Singh, Purushottam *et. al.,* (1991–92): Excavation at Imlidih Khurd, *Puratattava*, Bulletin of the Indian Archaeological Society, New Delhi, No. 22: 120–22.

Singh, Purushottam (1992–93): Archaeological Excavations at Imlidih Khurd-1992, *Pragdhara* No. 3, pp. 21–36.

Singh, Purushottam (1993–94): Further Excavations at Imlidih Khurd-1993, *Pragdhara* No. 4, pp. 41–48.

Singh, Purushottam (1994): *Excavations at Narhan (1984–89)*, Banaras Hindu University and B.R. Publishing Corporation, New Delhi.

Singh, Purushottam *et. al.,* (1994–95): Protohistoric Investigations in Ballia District Uttar Pradesh, *Pragdhara* No. 5: 21–36.

Singh, Purushottam *et. al.,* (1995–96): Excavation at Waina, District Ballia, *Pragdhara* No. 6: 41–61.

Singh, Purushottam, (1996): *Prelude to Urbanization in the Sarayupar Plain*, Presidential Address (Section V), Indian History Congress, Fifty Seventh Session, Chennai, 1996.

Singh Purushottam *et. al.,* (1997–98): Trial Excavations at Bhunadih, District Ballia (Uttar Pradesh), *Pragdhara*, No. 8: 11–29.

Singh, P. *et. al.* (2000): Excavations at Agiabir, district Mirzapur (Uttar Pradesh), *Pragdhara*, No. 10, pp. 31–68.

van Zeist, W. and H.E. Wright, (1963): Preliminary pollen studies at Lake Zeribar, Zagros Mountains, South-West Iran, *Science*, N.Y. 140, pp. 65–69.

van Zeist, W. (1969): Reflections and Prehistoric environment in the Near East: In Ucko, Peter J. and G.W. Dimbleby (eds.): *The Domestication and Exploitation of Plants and Animals*, London: Duckworth, pp. 35–46.

Chapter 12

A Model for Monsoon-based Protohistoric Subsistence Economy in the Semi-arid Region of South India

P.P. JOGLEKAR

Introduction: Why is a model necessary?

The semi-arid region of the Deccan witnessed the rise of several Protohistoric (Chalcolithic) cultures during the 2nd millennium BC (Dhavalikar, 1997). It has been postulated that these people were the first agriculturalists (farmers) in this part of India (Dhavalikar, 1988). These Chalcolithic cultures of (modern) Maharashtra and the Southern Neolithic cultures in neighbouring (modern) Karnataka have been extensively studied (Allchin, 1963; Paddayya, 1973,1993,1995). The settlement pattern, material culture, and particularly the faunal and botanical components of archaeological remains have been examined in detail (see Paddayya, 1975; Gogte and Kshirsagar, 1988; Thomas and Joglekar, 1994; Kajale, 1991; Fuller, 2000). The Chalcolithic culture sites in Maharashtra showed a decline during late 1st millennium BC and various postulates have been put forth explaining the decline in relation to climate change (see Shinde, 1994; Dhavalikar, 1997). However, these suggestions, especially the impact of monsoon rain failure and consequent rise in aridity has remained at the level of untested hypotheses. This is because earlier investigations had an implicit assumption that the people were primarily agriculture-oriented. The role of animal husbandry, hunting and gathering was considered to be secondary and restricted to a support system. These studies were not based on any working model of the agro-pastoral subsistence, which is a critical necessity in semi-arid regions. Lack of quantified faunal and botanical data (Fuller, 1999) in relation to the excavated areas and their correlation to the artefact remains is perhaps one of the major causes why such a model could not be proposed earlier. A model is particularly vital to any explanation of rise and fall of the Protohistoric cultures in this part of India as the semi-arid region has a marginal resource base (Bennett, 1996). The climate in arid and semi-arid parts places severe restrictions on the adaptation of biological systems (humans, plants and animals). All the living beings here face even more demanding challenges since the region has extremely variable monsoon precipitation (Abrol and Sharma, 1996; Singh and Sontakke, 1996). Various risks involved in monsoon-based agriculture (even accounting for modern post-independence irrigation facilities) have a significant effect on the cropping and land-use pattern (Bishwas, 1996). Though crops such as *jowar* (sorghum—*Sorghum vulgare*) and *bajra* (*Pennisetum* sp.) are adapted to the semi-arid climate, risk of damage due to weeds, insect and microbial attacks is high (Dubey and Yadav, 1980; Nagarajan, 1983; Chundurwar *et al.*, 1984; Mayee, 1996; Ramana Rao and Srinivasa Rao, 1996). The region in general has shallow and medium deep black soils that support a crop for 90–150 days in a normal monsoon precipitation regime (Sehgal and Mandal, 1996). Rainfall here is the principal limiting factor and the crop yield is affected by relatively minor climate variations (Bishwas and Khambete, 1979). At present, even in areas that receive a good amount of rainfall in semi-arid regions, the crop yields are highly variable (Khan, 1961; Kalyankar and Potekar, 1994). The cattle and sheep-goat pastoralism is equally vulnerable to damages of various kinds such as deaths due to diseases and overheating of animals in summer (Bala and Sidhu, 1981). The problem of long-term adaptation to the variable climate and rainfall is acute in such semi-arid regions (Fig. 1) since there is very little scope for recovery after harsh droughts (famines). Hence, it was felt that a formal model is necessary that can explain the dynamics of existence of Protohistoric cultures taking into account the role of monsoon-based agriculture as well as animal husbandry. Such a model would help in interpreting archaeological data, especially that dealing with the spatial and temporal aspects of subsistence strategies.

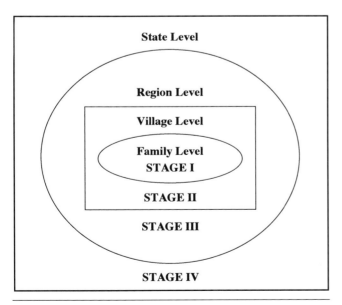

Fig. 1. Stages of modelling the subsistence in the semi-arid region.

The Scope and the Structure of the Model

Simulations are a new means of doing research in which the study is conceived as a game with a specific model in hand. Simulations and model building are assuming increasing importance in social science research since the 1950s (Bailey, 1982: 330). It is especially important for archaeology as an experimental tool since in archaeology the objects/events to be studied cannot be observed directly. Simulation involves the formulation of a dynamic model, i.e., a model concerned with change through time The idea of simulation modelling is simple: take any model in any form of some process, write a computer programme for it, run the model repeatedly and evaluate the results. Although in actual practice it is not as simple as that mentioned earlier (Aldenderfer, 1991: 194). Simulation study requires visualizing a complex task of setting rules of action and their results in terms of quantities. One starts with some initial information (data or conditions) and with repeated application of the model, generally through a computer, reaches various states of the system designed by the model. Simulation models can be used to predict the outcome of a process, explain the behaviour of a modelled system through time, experiment with different conditions of model behaviour and explore the model responses (Aldenderfer, 1991:205). This paper deals with formation of a model to explain the existence and then consequent decline of the Protohistoric cultures. These cultures existed in a semi-arid region where monsoon rainfall is marginal and highly variable.

One model of energy requirements for Pre-Neolithic and Neolithic people using various modes of production/food procurement strategies is available (Ulijaszek, 1992). Dahl and Hjort (1976) have simulated the structure of cattle herds to explain Neolithic cattle pastoralism in Europe. Flam (1976) has developed a model to examine settlement, subsistence and population of the Indus Valley Civilization. The model of subsistence designed for the Protohistoric economy is based on a number of assumptions. These are vital to the understanding of the possible deviations, if one or a few of these prove to be wrong/invalid. The model needs to cover various aspects of life under normal as well as stressful conditions. Though basic subsistence strategies originate at the lowest level of a family/clan, social and political ambience at a village, region and state level are bound to affect choice of the food procurement approaches. For instance, if several family/clan units were sharing their resources at the village level they would be spending a part of their output (energy) towards creations of mass utility facilities (such as tanks and temples). In this case, their decision calculus would include specific resource utilization schedules in order to meet the demands of surplus output. Similarly, if the villages/hamlets were operating under a control of state/feudal authority, they would need to adjust their economic activities so that even after contributing to the 'taxes' they would be in a position to survive. Therefore, this model building intends to construct four sub-modules at four specific levels of social organization (Fig. 1).

At the lowest level (Stage I) the sub-module 'family level' incorporates a family/clan engaged in subsistence for simple survival under normal (usual situations of climate and rainfall) and stressful life conditions (such as famine). At this level, the model assumes that there was no intra-family disparity of any kind (customary, cultural and calculated) though poor and ultra-poor family structures all over the modern world are known to have these as a common feature (Gill, 1992: 52). At the second level (Stage II) social and political aspects of resource sharing and/or exploitation dynamics among village families/clans is incorporated. In this stage the model attempts to weigh various factors governing choice of cooperation and sharing of resources so that agricultural and pastoral outputs could be optimized to the need of the community (village/hamlet). A formal stage III model integrates various options of subsistence strategies that would have been thought over at a regional level, if any such regional social/political authority existed during the Protohistoric period. Similarly, Stage IV examines theoretical issues and various factors that govern subsistence policies in a situation where a 'state' or an agent of the state like a feudal lord collects (either contributed willingly by the producers or extracted forcefully) a portion of the agro-pastoral outputs.

In order to simulate the conditions of rural life and subsistence-based activities during the Protohistoric period, empirical data is necessary. This data is essentially of modern origin and therefore it is vital that its utility and drawbacks are discussed in detail. Agricultural statistics and the census reports for three districts in the Deccan that were governed by British East India Company officials (Sholapur, Ahmadnagar and Poona) have been extensively used here. These numbers are thought to be useful indicators of pre-industrial agricultural conditions since research on improving food crop production began only during the 1920s (Guha, 1985: 195). Due to the nature of the information available to us about agriculture, pastoralism and hunting-gathering modes of life in the past; and climatic conditions during the Protohistoric period in India (c. 2000–600 BC) constraints and limitations operate on the interpretations of economic activities. This paper aims to deal with only Stage I of the larger modelling exercise where a monsoon-based Protohistoric subsistence economy in the semi-arid region of south India is being simulated. Yet all such constraints and the limitations of other modelling stages are discussed in the next section.

■ General Condition of Rural Economy

To understand the economics of alternative food procurement strategies (such as animal husbandry and agriculture), data used belongs to the pre-modern medieval period (c. AD 1000 – AD 1800) and the British colonial period. There are two reasons for using information chronologically so much detached from the Protohistoric period, for which this modelling exercise is being done. Firstly, though fragmentary, a good amount of authentic information is available for pre-modern medieval Indian society due to works of various historians (e.g., Chattopadhyaya, 1990; Jha, 1993; Singh., 1996; Habib, 1999). We know about various aspects of the agro-pastoral life of the rural masses in British India primarily due to meticulously kept records of British administrators since the late 18th century and subsequent compilations of agricultural statistics by British policy makers (Guha, 1992: 3). In Indian agriculture, technological changes for commercial exploitation and for getting surplus produce began to show their impact fairly late. Secondly, for the Protohistoric period, no direct archaeological data of any nature is available. Murty and Sontheimer (1980) have pointed to continuity of a sedentary, village-based dairy-cum-agricultural system in southern Deccan. Therefore, the general status of the agrarian society during the pre-modern medieval period and the British period is assumed to have been the same during the Protohistoric period.

■ Mode of Production

We are visualizing the Protohistoric settlements (the equivalent of the villages of later date) as places of permanent/semi-permanent habitations where a good portion of the population is engaged in agricultural operations. It is assumed that the mode of production is essentially for maintenance of life supporting activities. A number of contrasting theories are available to interpret the status and the nature of Indian rural folks during the pre-modern medieval period. For example, the 'Asiatic Mode of Production' (AMP) is a Marxist notion that looks at medieval Indian agriculture-based villages as centres of internal as well as external exploitation by the State (Kulke, 1995: 1–2). Particularly in the case of a pre-modern medieval context, the emergence and formation of the feudal State is discussed which essentially revolves around the idea of wealth generated either due to agricultural surplus or its forceful extraction from the peasantry (see Sharma, 1995; Jha, 1993). Although at the Protohistoric settlements in the semi-arid regions of the Deccan, some kind of power structure in form of chiefdoms has been suggested (e.g. Shinde, 1994: 176), the available archaeological record is inadequate to support the existence of any such social/economic power structure. Hence, we assume for the Protohistoric period that no such 'internal' or 'external' extractors of produce operated. In other words, it is assumed that they did not have to go for 'extra' (surplus) produce while choosing between various food procurement strategies. Like it has been suggested about the freedom of the medieval peasants (Sharma, 1995: 52) to decide what to cultivate, this model considers that each of the Protohistoric peasant units had full autonomy over means of production. Such assumption is necessary since the choice of crop and the amount of land to be brought under cultivation would make a difference to the total economic status of the Protohistoric peasant unit over a period of time. If the production from the land under cultivation had declined, the peasant unit might shift to a poorer quality land patch. This patch though poor in its yield capacity would have produced more during the first few years before showing signs of decline. During the British period it was observed that a land declined in yield after 20–30 years of continuous usage where no additional/deliberate efforts were made to replenish the lost nutrients (Guha, 1992: 42–43).

■ Community Efforts

For building of any material structures and creation of social-political and religious institutions, community efforts are necessary. Their maintenance also involves an energy cost in terms of days of labour contributed towards community work. The evidence at several Protohistoric settlements in the semi-arid regions of the Deccan (Maharashtra and Karnataka) indicates that almost no or very little communal participatory works (public works in modern terms) were carried out. For example, the houses built at the Chalcolithic sites of Inamgaon, Kaothe and Walki in the Bhima Basin (Dhavalikar, 1997) are made of poor quality construction material. Though Dhavalikar (1997: 173) suggests that some sort of public architecture might have existed at Inamgaon and Daimabad, it seems that this is unlikely. It is more likely that the early farmers in the Deccan did not have an energy budget adequate enough to even construct simple dwelling structures. Hence, it is safely assumed that they were either not much interested in public works or they could not afford to do so

The Protohistoric people of the Deccan might have been engaged in community economic activities of some sort. For instance, it is possible to use pastures as common resources as they were used in many parts of British India during the colonial period. However, community efforts in animal husbandry and practising agriculture for achieving common economic goals has some long-term disadvantages. In the short-term such community cooperation might be to the advantage of all the participants. By applying 'n person game theory' to the situation we can see what the trade-offs are of both short-term and long-term cooperative community activities. Assume that there are only two options of adding a resource unit (such as a cow, a piece of agricultural equipment, working bullock, etc.) to that of the other player in this game. The players

(common people) face two contrasting outcomes and this is the 'tragedy of the commons' (Masters, 1983).

In the short-term situation (Table 1), adding or pooling of resources is beneficial, but the benefit of pooling is lost in a long-term situation (Table 2), due to increase in the cost of existing resource units as compared to their benefits, adding more units actually leads to mutual loss.

Table 1. Short-term situation

Player B	Player A	
	Don't add 1 unit	Add 1 unit
Don't add 1 unit	A= -1 A= +1 Mutual harm	B= -1 B= -1 Nepotism
Add 1 unit	A= -1 A= +1 Sociality	B= +1 B= +1 Mutual benefit

Table 2. Long-term situation

Player B	Player A	
	Don't add 1 unit	Add 1 unit
Don't add 1 unit	A= X-1 A= X +1 Mutual benefit	B= X-1 B= X-1 Nepotism
Add 1 unit	A= X-1 A= X+1 Sociality	B= X+1 B= X+1 Mutual harm

As has been suggested by Masters (1983) one of the solutions is to delegate the authority (power) to choose among the options in such a way that pooling of common resources beyond a specific point could be avoided. However, such a situation calls for the creation of a 'power centre'. Emergence of such a centre or authority can be either formal or at a subtle level, one person can exercise such power to stop the overexploitation of resources, thus prolonging their usefulness. It is not possible to know whether Protohistoric people would have operated in such a way that they understood the relative cost-benefit ratios in implementing various subsistence options. As it has been pointed out by Randhawa (1980) traditional Indian agriculture and/or pastoralism-related activities were done without any idea of such ratios. In this condition the model assumes that the peasant units (families) would try to gain as much benefit as possible and avoid resource sharing at the lowest level.

■ The Technology Base

Technology is one of the most fundamental factors that determines the nature and quality of agro-based rural life in Asia in general and India in particular. In South Asia 'low profile' labour-intensive tillage equipment and tools are still in common usage. In this context it is necessary to look at the idea of 'indigenous' knowledge-base (see e.g., Kurin, 1983), or traditional technology-base in detail. It has been shown that even in modern (post-colonial) Indian villages, peasants neither possess 'systematic' indigenous knowledge nor the 'green consciousness' that is generally attributed to the so-called less developed peasants in Asia (Gupta, 1999: 265). For purposes of making a model, it is assumed that there were very minute innovative technological changes during the rotohistoric period. If there have been any, the model assumes that these changes were not aimed at commercial exploitation of the resources.

■ Choice Calculus

Though this model is about simulation of subsistence activities of the early farming communities that were present in the semi-arid tracts of the Deccan, there is evidence to suggest that even after becoming plant food producers these people continued to practise hunting (Joglekar, 1991). They might also have been engaged in collecting food such as shellfish, wild fruits and tubers. Thus in the model it is necessary to compare various options of acquiring food

since each of the sub-systems of the agro-pastoral mode of life have variable yields (Table 3). Possible choices and their various combinations could then be utilized to compute the paths of least energy expenditure. It is assumed that people would follow such a choice(s) that would involve less amount of energy and yet give higher returns in terms of food. Therefore, it is essential to discuss comparative cost-benefit economics not only of agriculture and pastoralism, but also of a gathering-foraging mode of subsistence.

Table 3. Output from dry land agricultural sub-systems (after Bhati, 1995)

Sub-system	Benefit-cost ratio
Agro-forestry	1.69
Silvi-pastoral	1.66
Agro-pastoral	1.87
Agro-horticultural	1.46
Crop production	1.24

While considering options of foraging and hunting it is natural that humans consider various logistical, nutritional and local ecological circumstances (Blumenschine, 1992). Not only does the decision to forage in a particular area depend on several such considerations, but so does the type of equipment (technology). A trade-off between risks involved and the net return (in terms of energy or food yield per participant) defines various activities at the settlement site and in the foraging area. Malhotra *et al.*, (1983), while studying a few non-pastoral nomadic communities from Maharashtra, observed that the prey choice operates depending on usefulness or harmfulness of the food species. On pastoral land, it has been shown (Crotty, 1980) that beyond a certain point (A) net output has no further improvement (Fig. 2). In fact, as input increases, the output may reach zero and subsequently a negative value (loss).

In India (perhaps all of South Asia), cows/bulls were kept traditionally by wisdom obtained over centuries without any formal risk

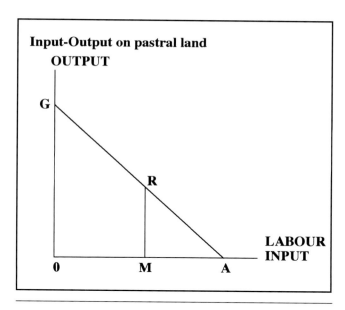

Input-Output on pastral land

OUTPUT

LABOUR INPUT

Fig. 2. Input-output calculus on a pastoral use of land.

analysis and cost-benefit considerations of the various options involved (Randhawa, 1980). Herding of domestic animals like cattle, buffalo, sheep and goat involves using their lifecycles to obtain products such as meat, hide, milk, blood and manure. Due to the biological adaptations of each of these animals, there are advantages and disadvantages in keeping them in a semi-arid region. For example, in warm-humid conditions the buffalo is an ideal animal for draft purposes. As has been observed in Bangladesh and eastern India, it can do good work though it is slower than a working bullock. As compared to the work output of oxen (4.8–6.4 km per hour) the working buffaloes give less (3.2 km per hour) work output (Fahimuddin, 1975). Mishra (1978) has provided comparative economics of agriculture and animal husbandry. It is clear that livestock keeping is more labour-intensive and produces more food per unit input of resources (Table 4).

Estimates of the economics of keeping cattle and buffalo in an Indian situation where these animals are kept mainly as a family-based subsistence enterprise (Shiva, 1995). These animals perform a very important function of providing manure for agriculture in an 'internal

Table 4. Input-output in agriculture and livestock keeping (after Mishra 1978)

		Agriculture	Livestock
Input	Land (acres)	0001.00	000001.00
	Labour (Man days)	0042.00	00297.00
	Capital (Rs.)	0024.40	00031.65
Output	Food (Kg)	4481.40	30584.65
	Calories (million)	0001.55	00002.48
	Protein (Kg)	0037.43	00100.00

input farming system' (Shiva, 1989). It has been found that Indian cattle and buffalo are less productive than those kept in the US on a commercial basis. However, they generate a supply of manure and employment for non-commercial cattle/buffalo keepers (Table 5). These estimates obtained for traditional agro-pastoral economics of rural India are useful to understand the energy balance and for simulation of output/input balance. This perhaps is the best model of traditional Indian rural society where resources are optimized by recycling them in the low-profile subsistence agro-pastoral system in South Asia (Fig. 3).

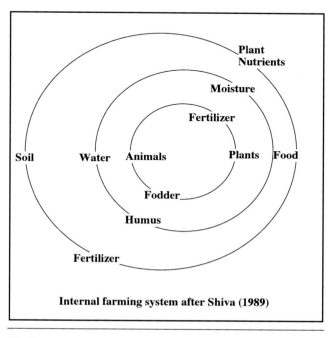

Internal farming system after Shiva (1989)

Fig. 3. Resource cycles in the internal farming system.

Table 5. Input-output in non-commercial cattle/buffalo keeping (after Shiva, 1995)

Input	Energy (1012 cal)	122.2
	Proteins (109 g)	35.4
	Edible matter (1010 Kg)	40.68
Output	Work (1012 cal)	6.50
	Proteins (109 g)	0.99
	Manure matter (1012 cal)	16.16

Core of Stage I Model

The core of the model includes the origin of a settlement, when people would have started claiming arable lands and making them actually suitable for cultivation. Therefore, it is necessary to examine various limiting factors that affect production. An implicit assumption made while interpreting the increase in population of early farmers during the Protohistoric period is that of unlimited arable land. Such cultivable land may have been present around the Chalcolithic sites in the Deccan. However, it is important to note simply an increase in cultivable area does not lead to an increase in output. The real limiting factor may be the manpower and/or animal work power to do the heavy work of clearing the natural growth on such good patches of land. The middle 19th century condition of a semi-arid monsoon-based region of the Deccan is a good case to illustrate the relation between land, people and net returns when technology remains unchanged. Official statistics of land under cultivation, bullocks and number of ploughs in the Deccan (Khandesh, Ahmadnagar, Poona, Sholapur, Belgaum and Dharwar districts during the British regime) in 1846 and 1876 (Table 6) indicate that the rate of change was marginal. Also the human population was changing slowly during the same period (1.2–1.6% per year). This was in spite of marked efforts of the peasants and British revenue administrators in some of the districts to increase agricultural output. Between 1846 and 1872 (census year) large areas were progressively being brought under cultivation. Most of these were classified earlier as non-arable. But agricultural output increased less rapidly than the rate at which new lands were brought under cultivation (Guha, 1985: 62).

Table 6. Agriculture and population in colonial Deccan (after Guha 1985)

			% change/year
Year	1846	1872/1876	
Bullocks (000)	1045	1270	0.80
Ploughs (000)	0237	0328	0.91
Humans (000)	3509	5300	1.60

Guha (1985: 69) referring to available revenue records has shown that on marginal land (poor quality) cropping of *jowar* and *bajra* was a labour-intensive and hence very expensive operation. Agricultural records of the area under cultivation of crops (cereals and pulses) was 16,236 thousands of acres in 1885–90, while it was 16,070 thousands of acres in the Deccan (Guha, 1985: 88). The area fluctuated but essentially no major change was seen in about fifty years though human population was growing steadily.

The quality of soil is an important factor that determines the net returns a peasant can get by cultivation. Yield values used for computing the net return for rain-fed food crops (Table 7) is based on the report of H.D. Robertson prepared in 1821 by observing bajra yields in Deccan district under British administration (quoted by Guha, 1992: 42).

Table 7. Yield variation of bajra observed in 1821 (converted figures)

Soil Class	Yield (Kg/acre)
First class black soil	470.38
Second class black soil	313.44
Third class black soil	134.26
Actual area under cultivation	294.84

To simulate the conditions of early farmers during the Protohistoric period it is essential to consider a family/clan unit of peasants using animal (bullock) power to do heavy agricultural work. As the British administrators have considered, one plough has been equated to one peasant unit. The district gazetteers of two districts (Ahmadnagar and Poona) published by the Bombay Presidency in late 19th century provide us with numbers necessary to visualize a peasant unit in quantitative terms (Table 8).

Table 8. Statistics of population engaged in agriculture in 1882–83 (modified)

District	Ahmadnagar	Poona
Population	0486248	511943
Bullocks	0252602	227619
Cows	0195210	144949
Buffaloes	0046497	052730
Sheep-goats	0456625	289688
Ploughs with 2 bullocks	023941	026722
Ploughs with 4 bullocks	040739	025908
Holding (1–30 acres)%	91.68	90.49
Holding (> 31 acres) %	08.32	09.51
Bullocks per plough	03.90	04.32
Buffaloes per plough	0.7188	01.00
Sheep-goats per plough	5.5042	7.0597
People/plough	7.1578	9.7272
% below 15 years	39.55	39.16
% above 15 years	60.45	60.84

An estimate of the agricultural and pastoral outputs is necessary to build a quantitative base upon which the model of early subsistence can be designed. We have some modern estimates of agricultural production of *jowar* and *bajra* (Table 9).

This model utilizes both the estimates of crop yields recorded in pre-Independence India and the modern period (before introduction of high-yielding varieties in the 1960s). The Protohistoric sites simulated have a few peasant units engaged in primitive agriculture and associated animal husbandry. Each unit would have 8–10 people (half of them adults) working entirely to the benefit of their own unit. Each unit exercises its

Table 9. Yields of jowar and bajra in monsoon-fed semi-arid region (from Joshi, 1996)

Bajra	Sub-region	Yield (Kg/ha)	Rainfall (mm)
Gujarat	4	801	735
Maharashtra	2	559	988
Maharashtra	3	551	602

Jowar	Sub-region	Yield (Kg/ha)	Rainfall (mm)
Karnataka	1	650	688
Karnataka	2	1149	684
Maharashtra	2	655	988
Maharashtra	3	562	602
M.P. (Nimar)	12	737	874

choice of subsistence mode on 'as they face situation' basis and not by doing any formal risk analysis. Salient features of this model peasant unit (Fig. 4) acting at Stage I are:

The unit is living in normal monsoon-based climate conditions. Choices and actions of the people during stress conditions (famine/flood) are not considered here in Stage I.

The peasant units are using their subsistence options at will and these do not include any

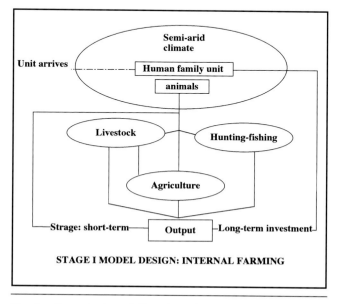

STAGE I MODEL DESIGN: INTERNAL FARMING

Fig. 4. Stage I of modelling the subsistence in the semi-arid region.

contribution of forced labour and bonded labour of any sort.

A family/peasant unit comprises 8–10 members and intra-family exploitation or 'unjust' utilization of the work capacity is not considered.

Climate is highly variable and monsoon rainfall shows cyclical trend as seen today.

Each family/peasant unit has 2–4 working oxen, a few cows, besides some sheep-goats and fowls.

Contribution of non-adult members of the family by means of foraging-collecting roots, tubers and shells towards food supplies is minimal.

The family/peasant unit is completely recycling its output. If necessary, the surplus is stored for future consumption and not for exchange in lieu of other items.

The Stage I model thus designed would be tested by computer simulations of variable climate and population conditions. Archaeological record, available at present—faunal, botanical and artefact remains (e.g., storage bins, house plans and their remains) would be examined using mathematical techniques of spatial analysis. Energy balance of the peasant unit shall form the basis for testing the model. Using the modern cyclical trends in monsoon precipitation, yields of *bajra* and *jowar* would be simulated. Assuming that a part of the unit is engaged in keeping domestic animals and a part of the year is spent in deliberate hunting and fishing, energy obtained and spent shall be computed. The model thus designed shall be applied to test the following hypothesis. Continuous use of soil and uncontrolled animal breeding over a long period of time by a group of peasant families led to a decline of productivity. Though climate and rainfall would have been variable, unplanned human actions aimed at short-term gains were responsible for overall degeneration of Protohistoric (Neolithic-Chalcolithic) cultures in the semi-arid regions in south India.

References

Abrol, Y.P. and A. Sharma (1996): Climate Variability and Indian Agriculture, *Science and Culture*, 62(7–8): 196–204.

Aldenderfer, M. (1991): The Analytical Engine: Computer Simulation and Archaeological Research. In Shaffer, M.B. (ed.): *Archaeological Method and Theory* 3, The University of Arizona Press, Tucson, pp. 195–247.

Allchin, F.R. (1963): *Neolithic Cattle Keepers of South India: a Study of the Deccan Ashmounds*, Cambridge University Press, Cambridge, 188 pp.

Bailey, K.D. (1982): *Methods of Social Research*, Free Press, New York, 553 pp.

Bala, A.K. and N.S. Sidhu (1991): Studies on Disease Resistance vis-à-vis Susceptibility in Cattle and Buffalo, *Indian Journal of Animal Health*, 20(1): 25–29.

Bennett, J.W. (1996): *Human Ecology as Human Behavior*, N.J., Transition Publishers, New Brunswick, 387 pp.

Bhati, T.K. (1995): Integrated Land Use System for Dry lands. In *Land Resources and Their Management for Sustainability in Arid Regions*, Scientific Publishers, Jodhpur, pp. 175–185.

Bishwas, B.C. (1996): Link Between Climate Variability and Agricultural Production. In Abrol, Y.P., Sulochana Gadgil and G.B. Pant (eds.): *Climate Variability and Agriculture*, Narosa Publishing, New Delhi, pp. 188–203.

Bishwas, B.C. and H.N. Khambete (1979): Distribution of Short Period Rainfall Over Dry Farming Tracts of Maharashtra, *Journal of Maharashtra Agricultural Universities*, 4: 145–156.

Blumenschine, R.J. (1992): Hominid Carnivory and Foraging Strategies and the Socio-economic Function of Early Archaeological Sites. In Whiten, A. and E.M. Widdowson (eds.): *Foraging Strategies and Natural Diet of Monkeys, Apes and Humans*, Clarendon Press, Oxford, pp. 211–221.

Chattopadhyaya, B. (1990): *Aspects of Rural Settlements and Rural Society in Early Medieval India*, Calcutta.

Chundurwar, R.D., R.R. Karanjkar, S.P. Kalyankar and Mir Ashfaq Ali (1984): Effect of Weather Parameters on the Incidence of Sorghum Shoot fly at Parbhani, *Sorghum Newsletter*, 27: 83.

Crotty, R. (1980): *Cattle, Economics and Development*, Commonwealth Agricultural Bureau, London, 253 pp.

Dahl, G. and A. Hjort (1976): *Having Herds*. University of Stockholm, Stockholm, 335 pp.

Dhavalikar, M.K. (1988): *The First Farmers of the Deccan*. Ravish, Pune, 80 pp.

Dhavalikar, M.K. (1997): *Indian Protohistory*. Books and Books, New Delhi, 329 pp.

Dubey, R.C. and T.S. Yadav (1980): Sorghum Shootfly Incidence in Relation to Temperature and Humidity. *Indian Journal of Entomology*, 41: 273–274.

Fahimuddin, M. (1975): *Domestic Water Buffalo*, Oxford and IBH, New Delhi.

Flam, L. (1976): Settlement, Subsistence and Population: A Dynamic Approach to the Development of the Indus Valley Civilization. In Kennnedy, K.A.R. and G.L. Possehl (eds.): *Ecological Backgrounds of South Asian Prehistory*, Cornell University Press, New York, pp. 76–93.

Fuller, D.Q. (1999): *The Emergence of Agricultural Societies in South India*. Unpublished Ph. D. thesis , Cambridge University.

Fuller, D.Q. (2000): Fifty Years of Archaeobotanical Studies in India: Laying a Solid Foundation. In Settar, S. and R. Korisettar (eds.): *Indian Archaeology in Retrospect, volume III. Interactive Disciplines*, Manohar, New Delhi, pp. 247–363.

Gill, G.J. (1992): *Seasonality and Agriculture in the Developing World*, Foundation Books, New Delhi, 343 pp.

Gogte, V.D. and A.A. Kshirsagar (1988): Chalcolithic Diet: Trace Elemental Analysis of Human Bones. In Dhavalikar, M.K., H.D. Sankalia and Z.D. Ansari (eds.): *Excavations at Inamgaon* Vol. I(ii), Deccan College, Pune, pp. 991–1000.

Guha, S. (1985): The Agrarian Economy of Bombay Deccan—1818–1941, Oxford University Press, Delhi, 215 pp.

Guha, S. (1992): Introduction. In Sumit Guha (ed.): *Growth, Stagnation or Decline? Agricultural Productivity in British India*, Oxford University Press, Delhi, pp. 1–62.

Gupta, Akhil (1999): *Postcolonial Developments, Agriculture in the Making of Modern India*, Oxford University Press, New Delhi, 416 pp.

Habib, Irfan (1999): *The Agrarian System of Mughal India, 1556–1707*. 2nd Edition. Oxford University Press, Delhi, 547 pp.

Jha, D.N. (1993): *Economy and Society in Early India. Issues and Paradigms*. Munshiram Manoharlal, New Delhi, 181 pp.

Joglekar, P.P. (1991): *A Biometric Approach to the Faunal Remains in Western India with Special Reference to Kaothe and Walki*, Unpublished Ph.D. thesis, University of Pune, Pune.

Joshi, S.N. (1996): Yield Gap Analysis in Ago-climatic Sub-regions. In Basu, D.N. and S.P. Kashyap (eds.): *Agro-climatic Regional Planning in India*, Concept Publishing, New Delhi, pp. 101–142.

Kajale, M.D. (1991): Current Status of Indian Palaeoethnobotany: Introduced and Indigenous Food Plants with a Discussion on the Historical and Evolutionary Development of Indian Agriculture and Agricultural System in General. In Renfrew J. M. (ed.): *New Light on Early Farming Recent Developments in Palaeoethnobotany*, Edinburgh University Press, Edinburgh, pp. 155–189.

Kalyankar, S.P. and E.M. Potekar (1994): An Economic Analysis of Farm Business in Assured Rainfall Zone of Maharashtra, *Journal of the Maharashtra Agricultural Universities*, 19(2): 269–272.

Khan, M.L. (1961): Variability of Rainfall and Bearing on Agriculture in the Arid and Semi-arid Zones of West Pakistan, *Pakistan Geographical Review*, 16: 35–50.

Kulke, H. (1995): Introduction: The Study of the State in Pre-modern India. In Hermann Kulke (ed.), *The State in India 1100–1700*, Delhi: Oxford University Press: 1–47.

Kurin, Richard (1983): Indigenous Agronomics and Agricultural Development in the Indus Basin. *Human Organization* 42: 283–294.

Malhotra, K.C., S.B. Khomne and Madhav Gadgil (1983): Hunting Strategies among Three Non-pastoral Nomadic Groups of Maharashtra. *Man in India* 63(1): 21-39.

Masters, R. (1983): Commentary on D.J. Ortner's Paper. In D.J. Ortner (ed.): *How Humans Adapt? A Biocultural Odyssey*, Smithsonian Institution Press, Washington D.C., pp. 149–159.

Mayee, P. (1996): Incidence of Fungal Diseases in Relation to Variation in Various Climatic Factors. In Y.P. Abrol, Sulochana Gadgil and G.B. Pant (eds.): *Climate Variability and Agriculture*, Narosa Publishing, New Delhi, pp. 295–305.

Mishra, S.N. (1978): *Livestock Planning in India*, Studies in Economic Development and Planning, Delhi.

Murty, M.L.K. and G.D. Sontheimer (1980): Prehistoric Background to Pastoralism in the Southern Deccan in the Light of Oral Traditions and Cults of Some Pastoral Communities. *Anthropos*. 75: 163–184.

Nagarajan, S. (1983): *Plant Disease Epidemiology*, Oxford and IBH, New Delhi.

Paddayya, K. (1973): *Investigations into the Neolithic Culture of the Shorapur Doab*. E.J. Brills, Leiden, 137 pp.

Paddayya, K. (1975): The Faunal Background of the Neolithic Culture of South India. In Clason, A.T. (ed.): *Archaeozoological Studies*, Elsevier, Amsterdam, pp. 329–334.

Paddayya, K. (1993): Ashmound Investigations at Budihal, Gulbarga District, Karnataka. *Man and Environment* XVIII (1): 57–88.

Paddayya, K. (1995): Further Field Investigations at Budihal, *Bulletin of the Deccan College Post-Graduate and Research Institute*, 53: 277–322.

Ramana Rao, B.V. and M. Srinivasa Rao (1996): Whether Effects on Insect Pests. In Abrol, Y.P., Sulochana Gadgil and G.B. Pant (eds.): *Climate Variability and Agriculture*, Narosa Publishing, New Delhi, pp. 281–294.

Randhawa, M.S. (1980): *A History of Agriculture in India*, Vol. I., Indian Council of Agricultural Research, New Delhi, pp. 541.

Sehgal, J. and D.K. Mandal (1996): Agro-ecological Regions of India and Climatic Change. In Y.P. Abrol, Sulochana Gadgil and G.B. Pant (eds.): *Climate Variability and Agriculture*, Narosa Publishing, New Delhi, pp. 204–222.

Sharma, R.S. (1995): How Feudal was Indian Feudalism? Hermann Kulke: *The State in India* In Hermann Kulke (ed.): The state in India 1100–1700, Oxford University Press, Delhi.

Shinde, V.S. (1994): The Deccan Chalcolithic: A Recent Perspective, *Man and Environment*, 19 (1–2): 169–178.

Shiva, V. (1989): *Staying Alive*: Zed Books, London, 234 pp.

Shiva, V. (1995): *Monocultures of the Mind*: Third World Network, Penang, 184 pp.

Singh, Chetan (1996): Forest. Pastoralists and Agrarian Society in Mughal India, in Arnold, D. and R. Guha (eds.): *Nature, Culture, Imperialism*, Oxford University Press, Delhi, pp. 21–48.

Singh, N. and N.A. Sontakke (1996): Climate Variability Over Pakistan and its Relationship to Variations Over Indian Region, in Abrol, Y.P., S. Gadgil and G.B. Pant (eds.): *Climate Variability and Agriculture*, Narosa Publishing, New Delhi, pp. 69–97.

Thomas, P.K. and P.P. Joglekar (1994): Holocene faunal studies, *Man and Environment* 19(1–2): 179–203.

Ulijaszek, S.J. (1992): Human Dietary Change. In Whiten A. and E.M. Widdowson (eds.): *Foraging Strategies and Natural Diet of Monkeys, Apes and Humans*, Clarendon Press, Oxford: 111–119.

Chapter 13

The Nile Floods and Indian Monsoon

MADHUKAR K. DHAVALIKAR

The monsoon plays such an important role in the life of Indians that their entire life cycle revolves around it. It is rightly said that agriculture in India is a gamble with nature, for every third year is a bad year and every fourth, a famine. Even at present, with tremendous advance in technology, the annual financial budget of the Government of India is totally dependent on the monsoon. One can then imagine what must have happened in the ancient past when technology was primitive and man was fully at the mercy of nature. It is interesting to examine how the monsoon has been shaping the destiny of India during the last five thousand years. It would certainly have been most interesting had we been able to have at our command adequate climatic data from the ancient past. But the reconstruction of past environment is still in its infancy in India, and we have therefore to search for reliable sources of information in this respect.

It is now increasingly being recognized that environment plays a very dominant role in the formation of human cultures. Perhaps that is why the adherents of New or Processual Archaeology define culture as 'an extra somatic means of adaptive system that is employed in the integration of a society with its environment and with other sociocultural systems' (Binford, 1972: 20). In history too a pioneering contribution has been made by the proponents of the *Annales* School of France which was founded by Lucien Febvre and Marc Block. It flourished under its most ardent advocate Fernand Braudel. For him, the history of man is nothing but the history of his relationship with environment—which is a great constraint in the development of human cultures.

The environment is a hindrance beyond which man and his experiences cannot go. For centuries man has been a prisoner of climate, of the animal population, of a particular agriculture, of a whole slowly established balance from which he cannot escape without the risk of everything being upset (Braudel, 1980: 30).

Monsoon constitutes the most important component of environment in India, but we have absolutely no knowledge about its pattern in antiquity. The only exception is the *Arthasastra* of Kautilya (4th – 3rd century BC) which gives information about the monsoon rainfall in some parts of the country (*Arthasastra* II, 24: 116). Incidentally it may be mentioned that Kautilya, who is identified with Chanakya, the prime minister of Chandragupta Maurya (323–300 BC), is the earliest source who introduced the measurement of rainfall (II, 5: 58). He mentions a rain gauge (*drona*) to measure rainfall. This is the earliest attempt to measure rainfall in human history. There are, however, many references to famines in inscriptions and literature. Thus Megasthenes, the ambassador of Seleucus Nicator, the Indo-Greek king, to the court of Chandragupta Maurya, states in his *Indica* (Ch. I, p. 35) that 'Famine has never visited India.' This indicates that there was no famine in India during his brief stay. Ironically enough, there was such a severe famine during the last years of Chandragupta Maurya (c. 300 BC) that, according to a Jain legend, he had to migrate to south India.

In all probability the famines recorded in the Sohagaura and Mahasthan inscriptions, which belong to the Mauryan period, were the same as that referred to in the Jain legend.

But what about the monsoon records? Surprisingly enough, we can obtain information about monsoon in the ancient past from the Nile flood data. It may rightly be asked what the Nile has got to do with India? The explanation can be found in Gilbert Walker's study (1986) of monsoon in India and northeast Africa. He was the Director General of the meterology department of Government of India. He studied the rainfall pattern of India and northeast Africa from 1840–1910 and concluded that there is a great correspondence, almost one to one, between the two (*Imperial Gazetteer of India*, 1908: 1, 37).

Gilbert Walker's conclusion is based on the fact that the Southwest Monsoon which causes summer rainfall in India is the same that causes rainfall in northeast Africa. It is because of this rainfall that rivers in Ethiopia, particularly the Blue Nile, get flooded. The Blue Nile joins the White Nile near Khartoum which is close to the southern border of Egypt. The White Nile originates in central Africa and is rarely flooded. What is more, Egypt is almost a desert with hardly any rainfall worth the name. The obvious conclusion then is that the flooding of the Nile in Egypt is the result of the Southwest Monsoon which comes from India. This has been very explicitly stated by Walker (ibid).

It is now fully established that years of drought in western and northwestern India are almost invariably years of low Nile flood. The relation is further confirmed by the fact that years of heavier rain than usual in western India are also years of Nile flood.

It is significant that in the last century north Africa and India experienced common intervals of drought and in the present century both have experienced weak monsoons between the 1960s and the 1970s (Kutzback, 1987: 289). Even

Alexander's historians seem to have been aware of this phenomenon. Nearchus' account is not available but has partly survived in Arrian's *Indica* which belongs to the 1st century AD. It seems that while crossing the Indus, Alexander had to face the Southwest Monsoon sometime in September 326 BC. He states that these winds blow every year in the summer from the sea to the land and make navigation difficult (*Indica* Ch. 21–22). Similarly while returning in autumn, Alexander was again troubled by unfavourable winds. It is further observed that the sea is fit for navigation in winter.

The testimony of Nearchus is of great importance because it suggests that there was heavy rain when Alexander came and the Indus was in flood. But much more important is Arrian's observation which suggests the connection between the Indian monsoon and the Nile floods.

India is visited by rain in summer, especially the mountains and from these the rivers flow swollen and muddy. In the summer also the plains of India are visited by rain, so that a great part of them are covered by pools; and Alexander's army had to avoid the Acesines river (Chenab, a tributary of the Indus) in the middle of the summer because the water overflowed the plains. Wherefore from this it is possible to conjecture the cause of the similar condition of the Nile, because it is probable that mountains in Ethiopia are visited by rain in summer, and the Nile being filled by them overflows its banks into the Egyptian country (*Indica Ch.6*).

There is a hint of the seasonality of monsoon in the Babylonian documents as well. A few of them refer to the preparation for voyage to Dilmun and Makan and specify the month of the year. The merchants are advised that in the months of February-March (the Sumerian month of *se-kin-kud*) the wind is most favourable for sailing. This document is dated 2031 BC whereas there is another one more or less of the same period which shows that July-August was most suitable. There is yet another record which supports this contention (Warren, 1987: 139–140).

The unique feature of the Nile flood data can be gauged from the fact that it is available from a very early period, the earliest going back to 2900 BC (Bell, 1970). The record is sporadic; there are sixty-three records from the Early Dynastic (3050– 2700 BC) to the Old Kingdom (2700–2215 BC) and twenty-eight for the Middle Kingdom (2040–1715 BC), but no systematic data is available for the New Kingdom (1570 – 1070 BC) (Butzer, 1976: 28–29). However, from AD 622 to AD 1521 this data is available for almost every year (Toussoun, 1925) when the annual flood level was recorded by a gauge-Nilometer-which was invented by an Egyptian sultan who installed it on the island of Roda in the Nile. The data is also available in the form of flood level marks on temples, foundation levels of buildings, steps leading to the river and so on. The earliest record (2900 BC) is evident from the marking on a stone slab known as the Palermo Stone which is housed in the Palermo Museum in Italy. As early as 1895, M. LeGrain noticed at Karnak a series of forty markings on the walls of the great temple, starting from 800 BC. The rate of the rise in the river bed has been calculated to 5.2 inches per hundred years (Quinn, 1992).

Three types of records have been noticed. The first marked the water level on cliffs on the river banks, the second was the flight of steps on the banks, and the third consisted of bringing the water to a well through conduits and marking the level on a central column. This last method was more accurate. The Roda Nilometer consists of a square well connected to the Nile by three conduits, and in the centre of the well is an octagonal column on which the flood levels are recorded (Biswas, 1970: 15).

Eminent scientists have studied the Nile flood data and have calculated with great precision actual flood levels with due consideration for the thickness of silt deposition every year (Quinn, 1992). We can therefore use the data for finding out the behaviour of monsoon rainfall in India. We can reasonably hypothesize that when the Nile had a good flood, India enjoyed a good rainfall, and conversely when the Nile level was low, India suffered droughts. It is, however, necessary to test the veracity of this proposition. The only recorded instance in India which we can verify is the great famine, known as the 'Durgadevi' famine which lasted for twelve years from AD 1396–AD 1408. It was so severe that even now it lingers in people's memory. It was during this famine that one Damajipant, an officer in charge of the Royal Granary of the Sultan of Bidar (now in Karnataka), is said to have distributed the grain from the granary to the famine-stricken people without the permission of the Sultan. There are many references to it in Marathi literature. Even official documents describe how a vast majority of the populace deserted their habitations and became pastoralists and hence their land remained uncultivated (Guha, 1999: 30 ff). The Durgadevi famine receives corroboration from the Nile flood data. There are five years—1399, 1401, 1402, 1403 and 1408—when extremely low levels were recorded (Quinn, 1992: Table 6.6) indicating poor monsoon in Ethiopia and if we accept Gilbert Walker's observation (1986), then it can be averred that there is almost a perfect fit between the Indian monsoon and Nile flood levels.

The 'Durgadi' famine referred to by the great Maratha historian, V.K. Rajwade (1905) has been described in a document he obtained from one Deshpande of Wai (Maharashtra). It is sometimes confused with the Durgadevi famine, but must have been different as it is stated to have occurred during AD 1468–AD 1475. It is noteworthy that even this famine receives corroboration from the Nile recordings which show that the levels were quite low from AD 1468 to AD 1474 (Quinn, 1992).

The Nile flood data also allows us to fix the date of an earlier devastating famine for which we have literary evidence. Dandin, a great writer of Sanskrit prose, gives a heart-rending description of an extremely severe famine in his *Dasa-kumara-charita* which is assigned to AD 700 (Gupta, 1971). He is said to have lived in the Pallava court during the reign of Parameshwaravarman I (AD 670–AD 700) and/or Narasimhavarman II Rajasimha (AD 700–AD 728) (Gupta, 1971: l8–19). In one of the sub-stories in this work, there is a pathetic account of a severe drought which lasted for twelve years (Kale, 1917: Ch. VI, 117). When did this famine occur? Although it is a piece of literature, it appears from the description that the author was witness to the miseries of the people. From the Nile flood record, it is seen that the river levels were extremely low from AD 683 to AD 696, and also from AD 770–AD 782. Since Dandin lived during the latter half of the 7th century AD, we may not be far off the mark if we assign the famine to AD 683–AD 696 to the time of Parameshwaravarman I (AD 670–AD 700). The work was probably composed when Narasimhavarman II (AD 700–AD 728) was ruling.

The Nile flood data is of great help to us for the period from AD 621 to AD 1520. A casual glance at the record shows that between AD 600 to AD 1000, there were 128 years of low levels of the river, and sometimes so low that even ice was seen floating on it. We may not therefore be wrong if we postulate poor monsoon rainfall in India during these years. This is extremely likely because at many sites in the country the excavation evidence clearly shows a hiatus after the post-Gupta epoch from the 6th century AD. This happened at about 60% of the sites which were abandoned and many of them came to be occupied only in the medieval period from AD 1400–AD 1800 (Dhavalikar, 1999). There is corroborative literary evidence too. The *Devi-Mahatmya* section of the *Markandeya Purana* describes a devastating drought which lasted for a hundred years. To alleviate the sufferings of the people, the goddess Durga caused rain to fall and covered the earth with vegetation. Hence she came to be known as *Sakambhari* (Dhavalikar 1986). This goddess has been described in some of the works as the headless goddess (*chhinna-masta*) and can be identified with what are generally known as the images of Lajja-Gauri who is shown without head and with legs stretched apart (Dhere, 1978).

The widespread desertion of habitation sites in the post-Gupta era is supported by the evidence furnished by the celebrated Chinese traveller, Hieun Tsang. He was in India from AD 629–AD 641, and travelled over a large part of the country. He records that at many places in north India cities were in ruins, and in the south, town after town was abandoned (Watters, 1904–05: II, 25, 63). It is interesting to note that Varaha-mihira, (6th century AD), the celebrated Indian astronomer, had predicted all round decline caused by drought and consequent famines (*Brihat-samhita*: Ch. 17). But the evidence from excavations indicates that in northwest and western India, many sites were abandoned after the Kushan period in the 3rd century AD and were occupied again after a lapse of over a thousand years in the medieval period (Dhavalikar, 1999). In Europe too decline set in from the 4th century AD (Lamb, 1977: II 424). There, the climate became intensely cold, and climatologists tell us that whenever temperate lands have a cold phase, tropical countries like India suffer drought (Sulman, 1982: II, 118). Thus drastic change of climate from the 4th century AD was an important cause of the decline that set in India. In addition, the cessation of Roman trade was also an important factor.

As we go backwards, the Nile data becomes rather sporadic, but whatever is available, shows that the climate was far more favourable. The period from the 5th century BC to 5th century AD all over the Old World is marked by the rise of empires: Maurya, Satavahana, Kushan and Gupta in India; Achemenid and Sassanian in West Asia; Macedonian and Roman in Europe and the Han in

China. One single factor responsible for them all was the favourable climate; a warm phase in the temperate zone. According to scientists when Europe enjoys a warmer climate, India receives a good rainfall (Currie and Fairbridge, 1985: 111). It was during this period that the existence of monsoon was recognized by Hippalus, a Greek navigator. The Nile flood data for the period between 300 BC and AD 400 points to very high flood levels (Evans, 1990: Fig. 5) suggesting a good monsoon in India. This is further corroborated by the archaeological evidence from many town sites of the Early Historical period which were the capitals of republics (*Mahajanapadas*). Many of them had fortification walls of which the one along the river was originally an embankment built in the 5th–4th century BC as a measure of protection from river floods (Mate, 1969–70). Dams were built and canals were dug as at Girnar and other sites, and wells were dug on a large scale. A new class of hydrolic engineers (*odayantrikas*) came into being during this period (*Luders' List* No. 1137). According to some even *Shaduf* and the Persian wheel was introduced in this period (Joshi, 1971).

The Nile data is scarce for the still earlier period but is significant as it is corroborative. From 3250 BC to 2750 BC there were high floods and India probably enjoyed a good rainfall. This phase witnessed the rise of the early Harappan culture. The Nile levels were very low from 2200 BC to 1750 BC and it is exactly this phase which marks the decline of the Harappan Civilization. Gurdip Singh's palynological evidence (1971) too corroborates it. Further between 1250 BC and 750 BC there were many years of very low levels. In India most of the Chalcolithic cultures died out after 1200 BC which is noteworthy. In Egyptian history this period has been referred to as the 'centuries of darkness' (James *et al.*, 1991). The climate probably began to improve from the 5th–4th century BC which marks a period of prosperity all over the Old World.

The foregoing discussion of Nile floods data amply brings into sharp focus its importance as a source for reconstructing the behaviour of monsoon in the ancient past. The close correspondence between the two underlines how the study can prove to be immensely fruitful for students of Indian history. In India, the sources for reconstructing past environment are extremely scarce and one can therefore imagine the crucial importance of the Nile flood data. Another important feature of this record is that the fluctuations in the Nile levels and the corresponding changes in the pattern of monsoon correlates with culture shifts in all the 5000-year-old history of India. I have undertaken a detailed study of this invaluable source for reconstructing the past behaviour of the monsoon and the contemporary culture-historical changes. There is little doubt that this study of cultural ecology goes a long way in explaining cultural efflorescence and hiatuses in Indian history (Dhavalikar, 2002).

References

Bell, B. (1970): The Oldest Records of the Nile floods, *Geogr. Jr.* 136: 569–73.

Bell, B. (1973): The Dark Age in History: The First Dark Age in Egypt, *AJA* 75: 1–26.

Bell, B. (1974): Climate and the History of Egypt—The Middle Kingdom, *AJA* 79: 223–269.

Binford, L.R. (1972): *An Archaeological Perspective*, Seminar Press, New York.

Biswas, A.K. (1970): *History of Hydrology*, North Holland Pub. Co., Amsterdam.

Braudel, Fernand (1980): *On History*, University of Chicago Press, Chicago.

Butzer, K.W. (1976): *Early Hydrolic Civilization of Egypt: A Study in Cultural Ecology*, University of Chicago Press, Chicago.

Currie G. and R.W. Fairbridge (1985): Periodic 18.6 year and cyclic 11 year induced drought in northeast China and some global implications, *Quarternary Science Review* 4: 109–34.

Dhavalikar, M.K. (1987): Sakambhari - The Headless Goddess, *Annals of Bhandarkar Oriental Res. Inst.* 68: 281–93.

Dhavalikar, M.K. (1999): The Golden Age and After: Perspectives in Historical Archaeology, *Presidential address, Indian History Congress* (60th Session), Calicut.

Dhavalikar, M.K. (2002): *Environment and Culture: A Historical Perspective,* Bhandarkar Oriental Research Institute, Pune.

Dhere, R.C. (1978): *Laijagauri,* Srividya Prakashan, (in Marathi), Pune.

Evans, T.E. (1990): History of Nile Floods. In P.P. Powell and J.A. Allen (eds.): *The Nile—Resource Evaluation, Resource Management and Legal Issues,* Progs. of Conference. London SOAS and Royal Geographical Society, pp. 5–21.

Guha, S. (1999): *Environment and Ethnicity in India 1200–1991.* Cambridge University Press, Cambridge.

Gupta, D.K. (1971): *Society and Culture in the Time of Dandin,* Mehrchand Lachmandas, Delhi.

Imperial Gazetteer of India (1908): *The Indian Empire,* 3 vols., The Clarendon Press, Oxford.

James, P. et al., (1991): *Centuries of Darkness,* Jonathan Cape, London.

Joshi, M.C. (1971): An early inscriptional reference to Persian Wheel. In *K.A. Nilakantha Sastri* Felicitation Committee, Madras, pp. 214–217.

Kale, M.R. (ed. & tr.) (1917): *The Dasa-kumara-charitam of Dandin.* The Oriental Pub. Co., Bombay.

Kutzback, I.E. (1987): The Changing Pulse of the Monsoon. In J.S. Fein and P. Stephens (eds.): *Monsoon.* John Wiley, New York.

Lamb, H.H. (1972): *Climate—Past. Present and Future,* Vol. 1, *Fundamentals and Climate Now,* Methuen, London.

Lamb, H.H. (1977): *Climate—Past. Present and Future,* Vol. 2, *Climatic History and the Future,* Methuen: London.

Luder, H. (1909–10): *Luder's List of Brahmi Inscriptions—Epigraphia Indica 10,* Archaeological Survey of India, New Delhi.

Mate, M.S. (1969–70): Early Historical Fortifications in the Ganga Valley. *Puratattva* 3: 58–69.

Quinn, W.N. (1992): A Study of Southern Oscillation-related activity for AD 622–1900 incorporating Nile river flood data. In H.F. Diaz and Vera Markgraph (eds.): *E-Nino: Historical and Palaeo-climatic aspects of Southern Oscillation,* Cambridge University Press, Cambridge, pp. 119–49.

Rajwade, V.K. (1905): Durgadevi Famine (in Marathi), *Saraswati Mandir,* Saka 1827: 1–10.

Singh, Gurdip (1971): The Indus Valley Culture as seen in the context of Post-Glacial Climatic and Ecological Studies in northwestern India, *Archaeology and Physical Anthropology of Oceania* 6: 177–89.

Sulman, F.G. (1982): *Short and Long Term Changes in Climate,* 2 vols, CRC Press, Boca Raton, Flo.

Toussoun, O. (1925): *Memoires sur Histoire du Nil.* Impremerie de l'Institut Francaise de Archaeologie Orientale, Cairo.

Walker, Gilbert (1986): *Long Range Forecasting of Monsoon Rainfall: Collected Papers of Gilbert Walker,* Indian Meteorological Society, New Delhi: pp. 16–22.

Warren, B. A. (1987): Ancient and Mediaeval Records of the Monsoon Winds and Currents of the Indian Ocean. In Fein & Stephens (eds.): *Monsoon,* pp. 137–158.

Watters, T. (1904–05): *On Yuan Chwang's Travels in India,* 2 vols., Royal Asiatic Society, London.

Beautiful blue houses in Jodhpur, India.
(Photo by Takeshi Takeda)

The cow is both a sacred animal and the symbol of monsoon Asia. A boy and his cattle carrying the crop that has been harvested in Western India.
(Photo by Takeshi Takeda)

Part III

Holocene Climate History
and the Harappan Civilization

Typical landscape of Rajasthan, India.
(Photo by Takeshi Takeda)

Chapter 14

Influence of Monsoon in the Regional Diversity of Harappa Culture in Gujarat

VISHWAS H. SONAWANE

Introduction

South Asia is regarded as the cradle of human civilization. Many of the ancient cultures that flourished and survived in this part of the subcontinent were mainly agricultural economies. An agricultural community depends heavily on the monsoon, which plays an important role in the development, growth and sustenance of an agrarian society. Agriculture was the mainstay in the economy of the Harappan people, who were one of the earliest settled communities in the Indian subcontinent covering an area of about one million km2. One of the important aspects of this civilization is its apparent mobility and proliferation over a much wider area, greater than that of the contemporary civilizations of Egypt and Mesopotamia put together. The area covered by this culture consists of a number of ecozones with diverse environmental variables and economic incentives (Chitalwala, 1989). Within these habitats, which are defined by geological and geomorphological features, soils, climate, plants and animals, with their specific landforms, resources and constraints, adaptive strategies were developed which were aimed not only for survival but for maximum exploitation of the resources and optimized land use. The enterprising Harappans thus demonstrated their skill in selecting suitable environmental niches during their multidirectional expansion. These factors and accessibility of natural resources were largely responsible for the growth, expansion and long survival of the Harappa culture.

The Indus valley, like Mesopotamia, is by and large devoid of basic raw materials with which to meet some of its basic needs. However, the Harappans could evolve a magnificent civilization mainly because of their expansionist tendency towards farflung resource areas and their ability to mobilize economic potentials as the Mesopotamians did by exploiting the rich mineral resources of the nearby hilly regions of Iran and Anatolia. The discovery of a Harappan settlement at Shortugai, located in the Badakshan province of northern Afghanistan for mining lapis lazuli (Francfort, 1984), Balakot on the Somani Bay in Pakistan for marine gastropods (Dales and Kenoyer, 1977), Cholistan region in Bahawalpur district of Pakistan for copper (Mughal, 1980), limestone hills at Sukkur Rohri in upper Sind region of Pakistan for flint/chert (Allchin and Allchin, 1982: 196–197) and Manda in Jammu and Kashmir for timber (Thapar, 1985: 52) are some examples of their hunt for procuring desired raw material or finished products. The penetration of Harappans further south into Gujarat demonstrates their enterprising character. The discovery of more than 500 sites, associated with different levels of Harappan culture during the 3rd and 2nd millennium BC (Possehl, 1992: 120) perhaps indicates the dependence of the Central Indus plain on the outlying resource regions like Gujarat to sustain its specific needs

(Fig. 1). The climate regime and the drainage network are complex and vary within various regions of Gujarat. The spread of Harappa culture was therefore, governed by areas of attraction, depending upon the availability of resources and geographical and ecological factors conducive to their cultural dynamics. These factors partly explain not only the regional diversities in the manifestation of the Harappa culture in Gujarat but also the innate capacity of the Harappans to mobilize different subsistence systems by integrating them into their economic structure. In this peripheral zone, the Harappan tradition and material culture displays an integrated regional style, synthesizing with the local indigenous agricultural and pastoral conditions and even hunting and food-gathering communities, offering interesting data for understanding the process of cultural transformation.

Environmental Setting

Culture and environment are coterminous (White, 1959). In other words, culture is an adaptation to the existing environment. Environment is thus not only a determinant factor in the growth of any cultural process, but an 'item' with which the human groups interact. Its role in the evolution and development of human culture is immense. Thus it is imperative to take

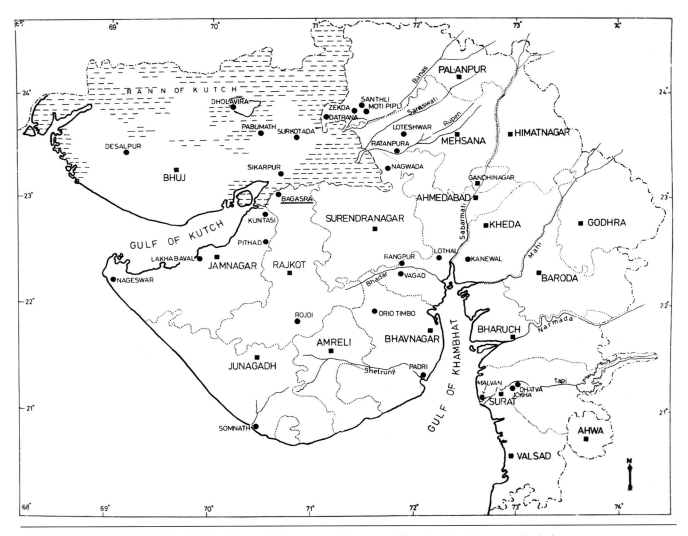

Fig. I. Map of Gujarat showing major excavated Harappan and Harappan affiliated chalcolithic sites (circles).

218

note of the environment of the region under review.

Whether prehistoric climate and its change throughout the last twelve millennia within the semi-arid and arid monsoon affected regions of the Indian subcontinent is a matter of great debate. Some agree for a more humid period between 3000 and 2000 BC followed by a strong increase in dryness (Singh, 1971), while others see no significant change (Flam, 1976). However, wide short-term and regional fluctuations, as observed today, must have been quite common. Therefore, a brief outline of the present-day environment will not be out of context here to appreciate the archaeological data related to the objectives of this paper.

On the western flank of Peninsular India lies the state of Gujarat. The present political state of Gujarat extends over an area of approximately 187,117 km^2 enclosed within 20°10'N to 25°50'N and 68°40'E to 70°40'E. The Arabian Sea coastline forms its western margin while the hills of southern Aravallis, the western rim of the Vindhyas and Satapudas besides the northern part of the Sahyadris (Western Ghats) form its eastern margins.

Physiographically, Gujarat constitutes three distinct zones: Kutch, the Saurashtra Peninsula or Kathiawad, and Mainland Gujarat. The mainland Gujarat is further conventionally subdivided into northern and southern segments by the Mahi river. The region between Mt. Abu and the Mahi river forms the arid zone of north Gujarat, while the region between the Mahi and Damanganga rivers forms the fertile alluvial tracts of south Gujarat.

Kutch is a vast desert-like expanse of mixed soils dotted with low hills, which break the monotony of the landscape. Inland, Kutch is bordered by desolate salt flats known as Ranns—the Great and the Little. The Ranns of Kutch were originally shallow bays connected to the sea. These bays have been gradually filled with silt and sand carried down from the minor rivers of the adjacent highlands of Rajasthan and the western Nara of Sind (Roy and Mehr, 1977).

The peninsula of Saurashtra (Kathiawad) forms a rocky tableland fringed by coastal plains. A major portion of Saurashtra is occupied by Deccan lava flows. The central part is made up of an undulating plain broken by hills and considerably dissected by various rivers (Bhadar, Kalubhar, Sukhabhadar and Bhogavo) flowing in all directions. Saurashtra has vast tracts of rich black cotton soil derived from weathered basalt, known for their moisture retaining capacities.

North Gujarat is a semi-arid, sandy plain, dotted with fossil sand dunes and attendant blowouts. The region extends in the north up to the southern Rajputana and gradually merges into the alluvial plains of Saurashtra towards the west, and central Gujarat further south. The region is drained by the Banas, Saraswati, Rupen, Sabarmati rivers and their tributaries of which the Sabarmati flows into the Gulf of Cambay while the rest disappear into the Little Rann of Kutch. None of these rivers, except the Sabarmati, are perennial and are therefore ill-suited for irrigation.

South Gujarat extends south towards Maharashtra and is bounded by the Satpura and Sahyadri ranges. The Gulf of Cambay (Khambhat) and the Arabian sea form its western boundary. It is an alluvial plain developed by the Tapi, Narmada and Mahi rivers, known for its fertile black cotton soil. However, the *Bhal* and *Bhalbaru* region of the Gulf of Cambay, occupying a strategic position between Saurashtra and north and south Gujarat, is characterized by sandy and silty soils, with brackish subsoil water. Present-day cultivators generally avoid this region due to its susceptibility to flooding during monsoons. However, this tract has given rise to rich pastoral grasslands (Bhan, 1989: 129).

Gujarat lies in the transitional zone between the monsoon climate of Konkan (coastal region of Maharashtra) in the south and the arid region of Rajasthan in the north. Climatic conditions vary greatly and Gujarat can be divided into three major climatic zones. Kutch, the western part of north Gujarat—along the eastern margin of the Little Rann of Kutch—and the northern fringes of Saurashtra form an arid zone. The entire south Gujarat has a sub-humid climate where annual rainfall ranges from 600 to 1500 mm, gradually decreasing from south to north. It has a relatively good vegetation cover and tropical dry forest can be seen in the hills near Godhra, Vadodara and Bharuch. The rest of Gujarat experiences a semi-arid climate with an annual rainfall ranging between 600 to 800 mm.

Although irrigation has been introduced in some regions recently, the agriculture of Gujarat is characterized by dry farming. Even now, 85% of its agriculture is rain-fed. The majority of crop production occurs during the monsoon season and the harvest is in the autumn (*kharif* crops). The black cotton soil (locally known as *Goradu* and *Bhata*, representing older and young alluvium, respectively) is labour-reducing. It swells with water during the monsoon season and later develops deep cracks as the dry season continues (Rissman, 1985: 164). Even today, in most parts of Gujarat, hoeing with blade harrows is practised as a substitute to ploughing. Annual ploughing to destroy weeds is not necessary because only one crop is grown each year (Patel, 1977: 39).

The subsistence crops of Gujarat are dominated by millets—*bajra* (pearl millet; *Pennisetum typhoides*) and *jowar* (sorghum; *Sorghum bicolor*). Pearl millet is cultivated under more acreage than any other crops (Patel, 1977: 47). It is the mainstay of the semi-arid and arid regions of Gujarat because it has a short maturation period of 85–90 days, the shortest among the common cereals. It is drought resistant and also provides much-needed fodder for cattle (Patel, 1977: 62). Sorghum replaces pearl millet in south Gujarat as the most popular food grain. It is a sturdy crop of deep and heavy soils and can be raised either in the summer (*kharif*) or the

winter (*rabi*) seasons. Again, like pearl millet, it is an excellent fodder for cattle.

The major food grain produced in the *rabi* season is wheat. Its production is restricted due to the dependence upon irrigation. However, it can also be grown without irrigation in the low-lying saline strips of land which border the Gulf of Cambay, and in the *Bhal* tracts (Patel, 1977: 98). The *Bhal* and *Bhalbaru* tracts are waterlogged during the monsoon and wheat can be planted in October. The residual humidity in the alluvial black cotton soil is sufficient to permit the maturation of this crop, even with no further moisture input (Rissman, 1985: 63). The cultivation of rice is confined to restricted areas of the Gulf of Cambay.

The saline wastelands of the region in general, and the western portion of north Gujarat covering the estuaries of the Rupen, Saraswati and Banas rivers in particular, favour excellent growth of many species of wild grasses. These are available immediately after the first monsoon showers and are exploited by the present-day pastoral communities of Gujarat for feeding their cattle (Bhan, 1994: 73). Gujarat is an important source for the famous *kankrej* breed of cattle, while sheep, goat, camel and buffaloes are also kept. Nomadic or semi-nomadic pastoral communities such as the *Rabari*, *Bharvad* and *Charan* make their living primarily out of breeding and/or herding of these animals. These communities are found distributed in pockets in Kutch, Saurashtra and north Gujarat, as these regions have stretches of open wastelands locally known as *padthar* in which a variety of grasses and other shrubs grow to support pastoral activities.

Early Inhabitants of Gujarat

Gujarat has been inhabited since the Lower Palaeolithic period, perhaps in the range of 500,000 years, except for a little gap between the Chalcolithic and Early Historic periods for about 1000 years. The current status of research in

Gujarat suggests that a vast and diverse set of interconnecting ecosystems has favoured the indigenous development of material culture by subsisting on the local environment. The sites of the Lower Palaeolithic period are located almost all over Gujarat, followed by the Middle and Upper Palaeolithic periods, though the latter two are relatively less known in terms of their distribution and chronology, besides tool typology. However, so far as the Mesolithic culture is concerned, Gujarat is considered as one of the prominent regions where a good deal of research has been carried out right from the inception of prehistoric studies in India. About four hundred sites of this culture have been reported from different parts of the state. The people who made and used microlithic tools—small stone implements in the form of blades, lunates, triangles, scrapers and other component parts of tool kits meant to be hafted on bone or wooden shafts as composite implements—were hunters and gatherers who appeared on the scene prior to the domestication of plants and animals. Important evidence from the sites of Loteshwar (Bhan, 1994: 74), Ratanpura, Kanewal and to some extent from Tarsang (Sonawane, 1996) indicate that some of these people managed herds of sheep and/or goat and perhaps cattle which complemented and added to the food resources they obtained by hunting and gathering. This is in conformity with the parallel evidence obtained from Bagor (Misra, 1973: 72–110) and Adamgarh (Joshi, 1978: 83) of northern and eastern neighbouring states. Though nothing is known about the domestication of plants, there are indirect indications of the use of either wild seeds or rudimentary cultivation in the form of grinding stones and ring stones found at some sites. A solitary radiocarbon date of a charred bone sample obtained from a Mesolithic deposit at Loteshwar positively suggests quite an early beginning of microlithic tradition in this part of the region around the 6th millennium BC (Sonawane, 1998). Even after the emergence of

the earliest farming and stock-raising Chalcolithic communities, microlith-using folk continued to live in Gujarat. A study of their material remains suggests a close interaction between these two communities resulting in a symbiotic interdependence (Possehl, 1976; 1980: 67–80; Sonawane, 1996).

Early Chalcolithic Phase in Gujarat

Existence of regional Chalcolithic communities in Gujarat prior to the arrival of Mature Harappans is well documented now. Excavations carried out at Loteshwar and Moti Pipli in north Gujarat, Prabhas Patan and Padri in Saurashtra, and Dholavira and Surkotada in Kutch has demonstrated that settled indigenous Chalcolithic communities flourished in different parts of Gujarat during the beginning of the 3rd millennium BC.

Saurashtra was the home for a few early village-farming communities spread into different pockets of this region. Pre-Prabhas culture, unearthed during the excavation at Prabhas Patan (Somnath) from the earliest levels, gives a date of c. 3000–2800 BC. The site is located right on the bank of Hiranya river and evidence of flooding could be seen from the lowest, Pre-Prabhas level. Very little is known about the early settlers of the site apart from their material inventory like the ceramics, with their independent indigenous origin, a few chalcedony blades, beads of faience and steatite, with a few segmented ones, and wall plasters with reed impression (Dhavalikar and Possehl, 1992). No other site with material culture similar to that of the Pre-Prabhas phase has been found in the region which can shed some light on the subsistence activity and settlement pattern of the Pre-Prabhas community[1] (Fig. 2).

Similar is the case with the earliest settlers of Padri. The site, situated in the coastal region of Bhavnagar district, at the lowest level was inhabited by an early indigenous Chalcolithic community living in square mud structures with a

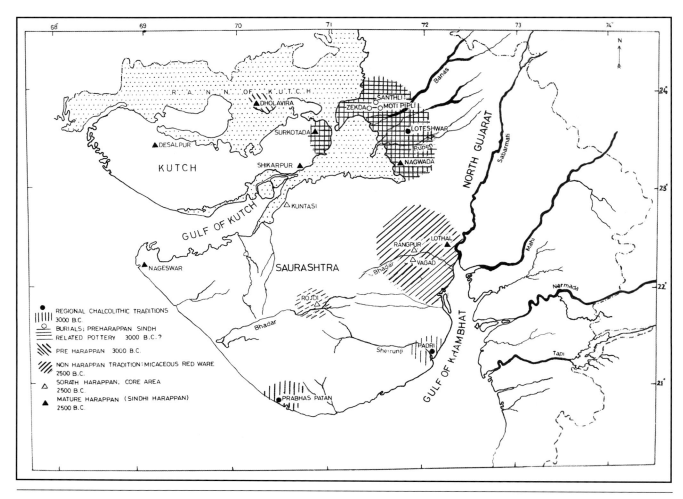

Fig. 2. Map of Gujarat showing concentration of regional Chalcolithic and Harappan culture sites.

material inventory like an indigenous ceramic tradition, christened as Padri Ware, a large number of steatite micro beads, a globular bead of carnelian and a few short blades of chert named as Padri culture and dating back to about the second half of the 4th millennium BC (Shinde, 1998). The site, according to the excavator, was not inhabited with the intention of utilizing the agricultural potentials, but for exploiting the sea resources. However, apart from Padri, no other sites belonging to this cultural phase has been discovered as yet.

Micaceous Red Ware is another early indigenous Chalcolithic ceramic, which is thought to pre-date the Mature Harappan phase in Saurashtra. The earliest levels at Lothal revealed the Micaceous Red Ware in association with the

Harappan pottery, though the ratio of Micaceous Red Ware to Harappan wares was increasing in the lower levels (Rao, 1979 and 1985). As it is well known that the lowest levels of Lothal could not be reached due to water-logging, the early occupation of Micaceous Red Ware, before the arrival of Mature Harappans, could not be established at the site. Since then, Micaceous Red Ware in an independent, self-contained context has not been discovered from any of the sites in this region. But it can be definitely said that the Micaceous Red Ware was not part of the Harappan ceramic inventory and should be considered as one of the early indigenous ceramic traditions of Saurashtra. The spread of Micaceous Red Ware is restricted to the Bhal region, the region between the Gulf of Cambay and the Nal depression and the adjoining area extending to its southwestern side (Sonawane and Ajithprasad, 1994; Herman and Krishnan, 1994).

Recent explorations carried out by Dimri (1999) in the region around Ahmedabad and Surendranagar districts yielded a large number of Micaceous Red Ware sites south of Sukhabhadar river, along the bank of Lilka river. Though this ware is mostly found associated with Rangpur Period IIA ceramics, the sheer quantity of Micaceous Red Ware recovered from this area suggests that the people using this pottery might have evolved in this region. The sites represented by Micaceous Red Ware in this region are not very large, ranging from 1–2 ha. Most of these settlements are located very close to the rivers showing their dependence on them.

From the evidence discussed above, it can be suggested that the sites in Saurashtra were small but permanent riverine settlements occupied round the year by the people who were basically cultivators, but also engaged themselves in herding their livestock in case of the third failed monsoon season, which is a common phenomenon in this region.

While Saurashtra was buzzing with activity during early 3rd millennium BC, north Gujarat did not lag far behind. Here explorations and excavations in the last three decades revealed that the region was occupied by a regional Chalcolithic culture, named as Anarta culture. The ceramic tradition, characterized by a Gritty Red Ware, Fine Red Ware, Burnished Red and Grey Ware, was the most predominant type in this region. That this ceramic tradition has an early origin was demonstrated from the excavation at Loteshwar, in Sami Taluka of Mehsana district. The ceramic assemblage of the Anarta culture continues and coexists with the subsequent Mature and Post-Urban Harappan phase.

The size of the Anarta culture sites is generally small, measuring less than 3 ha. Even in case of larger settlements, the artefact assemblage is found in small clusters, possibly separated

sufficiently in time. This feature is shared by the settlements in the subsequent phases of occupation. The structural remains from the early occupational sites are restricted to a few clay plasters with reed impression (Ajithprasad and Sonawane, 1993).

Equally important evidence regarding the early penetration of Harappans in north Gujarat has been brought forth recently from Nagwada and Santhli. The pottery found with the burials shows a close affiliation with the vessel forms recovered from Early Harappan levels at Kot Diji (Khan, 1965), Amri (Casal, 1964) and Balakot (Dales, 1974). Stratigraphically, the burials at Nagwada occur below the Mature Harappan levels (Hegde et al., 1988: 20).

However, Moti Pipli, another site in the same region, showed a substantial habitation deposit containing a similar type of pottery along with the Anarta ceramic, but without any evidence of burials (Majumdar and Sonawane, 1996–97). Subsequent exploration in the region revealed eight more sites of this category. Further excavation at Mathura revealed the Early Harappan ceramic in independent context, devoid of any associated pottery. The pattern of the Early Harappan settlements in this region is similar to that of the sites belonging to the Anarta tradition with small semi-permanent camp-sites located close to each other. Though no radiocarbon dates are available for this phase, the evidence gathered from these sites suggests that the early Harappans arrived in this region not later than 2800 BC (Majumdar, 1999).

The elements of the Early Harappan culture of Sind and Baluchistan extended southwards as far as north Gujarat at a time when this region was already inhabited by an indigenous Chalcolithic community. There was definite interaction between these communities, as they shared the same region for occupation and more often than not the settlement itself, and might have borrowed a lot of cultural traits from each other.

These sites are mostly located close to the low-lying marshy wastelands, which are not suitable for agriculture but contiguous land around such depressions forms good pasture land for cattle and sheep, as they support many types of grasses. The abundance of faunal remains from these sites, their small size, their location and absence of permanent architecture suggests that these were semi-permanent pastoral camps of nomads who followed a pattern of seasonal migration.

In Kutch region, impressive evidence of the formative stage of the Harappan culture has been brought to light from the excavation at Dholavira. A 60–70 cm thick deposit at the bottom of the total 12 m habitation sequence belongs to an Early Harappan phase beneath a typical Mature Harappan occupation (Bisht, 1991: 76). The pottery assemblage of this early phase is dominated by a wheel-made Red Ware of light or pinkish tones treated with a variety of slips, or decorated with incised horizontal grooves. There are examples of red slip, casual smearing with a dull brown pigment, a thick-coated buff paste or a thinly applied white paint. A white colour was often used as a background for simple decorative patterns in black. Little is known about forms, but deep dishes and jars have been reported. Some of the above features of surface decoration are comparable with the Anarta ware found in north Gujarat on the one hand, and the Early Harappan pottery found in the same region on the other. They show generic similarities with the Amri, Nal and Kot Diji complexes (Dhavalikar and Possehl, 1992: 77; Sonawane and Ajithprasad, 1994: 134).

The cemetery at Surkotada revealed a few secondary fractional and pot burials containing ceramics identical to that found from the burials at Nagwada. However, related ceramics could not be found from the habitation deposit.

From the available archaeological data, it is apparent that there is considerable cultural diversity in the early levels of most of the sites mentioned above. Although apart from pottery and a few other material remains very little is known about the origin and development of these early indigenous Chalcolithic communities, there is enough evidence to gauge the economic variability and adaptive capability of these people. The fluctuating monsoon and unavailability of perennial water source had forced them to adapt various alternative strategies for survival. These Chalcolithic traditions were so well rooted by 2500 BC that the Mature Harappans, once they arrived in Gujarat, assimilated these regional traits into their broad cultural spectrum. The interesting phenomenon of the Mature Harappan settlement in Gujarat is that they occupied those very regions which were earlier familiarized by the Early Chalcolithic communities and tried to adopt the strategies already evolved by them. These Chalcolithic communities thus played a vital role in the regional manifestation of Harappa culture in Gujarat.

Mature Harappan Phase in Gujarat

Kutch was the gateway for the Harappans migrating into Gujarat and witnessed the continuous flow of people from the early Harappan time itself. This was the region connecting Harappan people travelling from Sind to Saurashtra and north Gujarat. During the Mature Harappan period, there was a sudden increase in the movement of people and a rise in trade and industry, and this region rose into prominence.

The Mature Harappan settlements in Gujarat can be separated into two distinct categories: 1) sites with classical Harappan traits in their habitation deposit and 2) sites with some material inventory of Harappan culture, but generally devoid of classical traits. Possehl (1992) termed the first category of sites as 'Sindhi Harappan' and the second category of sites (especially from Saurashtra) as 'Sorath Harappan'. It is interesting to note that the second category of sites is predominated by the regional Chalcolithic traits, be it in Saurashtra or north Gujarat. The question arises as to whether they are Harappan sites in the real sense or regional Chalcolithic sites affiliated to Harappa culture. Since there is evidence of the early settlements of these regional Chalcolithic cultures in Gujarat, the hypothesis of their being indigenous Chalcolithic communities with Harappan affiliation seems more reasonable in the light of the present evidence. These Harappan affiliated sites seem to have played the role of supplier, catering to the needs of the Harappans, providing agricultural products, raw materials as well as labour for their varied industries. So one large urban settlement is normally supported by a large number of small satellite Harappan affiliated Chalcolithic settlements.

In Gujarat, out of more than five hundred sites with varying degrees of Harappan affiliation, only twenty-five settlements belong to the Mature/Urban (Sindhi Harappan) category. Kutch being the connecting link between Sind and Saurashtra and in the centre of activities, reveals fifteen of the twenty-five 'Sindhi Harappan' settlements. Location-wise, these sites can be broadly categorized as coastal settlements, situated either on the sea coast or on the margins of the Ranns, which are hypothesized as originally forming an arm of the Arabian sea. These settlements were engaged in specialized craft production as industrial/manufacturing centres, or served as trading-cum-administrative centres, or both. Almost all these sites are associated with the manufacture of specialized items of semi-precious stones, steatite, faience, chank shell, ivory and copper for the purpose of trade.

These trading centres were ably supported by the small Harappan affiliated settlements wherein one can see the regional diversity of the Harappans and their skill in adopting various environmental niches. Sites representing Harappan affiliated Chalcolithic settlements are more frequent in the region of Saurashtra, though their presence in north Gujarat is fairly well represented. Although their precise number is not known, they certainly outnumber here the 'Sindhi Harappan' sites. These settlements are generally small in size, though sites like Rojdi, with a site-size as big as 7 ha, are not uncommon. These sites are basically rural settlements based on farming and animal husbandry and with the exception of Kuntasi no other settlement shows any form of craft activity or industrial production.

Rojdi, Kuntasi and Sikarpur are the only sites in Gujarat from where detailed palaeobotanical evidence is available. In the absence of craft activities or industrial production, agriculture with herding seems to have been the mainstay of the subsistence system of the Chalcolithic settlements. Three plants are especially important in this regard; pearl millet, sorghum and finger millet (ragi; *Eleusine coracana*). At Rojdi, finger millet forms the most important plant right from the beginning of settlement (Weber and Vishnu-Mittre, 1989). Other plants, sharing the same hardy features with finger millet, such as foxtail and broomcorn millet (*Setaria italica* and *Panicum miliaceum* respectively), are also present in Rojdi A (c. 2500–2200 BC). These plants are drought-resistant, need little care, and do well in the uncertain climate of Saurashtra. Barley was also recovered from Rojdi A. Finger millet continued to be a significant cultivar during Rojdi B (c. 2200–2000 BC), but in this phase there was also a significant increase in the presence of *Chenopodium album*, a wild (possibly tended) plant (Weber and Vishnu-Mittre, 1989) harvested in the spring, prior to the monsoon. Though the presence of wheat and barley are documented in Gujarat, unlike in Sind and Punjab, their success as principal cereals has been questioned. Rice husks are present in Rangpur and Lothal, either used as binding material in mud plaster or surviving as impressions on sherds (Saraswat, 1992: 528). The millets continued in use into the Post-Urban phase at Oriyo Timbo (Rissman, 1985; Reddy, 1991: 80).

From the aforesaid data it is clear that the system of double cropping, i.e., *kharif* and *rabi* or monsoon and winter crops, was developed in Saurashtra as early as 2500 BC, at the beginning of the Mature Harappan phase in Gujarat. This development perhaps reflected a reaction of the presence of recurrent risk (Possehl, 1992: 135). Meadow refers to this development as the result of two agricultural revolutions, of which '...the 1st involved the establishment in the 6th millennium of the farming complex based principally on *rabi* (winter sown, spring harvested) crops of wheat and barley and on certain domestic bovids, including zebu, cattle, sheep, and goats. The second saw the addition by the early 2nd millennium of *kharif* (summer sown, fall harvested) cereals including sorghum, various millets and rice along with new domestic animals including the camel, horse and donkey' (Meadow, 1989: 16).

Surprisingly, most of the above elements were in fact part of the 3rd millennium BC subsistence system in Gujarat. Domestic animals present include the horse from Surkotada, Lothal, Kuntasi and Shikarpur; camels from Surkotada, Jekhada, Pabumath and Rojdi and onagers from Surkotada, Shikarpur, Kuntasi and Rojdi (Thomas and Joglekar, 1994: 185). Moreover, the chronological priority for this development of a summer cropping regime lies in Gujarat and not on the Kachi plain, where events seem to have lagged behind Gujarat for seven or eight centuries (Sonawane, 2000).

It is unlikely that the Mature Harappans, appearing in Gujarat around 2500 BC and unaware of the intricacies of the environmental fluctuations in this region, had authored the second evolution in the cropping pattern. The onus of this evolution lies on the regional Chalcolithic communities of Gujarat, who were more familiar with the local environment, the fluctuations of the monsoon and the soil types, to have evolved the summer cropping after centuries of trial and error. The Harappans seem to have merely borrowed the cropping strategy from these Chalcolithic people and adapted it in their subsistence regime.

Intensive surveys carried out by Chitalwala (1979), Momin (1979), Possehl (1980), Bhan (1986), Dimri (1999) and by the team of archaeologists from M.S. University of Baroda (Ajithprasad and Sonawane, 1993), particularly in Saurashtra and the western part of north Gujarat, reveal that most sites occupied during Rangpur Period IIB (Mature/Urban phase) continued during Period IIC (Post-Urban phase). However, there was a marked decline in the number of settlements occupied during Rangpur III, characterized by the Lustrous Red Ware (Bhan, 1992). The recently discovered sites in north Gujarat offer a good example of a dispersed type of settlement pattern, in contrast to the linear-dendritic pattern in Saurashtra. The latter was mostly controlled by river systems, whereas in north Gujarat it was based on the locations of relict sand-dunes associated with inter-dunal ponds. While the majority of the Saurashtran sites were based on an agrarian economy, those of the western part of north Gujarat favoured pastoral activities, evidenced by thin cultural deposits and large frequencies of animal bones (Bhan, 1992: 175; Sonawane, 1994–95: 8). An important feature of these western north Gujarat settlements is that there are small clusters of two or three sites and the average distance between two such clusters is not more than 10 km.

Conclusion

Nature determines the route of development, while man determines the rate and the stage (Taylor, 1953: 14–15). Thus, the stage and the technological attainments of a society at any given time determine the relative influence of the environment at that period. Unlike the fertile tracts of the Indus valley, none of the rivers in Gujarat lend themselves to irrigation using ancient technology. Agriculture here is entirely dependent on dry farming techniques. As a result, alternative subsistence strategies were adapted by the Harappans and allied Chalcolithic communities to suit environmental settings. In order to compensate for scarcity and to minimize the risk of periodic rainfall fluctuation, one can hypothesize an extensive land-use pattern, which took advantage of total available subsistence potentials. The wide range of site locations not only highlights the adaptive skills of the enterprising Harappans but also throws light on how the regional Chalcolithic cultures engaged in agricultural and pastoral subsistence activities influenced the Harappan settlements in Gujarat.

Between the 4th and 2nd millennium BC in Gujarat there existed a mosaic of different adaptations such as hunting and gathering, pastoralism, agriculture and various specialized craft production strategies. Hence, we see not only regional diversity in the manifestations of the Harappa culture in Gujarat, but also the capacity of the Harappans to mobilize different subsistence systems by integrating them into their economic structure. In addition, the integration of the Harappa culture with indigenous counterparts offers an interesting situation in terms of the processes of cultural transformation. Regional diversity is now an accepted phenomenon of the Harappan, for which the role of the regional Chalcolithic traditions can be held responsible. Moreover, fresh data brought to light from recent studies have revealed that some of the non-Harappan indigenous settled communities emerged on the scene more than half a millennium prior to the beginning of the Mature Harappan phase in Gujarat.

Note

1 It is interesting to note that the pre-Prabhas ceramic was the dominant pottery type in the assemblage of Datrana, a site in Banaskantha District of north Gujarat. However, this was the only site which revealed this ceramic from the region of north Gujarat.

References

Ajithprasad, P. and V.H. Sonawane (1993): Harappa Culture in North Gujarat: A Regional Paradigm. Paper presented at the Conference on *The Harappans in Gujarat: Problems and Prospects*, Deccan College, Pune.

Allchin, B. and F.R. Allchin (1982): *The Rise of Civilization in India and Pakistan*, Cambridge University Press, Cambridge, 379 pp.

Bhan, K.K. (1986): Recent Exploration in Jamnagar District of Saurashtra, *Man and Environment* 10: 1–21.

Bhan, K.K. (1989): Late Harappan Settlements of Western India, with Specific Reference to Gujarat. In Kenoyer, J.M. (ed.): *Old Problems and New Perspectives in the Archaeology of South Asia*. University of Wisconsin, *Wisconsin Archaeological Reports* 2, Madison, pp. 219–43.

Bhan, K.K. (1992): Late Harappan Gujarat, *The Eastern Anthropologist* 45 (1–2): 173–92.

Bhan, K.K. (1994): Cultural Development of the Prehistoric Period in North Gujarat with Reference to Western India. *South Asian Studies* 10: 71–90.

Bisht, R.S. (1991): Dholavira: A New Horizon of the Indus Civilization, *Puratattva* 20: 71–82.

Casal, J.M. (1964): *Fouilles D'Amri*, Publication de la Commission des Fouilles Archaeologique, Fouilles de Pakistan.

Chitalwala, Y.M. (1979): Harappan and Post-Harappan Settlement Patterns in the Rajkot District of Saurashtra. In Agrawal, D.P. and D.K. Chakrabarti (eds.): *Essays in Indian Proto-History*, B.R. Publishing Corporation, Delhi, pp. 113–121.

Chitalwala, Y.M. (1989): Harappan Settlements in Kutch and Saurashtra: Problems and Perspectives, *Bulletin of the Deccan College Post-Graduate and Research Institute* 47–48: 339–42.

Dales, G.F. (1974): Excavations at Balakot, Pakistan 1973, *Journal of Field Archaeology* 1: 3–22.

Dales, G.F. and J.M. Kenoyer (1977): Shell Working at Ancient Balakot, Pakistan, *Expedition*, 19 (2): 13–19.

Dhavalikar, M.K. and G.L. Possehl (1992): The Pre-Harappan Period at Prabhas Patan and the Pre-Harappan Phase in Gujarat, *Man and Environment* 17 (1): 72–78.

Dimri, Kiran (1999): *Chalcolithic Settlements in Bhogava and Sukha Bhadar Valley*. Unpublished Ph.D. thesis, Baroda: M.S. University of Baroda.

Flam, L. (1976): Settlement, Subsistence and Population: A Dynamic Approach to the Development of the Indus Valley Civilization. In Kennedy, K.A.R. and G.L. Possehl (eds.): *Ecological Backgrounds of South Asia Prehistory*, Ithaca, pp. 76–93.

Francfort, H.-P. (1984): The Harappan Settlement of Shortugai. In Lal, B.B. and S.P. Gupta (eds.): *Frontiers of Indus Civilization*. Books and Books, Delhi, pp. 301–310.

Hegde, K.T.M. *et al.* (1988): Excavations at Nagwada 1956 and 1987: A Preliminary Report, *Man and Environment* 12: 55-65.

Herman, C.F. and K. Krishnan (1994): Micaceous Red Ware: A Gujarat Proto-Historic Cultural Complex or Just Ceramic? In Parpola, A. and P. Koskikallio (eds.): *South Asian Archaeology 1993*, Vol. I, Suomalainen Tiedaekatemia, Helsinki, pp. 225–43.

Joshi, R.V. (1978): *Stone Age Culture of Central India*, Deccan College, Poona.

Khan, F.A. (1965): Excavations at Kot Diji, *Pakistan Archaeology* 2: 11–85.

Majumdar, A. (1999): *A Ceramic Study of the Harappan Burials from North Gujarat*. Unpublished Ph.D. Dissertation, Baroda, M.S. University of Baroda.

Majumdar, A. and V.H. Sonawane (1996–97): Pre-Harappan Burial Pottery from Moti-Pipli: A New Dimension in the Cultural Assemblage of North Gujarat, *Pragdhara* 7: 11–17.

Meadow, R. (1989): Continuity and Change in the Agriculture of the Greater Indus Valley: The Palaeobotanical and Zooarchaeological Evidence. In Kenoyer, J.M. (ed.): *Old Problems and New Perspectives in the Archaeology of South Asia.* University of Wisconsin, *Wisconsin Archaeology Report* 2, Madison, pp. 61–74.

Misra, V.N. (1973): Bagor: A Late Mesolithic Settlement in North West India, *World Archaeology* 5 (1): 92–110.

Momin, K.N. (1979): *Archaeology of Kheda District (Gujarat) up to 1500 AD.* Unpublished Ph.D. Dissertation, Baroda, M.S. University of Baroda.

Mughal, M.R. (1980): New Archaeological Evidence from Bahawalpur, *Man and Environment* 4: 93–98.

Patel, G. (1977): *Gujarat's Agriculture.* Overseas Book Traders, Ahmedabad, 363 pp.

Possehl, G.L. (1976): Lothal: A Gateway Settlement of Harappan Civilization. In Kennedy, K.A.R. and G.L. Possehl (eds.): *Ecological Background of South Asian Prehistory,* Cornell University South Asian Programme, Ithaca, pp. 115–131.

Possehl, G.L. (1980): *Indus Civilization in Saurashtra.* B.R. Publication, New Delhi: pp. 264.

Possehl, G.L. (1992): The Harappan Civilization in Gujarat: The Sorath and Sindhi Harappans, *The Eastern Anthropologist,* 1–2: 117–54.

Rao, S.R. (1979): Lothal: A Harappan Port Town 1955–62, *Memoirs of the Archaeological Survey of India* 78 (I).

Rao, S.R. (1985): Lothal: A Harappan Port Town 1955–62, *Memoirs of the Archaeological Survey of India* 78 (II).

Reddy, S.N. (1991): Archaeobotanical Investigation at Oriyo Timbo (1989–90): A Post-Urban Site in Gujarat. *Man and Environment* 16 (1): 73–83.

Rissman, P.C. (1985): *Migratory Pastoralism in Western India in the Second Millennium BC: The Evidence from Oriyo Timbo (Chiroda),* Ph.D. Dissertation, University of Pennsylvania, Philadelphia.

Roy, B. and Mehr, S.S. (1977): Geomorphology of the Rann of Kutch and Climatic Changes. In Agrawal, D.P. and B.M. Pande (eds.): *Ecology and Archaeology of Western India,* Concept Publishing Co., Delhi, pp. 195-200.

Saraswat, K.S. (1992): Archaeobotanical Remains in Ancient Cultural and Socio-economic Dynamism of the Indian Subcontinent, *Palaeobotanist* 40: 514–45.

Shinde, V.S. (1998): Pre-Harappan Padri culture in Saurashtra: The Recent Discovery. *South Asian Studies* 14: 173–182.

Singh, G. (1971): The Indus Valley Culture, *Archaeology and Physical Anthropology in Oceania,* 6 (2): 177–189.

Sonawane, V.H. (1994–95): Harappan Settlements of Rupen Estuary, Gujarat, *Pragdhara* 5: 1–11.

Sonawane, V.H. (1996): Mesolithic Culture of Gujarat. Paper Presented at the National Seminar on The Mesolithic in India, Allahabad University, Allahabad.

Sonawane, V.H. (1998): Harappan and Harappan Affiliated Chalcolithic Settlements of Rupen, Saraswati and Banas River Valleys of North Gujarat. Paper Presented in the *National Seminar on River Valley Cultures,* Bhopal: IGMRS

Sonawane, V.H. (2000): Early Farming Communities of Gujarat, India, *Indo-Pacific Prehistory Bulletin* 19 (3).

Sonawane, V.H. and P. Ajithprasad (1994): Harappa Culture and Gujarat, *Man and Environment* 19 (1–2): 129–39.

Taylor, G. (ed.) (1953): *Geography in the Twentieth Century,* London, 661 pp.

Thapar, B.K. (1985): *Recent Archaeological Discoveries in India,* UNESCO, The Centre for East Asian Cultural Studies, Paris/Tokyo, 159 pp.

Thomas, P.K. and P.P. Joglekar (1994): Holocene Faunal Studies in India, *Man and Environment* 19 (1–2): 179–203.

Weber, S.A. and Vishnu-Mittre (1989): Palaeobotanical Research at Rojdi. In Possehl, G.L. and M.H. Raval (eds.): *Harappan Civilization at Rojdi,* Oxford and IBH, New Delhi, pp. 117–81.

White, L.A. (1959): *The Evolution of Culture,* McGraw Hill Book Company, New York, 378 pp.

Chapter 15

Effects of Ocean/Atmosphere on Climate and the Arabian Sea Environment for Harappan Seafaring

LAJWANTI SHAHANI

Introduction

The monsoon system prevalent in the Indian Ocean region is one of the major climate systems in the world and many facets of the ocean-atmosphere interaction are collectively responsible for the Indian monsoon. One of its special features is that despite being centred near 30° N, it becomes intense enough to draw in the intertropical convergence zone and the equatorial rains (Lamb, 1972). The biannual reversal of winds is a manifestation of the seasonal migration of the intertropical convergence zone. The Himalayan range and the Tibetan plateau also have an important role in this (Hahn and Manabe, 1975; Lamb, 1982; Overpeck *et al.*, 1996; Washington, 1981).

Our climate system has however been highly unsteady in the geological past. Evidence of this can be found in terrestrial as well as oceanic nature. Further, the palaeoclimate record is well matched with the record of sea level fluctuations as well as effects of the climatic changes, which are contained in the ocean floor (Fairbridge, 1960; Gupta and Fernandes, 1997; Lamb, 1982; Mehr, 1987). An attempt has been made here to interpret this data (meteorological and oceanographic) for the origins of seafaring. While archaeological evidence for Harappan seafaring activities is scant, with the data available to us we may still be able to conjecture the types of vessels that were used. On the other hand, coastal/port sites of the Bronze Age have been reported in India on the Makran coast and in Oman Peninsula. Archaeological evidence emerging from these sites can now be used to put together a fairly coherent picture. Data from the Oman Peninsula has also opened up new areas for studies including an ethnoarchaeological study based on contemporary traditional sea trade between India and the Persian Gulf (Shahani, 1996, 1997).

The Climate System

Variations in different aspects of the weather in different parts of the world are the manifestations of deviations in a single chain of events; the supply of heat from the sun warms the earth and drives the winds, which then redistribute this heat to all parts of the world. However, only some of the heat is conveyed by the winds; much of it is stored in the oceans, which transport this heat through the currents driven by winds all over the world (Lamb, 1982; Perry and Walker, 1976).

■ Solar Radiation Balance

The earth's surface receives virtually all its heat energy from the sun. The return flow of geothermal heat from the earth's interior is comparatively negligible. The mobile reserves of heat stored in the world oceans are an additional energy supply. The earth's surface as a whole possesses a large positive radiation balance, except in the polar regions, where there is a negative balance when the sun is at its greatest distance (Perry and Walker, 1976; Lamb, 1982). The annual radiation balance of the ocean surface is positive everywhere.

Radiation is received most where the skies are largely cloud-free: over land, in arid tracts of the tropics and sub-tropics (the passage of solar radiation through the atmosphere is impeded by cloud activity). There is little surface moisture here that can be evaporated for the provision of latent heat, which results in widespread subsidence of air in these regions. The subsidence inversion prevents thermals from ascending far enough for condensation to take place.

On the other hand, due to presence of clouds in most oceanic regions, insulation is considerably less than maximum. Solar radiation is either absorbed by the sea or reflected from its surface. The largest oceanic radiation balance is located in the northwestern part of the Arabian Sea. Cumulus (heap-like) clouds which are characteristic of trade-wind belts and extensive cloud-sheets associated with monsoon circulation form over the equatorial and monsoon belts in this region and are responsible for the wastage of a large amount of solar radiation, which is reflected off the cloud tops (*albedo*). Thus India receives more radiation in March than in August.

Heat and moisture are conveyed towards the poles through a single huge convection system of the atmosphere in slanting paths, and colder air from the poles is carried in the opposite direction with either clearer skies or low-level cumulus clouds. It is this unequal heating of different zones of the earth which sets the air in motion.

■ **Surface Wind Zones**

A single great circumpolar flow of upper westerlies circumvents the earth over each hemisphere—mainly over the middle latitudes—in wave-like meanders from southwest or west-southwest over the temperate zone of the northern hemisphere and northwest or west-northwest over the temperate zone of the southern hemisphere. This is known as the *circumpolar vortex* and is in fact the main flow of the atmosphere and carries most of the momentum of the wind in its meandering whirl (Perry and Walker, 1976).

Apart from the westerlies, the other surface wind zones are the polar winds which are cold Arctic and Antarctic easterlies; the trade winds between the sub-tropics and the equator which are not very strong except for a daytime sea-breeze, and are so named for their reliability by past merchant seafarers. The doldrums situated on the equator are mostly light variable winds. This last zone is the meteorological equator; the air-streams from the tropical zones of the two hemispheres meet here with a concentration of deep cumulonimbus activity. The air motions are nearly horizontal and are subject to the Coriolis effect (the influence of the earth's rotation on air currents flowing towards the equator). This is also called the intertropical convergence zone (hereafter—ITCZ) and it undergoes a greater seasonal movement between mean positions. This range is enhanced in the continental sectors.

At present the average position of the ITCZ/meteorological equator is situated 6° N of the equator (Gupta and Fernandes, 1997; Lamb, 1982). The greatest total rainfall is yielded by the equatorial rains, which are produced in the convergence zone between the wind systems of the two hemispheres.

The Monsoon System over the Indian Ocean Region

The word 'monsoon' is a corruption of the Persian and Arabic *'Mausam'* meaning season. In meteorology, the term monsoon stands for wind and pressure regimes, which reverse seasonally. In the early part of the warm season, monsoonal low pressure develops over the heated continents in association with waves in the westerlies. The precipitation belts—also known as the monsoon system—move north and south along with the wind circulation zones produce them.

One of the features of the Indian Southwest Monsoon is a belt of southwesterly winds, which develops in the lower atmosphere between the ITCZ and the equator and moves towards the northeast (Lamb, 1972; Perry and Walker, 1976; Subbaramayya and Ramanadham, 1981). During winter, the anticyclone (high pressure) and the massive build-up of cold air over Siberia drives winds from the northeast which are strong enough to cross the mountains into India, thus reversing the monsoon to a northeast pattern.

Sea Surface Topography of the Northern Indian Ocean

There is an upward slope from the western to the eastern part of the sea-surface of the Northern Indian Ocean—from 8°N—which is related more to salinity gradients (caused by climatic conditions in the overlying atmosphere) than to wind-stress. The western part, i.e., the Arabian Sea region has a dry and arid climate and correspondingly high rate of evaporation. In the Bay of Bengal on the east, the climate is more humid with copious precipitation. Surface salinity in the Bay is further affected by drainage from the various rivers.

Hydrographic Front

The topography of the northern Indian Ocean is characterized by two distinct surface water masses in the salinity gradients, i.e., the Arabian Sea which is highly saline (>34.5 ppt) and the Bay of Bengal, low saline (<34.5 ppt). These two extreme (hyper and low) salinity water masses 'are separated by an oblique boundary along the 34.5 ppt isohaline' (Gupta and Fernandes, 1997) located south of Sri Lanka. This hydrographic front is located in the Indian Ocean at 10°S and consists of sub-parallel isohalines.

Somali Current

The Somali Current is one of the most prominent current systems of the Indian Ocean and is a part of the ocean's dynamic response to the pattern of wind-stress over a large part of it (Lamb, 1972; Perry and Walker, 1976). This is noticed mainly along the Somali coast. During the Southwest Monsoon it is noticeable in a strong northeastward flow parallel to the coast in the upper 200 m or so of the ocean.

In response to the northeasterly trade winds the Somali Current reverses its flow to southwestward. During the Southwest Monsoon the Somali Current is considerably stronger. The Somali Current forms, within a month of the onset of the southwesterly monsoon winds, possibly due to proximity to the equator.

The Ekman Effect

In his theory of wind-driven currents, Ekman (in Perry and Walker, 1976) supposed wind-stress on the sea surface to be the driving force of surface currents, and frictional coupling between adjacent layers of water to be the only means by which water beneath the surface is set into motion. Ekman demonstrated that the total momentum of a current driven only by the wind (experiencing no acceleration) is directed in a right angle to the right angle of the wind itself, i.e., 180°. Thus, while motions in the upper ocean are a result of wind-stress, the Ekman layer is the body of water, which is set into motion by wind-action.

Effects of Mountains

Hahn and Manabe (1975) performed simulation experiments on the effects of the Himalayas on the Asian Monsoon circulation. These experiments show that thermal influence over orographic features is more significant here than the topographic influence, in the creation of large-scale features of the monsoon. The Himalayas block the flow of colder air from the north thus preventing its penetration farther south. They concluded with their no-mountain experiment that in the absence of the mountains the monsoon circulation would extend into the Asian continent deeper than it would do so in the control experiment with mountains. Clearly the Himalayas modulate the placement of the dominant features of the monsoon system.

Palaeoclimate and Changes in Sea Level

At the time of the Post-glacial Climatic Optimum of 6000–4000 cal. yrs. BP, the world temperature was 2–3°C higher than at present. There was a contemporaneous 'sub-pluvial' period between 7000–4400 cal. yrs. BP in North Africa and the Sahara seems to have become drier (Fairbridge, 1960; Kutzbach *et al.*, 1996; Lamb, 1982). On the other hand, northern Europe experienced very heavy rainfall for a while after the retreat of the ice.

Isostatic reactions and changes in the volumes of ocean water however, are due to expansion with the rise of temperature and reduction when the temperature falls. These changes work in the same sense as glacio-eustatism (Mehr, 1987). Due to the warmer ocean surfaces, the total amount of moisture in the atmosphere would be greater than at present, with greater total precipitation thus amounting to greater rainfall.

For Holocene sea-level rise, Kale and Rajguru (1985) believe that on the Indian coastline, the rate of progress was extremely rapid and this is evidenced in other parts of the world as well. The best example of rise in sea level from 6000 cal. yrs. BP lasting up to 2000 cal. yrs. BP, almost 6–7 m above the present level, according to Mehr (1987) may be found in the Ranns of Kutch which are remnants of a high sea. Further, there is

evidence of sea level fluctuations even within the Holocene (Gupta and Ghosh *in* Mehr, 1987) and these fluctuations may or may not be related to neo-tectonism. On the Gujarat and Saurashtra coastline the three segments—Mainland Gujarat, Saurashtra and the Ranns of Kutch show diversity. The Kutch region, in particular, falls in the high seismic zone (Grover and Sareen, 1993).

On the eastern board of the Arabian Sea, the northern Oman mountains (al Hajar range) seem to play a major role in blocking the monsoonal airflow, preventing it from entering Arabia (Orchard and Stanger, 1999) although the monsoon system does appear to dominate rainfall in coastal regions of southern Oman. The eastern Mediterranean airflow, which carries moist air into the northwestern Gulf area, plays a significant role in northern Oman, bringing winter rains across a broad frontal band. These rains account for more than 80% of the precipitation in northern Oman. In the Persian Gulf, eustatic and relative sea level changes have been well documented since early Holocene, as are climatic shifts (Carter *et al.*, 1999).

■ **Sea Cliffs**

A sea cliff have been described by Snead (1982: 202, 224), as 'an escarpment eroded by wave action'. This 'steep land surface facing a waterbody' if inactive, is called a dead cliff. Dead cliffs stand above present sea level and are not much affected by seasonal monsoon inundation. Most cliffs, are a transitional geomorphic feature and they fall partly in the shore and partly in the coastal zone (Ahmad, 1972). Cliffs occurring within the reach of the highest storm levels of the sea are shore features and their sections above this level are a part of the coastal zone.

Cliffed shores and coast occur with relative frequency on the western coast of India (Ahmad, 1972). In Gujarat, cliffed coast is seen mainly in the Gulf of Cambay near Daman and on the other side between Gogha and Verawal. These cliffs stand as high as 3 to 30 m (mostly between 15 and 30 m). Littoral concrete, extensive stretches of swamps associated with estuaries and creeks are prominent features in this area.

The Rann of Kutch, which is now only seasonally characterized by a shore, has some notable cliffs around some of its islands: Khadir Island, Bhanjada Bet, Kakinadia Bet and Jalandar Island. In the north of Khadir Island, I noticed a marked and bold line of cliffs. At one place it was possible to climb down a height of 4 m (no photographs or drawings were allowed as this marks the International Border). These dead cliffs run for about 25 km in length, about 80–160 m in height and lie within a width of 200–400 m from the shore. One opinion on these dead cliffs in the Rann is that along with well-defined channels of streams in the inland area, they 'suggest submergence in recent geological past' (Ahmad, 1972: 121).

I also noticed a similar process in the Gulf of Kutch near Bagasra (a Harappan site under excavation by the M.S. University of Baroda) where a 2 m high depression began about 200 m northwest of the site, opening toward the Gulf. This depression is carved out with 'gulleys' or well-defined channels and sustains dense vegetation of acacia. At present, during the Southwest Monsoon, the seawater inundates this depression and during heavy rainfall the land around the depression is submerged as well.

■ **Hydrographic Front**

Gupta (1999:99) is of the opinion that, 'the use of empirical relationships between radiolaria (oceanic microplanktons) and salinity, SSTs (sea surface temperatures) and solar insolation are a prerequisite for the synthesis of monsoon dynamics in the geological past.' To this end, Gupta and Fernandes (1997) have analysed sediment cores for radiolaria and found indications of palaeomonsoonal oscillation of the 10°S hydrographic front. They inferred that

higher freshwater flux in the warmer periods resulted in southward migration of the 10°S hydrographic front during the geological past.

Similarly, a lower freshwater flux during colder periods would have resulted in northward migration of the front. According to Dr. S.M. Gupta, around 5000 years ago this hydrographic front was located northward by a few degree (personal communication). Lamb's (1982: 119) suggestion that during the warmest post-glacial epoch between 7000 to 3000 BC, the meteorological equator was higher northward from its present location at 6°N, can be related here to Gupta's evidence of the hydrographic front.

In such a scenario, the ITCZ would pull in the equatorial rains as well as the prevailing Indian Summer Monsoon system, with rains reaching as far north as the Indus Valley region, Rajasthan and Gujarat, perhaps with rough weather. Along with the existing northward location of the hydrographic front, this would result in a higher location of the wind circulation pattern. The Somali Current, which is also affected by the winds, would then be driven northward towards Sind and Kutch. A higher freshwater flux due to higher rainfall would then cause the hydrographic front to migrate southward. Further, Gupta and Fernandes (1997) have suggested cyclicity in the migration of the 10°S hydrographic front, which corresponds to the earth's orbital eccentricity cycle.

Origins and Evolution of Seafaring

Any attempt to piece together a picture of the origins and evolution of seafaring, must of necessity, rely partly on imagination. Coupled however with common sense, logic and self-discipline this could lead to an approximation of the evolutionary process.

Apart from harnessing water for his needs, man was also bound by water, which many times cut him off from the rest of the world. So, did early man willfully set out on an adventure to navigate waters—or did he perhaps chance upon a floating tree-trunk?

A dugout canoe is then the second step from a floating tree-trunk. Initially, man may have used his bare hands to manoeuvre his craft, but quickly turned to ingenuity. A pole and finally a paddle and oar came into use for propulsion. Dugouts and canoes are still used by some cultures in Australia, South America and Africa. In some parts of the world, inflated skins of animals have been used for kayaks as noticed in some primitive cultures around the North Pole (Angelucci, 1970).

Another form of progress from the floating tree-trunk would have been the raft, made from a number of trunks tied together. Thus man had two types of watercrafts at his disposal—a dugout for individual journeys and rafts to carry more people or cargo. However, this development could only take place in those parts of the world where there was an abundance of trees. For the Old World one such example may be found in the Harappan Civilization. In places where no large trees were available, bundles of rushes, papyrus and even balsa were used to make the early boat (Heyerdahl, 1980).

The raft is the first link in the origins of shipbuilding and navigation. The second 'element' or link is without doubt the discovery of sail. Once man had a decently floatable vessel he could paddle to some distance—perhaps along calm waters in river estuaries or in bays—his observations would be that anything which floats moves in the direction of the wind, that is, downwind. How did man discover sails? By observing creatures of the water, like the swan or even the smallest water bug fluttering its wings to harness the power of the wind. At this stage the problem may well have been how to avoid sailing too far away from known land with no knowledge of how to get back.

Two developments may have taken place at this point: first, the discovery of wind-propelled navigation, and secondly, that of the flow of wind-driven currents in waters. It is practically impossible to get a rudderless boat to sail straight downwind; it will bow slightly into the wind and very slowly drift sideways as it moves forward. An inhibiting factor maybe the effect of earth's rotation (the Coriolis effect) as is seen with currents and also the Ekman effect.

The pioneering sailors, after recovering from their baptism by fire, would have quickly learned to have a measure of control over the process of navigation. Thus while sailing downwind in a known direction, a square of cloth or matting or even fur may have been stuck up on a pole to serve as early sail. As a result, early sails, by their very simplicity, were in fact capable of windward sailing. The evolution of shipbuilding and navigation therefore would depend much more on the development of windward working hull forms rather than the rigging. Having said so, we may add that the raft was not invented just once but hundreds, maybe thousands of times, in different places and at different times. This could also apply to the sail, the rudder and the oar.

From Mesopotamian texts we know that in 2500 BC, Sargon of Agade 'made ships from Dilmun, Magan and the faraway Meluhha moor at his piers'. With the consequent internal breakdown of the Akkadian political system long-distance trade dwindled for a while; it was revived again in the latter part of Ur III in Ibbi Sin's reign (Thapar, 1984). We find textual reference from Mesopotamia of sea trade brought by Harappans, various types of Harappan ships and some descriptions of Harappan vessels. Some references to the boat-types from Mesopotamian textual evidence, taken from Possehl (1996: 144–44) are presented here:

From *Gudea Inscriptions:* p. 141
    18: lu-KU-má-mé-luh-ha-ka. This is a reference to a man involved with a ship of Meluhhan type (Hackman 1958: 298.8).

Ur III economic texts with references to Meluhha:
(p.142)

42: má-gur$_8$-re IM-e tag$_4$-a dumu-mé-luh-ha. This is a list of rations dealt with in the context of a ship and a man called 'son of Meluhha' (Reisner 190: no. 154, column 6. Parpola *et al.* 1977: pp.142 f).

47: From *Old Babylonian references to Meluhha*: p.143

Enki and Wold Order (Benito 1969: p. 120):

125. I am Enki, may he look upon me.

126. Cause the trees to be cut down for (?) Dilmun-boats,

127. Cause the Magan-boat to reach (?) the horizon (?).

128. The 'Magilum' boat of Meluhha

129. May transport gold and silver,

130. May bring them to Nippur for Enlil, the [king] of all the lands.

49: *Enki and Ninhursag*, Ur version (Jacobson 1987: 189: p.143

...May the Land of Meluhha
load precious desirable sard,
mesu wood of the plains,
the best abba wood up into large ships!

From *Old Babylonian and later lexical and other references*: p.144

60: Gis-má-mé-luh-ha. A wooden ship of Meluhha
(Landsberger 1957: 174). Post-Old Babylonian.
(References 18, 42, 47, 49 and 60 figure in Possehl in 1996: 141-144).

Of special note here, are the 'Dilmun-boat', a 'Magan-boat' and a 'magilum' boat of Meluhha. Not only are the boats described as belonging to the three different lands, but we also have a particular type of boat from Meluhha shipping gold and silver. The word má is repeated thrice in the above references if one does not count the 'magilum' as well. Could the word má possibly mean 'boat'?

■ **Harappan boat-types**

A brief account of boat-types as seen from Harappan archaeological contexts is presented here. From Mohenjo Daro we have an incised potsherd depicting a rough sketch of a masted boat. Though the sketch of the boat is rather rough, certain points can be observed here: it has a sharply upturned stem and stern, a mast with a yard and a rudder. According to Mackay (1938) the mast may be a tripod and the second line with the yard is perhaps a furled sail.

The second example reported by Mackay (1938), also from Mohenjo Daro, was excavated in two pieces but found to form a rectangular whole. The depiction seems to be rough and incomplete and the work of an inexperienced seal maker. This boat also has a sharply upturned stem and stern along with a cabin at its centre. Fastened at each end of this cabin is seen a standard-like object bearing an emblem. According to Fairservis (1971) this boat does not appear to have a keel and is of a shallow draft. A steersman is noticed at one end of the boat (possibly stern with two rudders?). Further, this boat appears to be lashed together at both stem and stern perhaps indicating that it was made of reeds.

The third example, from Harappa, is on a square seal made on pottery instead of steatite. Here the stem and stern are raked, a big sail is seen hoisted above the stem, perhaps some sort of structure at the centre with two oars and an anchor hanging from above the stern. This appears to be a larger vessel, notably with a triangular anchor. A single Harappan script is shown on the top right of the seal, which makes it difficult to tell more about the centre of the boat structure.

From the excavations in the habitation area at Lothal, Rao (1979) has reported four miniature terracotta models of boats as well as a painted potsherd, which have been described and shown in the report. The first type is apparently 'heavy' with a pointed stem and broad stern. This model (Rao, 1979: pl. XCIA) has four holes: one near the stem, one each on the port side and starboard side and the last one toward the stern. The hole near the stern could be meant for a mast and the one at the stem for securing the sail/s or for towing. The two holes at the margins are apparently meant for 'pegs to rest the oars on'. Another similar model found broken is described as a 'heavy boat' with a thick broad stern. The excavator presumes the stem was pointed as it narrows towards that end.

A third type is a flat-bottomed barge-like boat with no provision for sails; this has been described as a 'light' boat and is wide at the centre with a pointed stem—the stern portion of the model is broken. Rao believes that this type of boat (barge) may have been used on the river and for unloading/loading vessels which could not enter the dock in phase IV (2000–1900 BC). Another model, which is long and narrow has been compared to the South Indian catamaran which is mostly used for fishing near the coast. Rao feels that this was perhaps a dugout. A potsherd of Micaceous Red Ware (Rao, 1979: pl. CLXXVA) with a painting of two boats, multiple oars and wavy lines beneath indicating water have been described in connection to the last boat model.

■ Harappan evidence from the Persian Gulf

Harappan artefacts have been reported from Mesopotamia, Bahrain and the Oman Penisula. Of these, pottery is the most important with specific information, which has helped archaeologists build a wide-ranging perspective. The black-slipped jar, a red micaceaous ware with a sharply everted rim and a narrow molded base has been reported from most of the sites (Cleuziou and Méry, n.d.). Very often, post-firing short inscriptions with four or five Harappan signs are seen on the shoulder or body (around the maximum diameter).

In Oman the presence of these jars has been chronologically dated with Umm an Nar context—c. 2500 BC to c. 2000 BC. An inscribed sherd of this ware was reported from Ras al Junayz and has been chronologically dated with the Wadi Suq context—2000 BC (Tosi, 1993: 374, Figure 30.8). Significantly, a black-slipped jar sherd with a similar inscription has also been recovered from Bagasra in Gujarat (Shahani 1996, 1997). Although no dates are yet available from this ongoing excavation of the M.S. University of Baroda, there is an unparalleled suggestion of contact between the port/coastal sites of Bagasra and Ras al Junayz. Perhaps it would not be fanciful to suggest that the two jars were used for trade of a particular commodity, further it would appear that these two specimens possibly originated from the same source of trade.

Another significant discovery was that of Harappan grave-goods at Shimal (de Cardi, 1989), which included a large globular jar that has been chronologically matched with Lothal A and Rangpur IIA. The pottery along with other burial goods (including a chert weight) throws light upon burial practices prevalent in the peninsula—possibly a burial of a resident Harappan trader.

Apart from pottery, associated Harappan artefacts from the Oman Peninsula include chert weights, which are reported from almost all sites, seals, metal objects and a small number of carnelian beads. The fairly widespread presence of Harappan weights points towards a strong influence of Harappan traders (Edens, 1993; Shahani, 1997).

As for West Asian imports in the Harappan context, Chakrabarti (1997) has considered two lapis lazuli objects: a bead and a pendant. He points out that the bead at Mohenjo Daro is common in Mesopotamia but singularly absent in the Harappan context. The pendant also seems to be the only instance of lapis inlay work in this region while the technique apparently goes back to the period of the Royal Graves at Ur. Apart from these, he argues for a 'Mesopotamian or generally West Asiatic influence' in items like 'Reserve Slip ware, gold beads with axial tubes, knobbed ware, segmented beads, designs like swastika, etc., on some seals'. Also falling in with imports, according to Chakrabarti (1997: 564–65), are stylistic treatments/motifs seen on seals, which recall the Mesopotamian Gilgamish/Enkidu myths.

Lothal played an important role in external trade and its importance lies in manufacturing and trade. Apart from being a bead-manufacturing centre it may have exported gemstone beads, shell inlays and ivory gamesmen to Sumerian cities. Lothal probably was also a centre for imports; copper ingots recovered here have been thought to be of Persian Gulf origin (Rao, 1979). A number of sealings found in the aisles of the warehouse at the site also indicate an import of sealed goods, although the sealings are of Harappan seals. A circular Persian Gulf seal found at Lothal is the only one of its kind in the Harappan region.

Kuntasi has revealed production of bichrome ceramic and also bead manufacture of carnelian, agate, chalcedony, jasper, steatite, faience, shells and lapis lazuli. The most noteworthy copper artefacts at Kuntasi are two finger rings with large double spirals, which have no parallel at any Harappan site. The spiral motif itself is a rare occurrence in the Harappan context. The two rings were located near the eastern gateway and the occupant/owner was presumably in charge of security of the settlement (Dhavalikar *et al.*, 1996). As for the Bichrome ware, Dhavalikar is of the opinion that the ware was imported from the Persian Gulf, whereas Chitalwala (1996) argues that this ware was exported to the Gulf (Qa'lat al Bahrain).

Conclusion

The fate of civilizations—past and present—whether in cold climes or in the drought-and food-ridden tropics has been determined by climate. Archaeologists as well as historians, geologists, oceanographers, biologists and others may find something in the evidence of the behaviour of climate, which may concern their own research. The climatic and oceanic conditions in the northern Indian Ocean region during the Bronze Age were (and still are) fairly conducive to seafaring, although perhaps still rough in the 4th millennium.

The broadest trough in the southern hemisphere westerlies lies in the Indian Ocean sector corresponding to the position of the greatest extent of the Antarctic ice surface, where the Antarctic continent stretches farthest northward, as well as the greater range of the ITCZ in this region (Lamb, 1972). It amounts to an eccentricity of the entire circumpolar vortex towards this side of the hemisphere. This great trough over the Indian Ocean is one of the persistent features of the southern hemispheric vortex; the summer heating of the Asian continent is a direct result of this uneven heating. The high mountain wall of the Himalayas and the Tibetan plateau (which bar the path of the northerly winds), along with this trough cause the equatorial rain system to move as far northward as the Indian subcontinent at 30°N during the summer (Washington, 1981; Hahn and Manabe, 1975; Subbaramayya and Ramanadham, 1981).

Glacials marked a fall in sea levels whereas Inter-glacial periods saw high sea levels. However, between 5000 to 3000 cal. yrs. BP, a marked regression in the sea level is indicated along with transgressions and still stands. The northward positioning of the hydrographic front at 5000 yrs. BP as suggested by Dr. S.M. Gupta (in personal communication) can be related to the end of the glacial climate before the freshwater flux from the rain-inundated rivers reduced salinity in the northern Indian Ocean.

Although scant, it is possible to glean some data on the Harappan boat from archaeological evidence. The square sails in their simplicity were ideal for downwind sailing and coupled with oars or rudders these navigational aids could easily have carried a vessel along the Gujarat and Makran coastline towards the Gulf, making use of the northeasterlies. For the return voyage, the southwesterlies would gently and easily push a square-sailed vessel towards the Makran coast. The Harappans were certainly sailing directly across the Arabian Sea, perhaps from Sutkagen dor to Ras al Junayz and Ras al Hadd. The square sails in Harappan depictions were suitable for windward navigations, which would also mean that they were aware of the monsoon winds on the northern Indian Ocean and were the first to use them for sailing with a partly coastal and partly open sea-route.

The Black-slipped jar has been identified as a maritime storage jar of the Harappans serving a similar purpose as the classical Mediterranean amphorae (Cleuziou and Méry, n.d.). Perhaps the jars, especially the inscribed ones were used for trade of a particular commodity. Further, it would appear that the two specimens noted from Ras al Junayz and Bagasra, possibly originated from the same source of trade. If this is so, we have evidence of one segment of the Bronze Age sea-route on the Arabian Sea.

In Gujarat, archaeological evidence has been reported from Lothal and Kuntasi for port and wharf structures. Although Kuntasi appears to be solely an industrial site (not residential) it may also have been an intermediate port for the West Asian trade as evidenced by architectural remains. The settlement was located close to the required resources; the main functions of this site were procuring raw materials from the hinterland and producing finished goods for export to Kutch, Sind and West Asia.

Port or berthing facilities are found mentioned in the Ur temple texts from Mesopotamia (of Sargon of Agade, for example). Egyptian murals also show unloading/loading activities at wharf-like structures. At Lothal, the excavator (S.R. Rao) has identified a dockyard. With the evidence of Harappan seafaring activities established, one must now look for possible boat-building activity. Gujarat had the raw material—wood—and an ideal coastline compared to Sind or Makran. Ethnographic data (Shahani, 1996, 1997) shows an open boat-building yard at Mandvi (in Kutch) along the inlet, but the Bronze Age climate (mentioned earlier in this paper) may have dictated a more protected location for such a yard. From our present status of research Lothal presents the best possibility for such activity. Lothal is the only Harappan site where we find terracotta models of different type of boats apart from a metrical scale and a compass. More focused study on this is required but for now this hypothesis stands open for discussion.

Acknowledgements

The author would like to thank the editors Prof. Yasuda and Dr. Shinde for inviting this paper. The author was greatly enriched by the discussions which followed the presentation of this paper at the second ALDP workshop and would like to express her gratitude for the same.

References

Ahmad, E. (1972): *Coastal Geomorphology of India*, Orient Longman, New Delhi.

Angelucci, E. (1970): *Encyclopaedia of ships*, The Hamlyn Publishing Group Limited, Middlesex.

Carter, R., H. Crawford, S. Mellaliew and D. Barett. (1999): The Kuwait-British Archaeological expedition to As-Sabiyah, Report on the First Season's Work. *Iraq*, Vol. LXI: 43–58.

Chakrabarti, D.K. (1997): *The External Trade During the Harappan Period: Evidence and Hypotheses*. In Allchin F.R. and D.K. Chakrabarti (eds.): *A Source Book of Indian Archaeology II*. Munshiram Monoharlal Publishers Pvt. Ltd., New Delhi, pp. 556–579.

Chitalwala, Y.M. (1996): In Dhavalikar *et al.* (ed.): *Kuntasi, a Harappan Emporium on West Coast*, Deccan College, Pune.

Cleuziou, S. and S. Méry (n.d.): In Between the Great Powers. In Wright R.P. (ed.). *Comparative and Inter-societal perspective: the Indus, Mesopotamia and Egypt*.

de Cardi, B. (1998). Harappan Finds From Tomb 6 at Shimal, Ras al Khaimah, United Arab Emirates. In Friefelt, K and P

Sorenson, (eds.): *South Asian Archaeology 1985*, Curzon Press, London, pp. 9–13.

Dhavalikar, M.K., M.R. Raval and Y.M. Chitalwala (eds.) (1996): *Kuntasi, a Harappan Emporium on West Coast*, Deccan College, Pune.

Edens, C. (1993): Indus-Arabian Interaction during the Bronze Age: A Review of Evidence. In G.L. Possehl (ed.): *Harappan Civilization, 2nd edition*, Oxford University Press and IBH Publishing, New Delhi, pp. 335–363.

Fairbridge, R.W. (1960): The changing level of the sea, *Scientific American*, 202, no. 5: 70–79.

Fairservis, W.A. (1971): *The Roots of Ancient India: The archaeology of early Indian Civilization*, George Allen and Unwin, London.

Grover, A.K. and B.K. Sareen (1993): Recent Geomorphic Changes Around Little Gulf of Kachchh, Gujarat, *Photonirvachak*, 21, no. 4: 209–215.

Gupta, S.M. and A.A. Fernandes (1997): Quaternary radiolarian faunal changes in the tropical Indian Ocean: Inferences to palaeomonsoonal oscillation of the 10°S hydrographic front, *Current Science*, Vol. 72, No. 12: 956–972.

Gupta, S.M. (1999): Radiolarian Monsoonal Index Pyloniid Group Responds to Astronomical Forcing in the Last ~500,000 Years: Evidence from the Central Indian Ocean. *Man and Environment* XXIV (1): 99–107.

Hahn, D.G. and S. Manabe (1975): The role of mountains in the south Asian monsoon circulation. *Journal of Atmospheric Science*, 32: 1515–41.

Heyerdahl, T. (1980): *The Tigris Expedition*, George Allen and Unwin, London.

Kale, V.S. and S.N. Rajguru (1985): Neogene and Quaternary transgressional and regressional history of the west coast of India, an overview, *Bulletin of Deccan College Research Institute*, 44: 153–165.

Kutzbach, J.G. Bonan, J. Foley and S.P. Harrison (1996): Vegetation and soil feedbacks on the response of the African Monsoon to orbital forcing in the early to middle Holocene, *Nature*, Vol. 384: 623–626.

Lamb, H.H. (1972): *Climate: Present, Past and Future,* Vol. I, Methuen and Co. Ltd, London.

Lamb, H.H. (1982): *Climate, history and the modern world*, Methuen and Co. Ltd, London.

Mackay, E. (1938): *Further Excavations at Mohenjo Daro*, Archaeological Survey of India, New Delhi.

Mehr, S.S. (1987): Quaternary Sea Level Changes: the Present Status vis-à-vis Records Along Coasts of India, *Indian Journal of Earth Science*, Vol. 14, No. 3–4: 235–251.

Orchard, J. and G. Stanger (1999): Al-Hajar Oasis Towns Again! *Iraq*, LXI: 89–119.

Overpeck, J., D. Anderson, S. Trumbore and W. Prell (1996): The southwest India Monsoon over the last 18,000 years, *Climate Dynamics*, 12: 213–225.

Perry, A.H. and J.M. Walker (1976): *The Ocean-Atmosphere System*, Longmans, Cardiff.

Possehl, G.L. (1996): Meluhha. In Julian Reade, J. (ed.): *The Indian Ocean in Antiquity*. Kegan Paul International and the British Museum, London, pp. 133–208.

Rao, S.R. (1979). *Lothal: A Harappan Port Town 1955–62*, Vol. I, Archaeological Survey of India, New Delhi.

Shahani, L. (1996): *Ethnography of the Dhow-trade between India and the Persian Gulf: Towards Understanding Harappan Sea-trade in this Region*. Unpublished M.A. Dissertation from the M.S. University of Baroda, Vadodara.

Shahani, L. (1997): Ethnoarchaeology of Harappan Sea trade—A Preliminary Study, *Man and Environment*, XXII (1): 9–18.

Snead, R.E. (1982): *Coastal Landforms and Surface Features*, Hutchinson Ross Publishing Company, Pennsylvania.

Subbaramayya, I. and R. Ramanadham (1981): On the onset of the Indian southwest monsoon and the monsoon general circulation. In Lighthill, J. and R.P. Pearce (eds.) *Monsoon Dynamics,* Cambridge University Press, Cambridge, pp. 213–220.

Thapar, B.K. (1984): Six Decades of Indus Studies. In B.B. Lal and S.P. Gupta. (eds.) *Frontiers of the Indus Civilization*, Books and Books, New Delhi, pp. 1–26.

Tosi, M. (1993): The Harappan Civilization beyond the Indian Sub-continent. In G.L. Possehl (ed.): *Harappan Civilization, 2nd edition*, Oxford University Press and IBH Publishing, New Delhi: pp. 365–378.

Washington, W.M. (1981): A review of GCM experiments on the Indian Monsoon. In Lighthill, J. and R.P. Pearce (eds.) *Monsoon Dynamic*, Cambridge University Press, Cambridge, pp. 111–130.

A parent and a child on an oxcart. The cow is regarded as a sacred animal in India and an important source of power. (Photo by Takeshi Takeda)

Securing water supply is vital to people living in the northern periphery of the Indian monsoon region. The well sometimes may have to be dug to a depth of 50 m. A well in Rajasthan, India. (Photo by Takeshi Takeda)

Chapter 16

Monsoon Behaviour and the Late Harappan Subsistence Pattern in Gujarat, Western India

KULDEEP KUMAR BHAN

Prehistoric Climate

In any attempt to understand the settlement pattern, subsistence strategies and culture change of a region, the climate and environment variables play a very crucial role. Both these variables are viewed as intimately linked with any cultural adaptation and/or change. Therefore, information on climate and environment is a prerequisite in any such study. Unfortunately, there is no consensus among South Asian archaeologists about the nature and interpretation of climate and environment during the Harappan period. The undertaking of palaeoclimatic studies in western India under the Asian Lake Drilling Programme by Professors Yoshinori Yasuda and Vasant Shinde is a welcome step in this direction.

The climate and environment of the Harappan period has long been debated in South Asian archaeology. The disagreement concerning the amount of rainfall and interpretation has ranged from greater rainfall, to lesser rainfall or the same rainfall as today. The earliest scholars including Marshall (1931), Dikshit (1938), Mackay (1943) and Piggott (1962) argued that prehistoric climate and environment were distinct from today. Most of them argued for a much wetter climate on the basis of a variety of archaeological evidence. This included the presence of laboriously constructed dams in south Baluchistan called *gabar bands*, the use of burnt instead of sun-dried bricks at Mohenjo Daro and the depiction on seals of animals which are commonly found in damp jungle country.

Later Singh's (1971, 1977) pollen sequence from salt lakes—Sambhar, Didwana and Pushkar of Rajasthan also suggested greater rainfall during the Harappan period, which was later supported by Bryson and Swain (1981: 135–45) on the basis of climatic data from other areas. These palaeoenvironmental reconstructions indicated there was increased rainfall (as compared today) in northwest South Asia during the Harappan phase, which decreased during the early parts of the 2nd millennium BC. Singh (1977) and others used this sequence as an explanation for the decline of the Indus valley tradition (Agrawal and Sood, 1982; Allchin and Allchin, 1997: 208). Vishnu-Mittre (1974) expressed doubts about the pollen indicators used by Singh for suggesting wet climate. Misra (1984: 461–489) also argued the testing of this climatic theory and the Harappan collapse. Shaffer and Lichtenstein on the basis of recalibrated radiocarbon determinations refuted Singh's climatic data. They suggested that subphase IVa (between 3000–1800 BC) should instead fall in subphase IVb—period of increased aridity (Shaffer and Lichtenstein, 1989: 120–121).

The hypothesis of increased aridity finds additional support in the recent study carried out by Enzel and his colleagues on the lacustrine record of Lunkaransar lake in the Thar desert. Their record suggests that the Lunkaransar lake rose abruptly around 5000 BC and it persisted with minor fluctuations for the following 1000 years, and then fell abruptly to the range of 10–40 cm of water at about 4200 BC. The final drying occurred between 2894 to 2643 BC (Enzel *et al.*, 1999: 127). However, several scholars have argued that the climate was similar to the present times (Fairservis, 1956; Hegde, 1977; Raikes and Dyson, 1961; Meadow, 1989; Rissman and Chitalwala, 1990; Possehl and Rawal, 1989). Kenoyer (1991a) notes that even if there were changes in rainfall, the area is so vast that it would not have affected all regions uniformly.

In Gujarat, a large number of Harappan and other contemporary settlements have revealed bones of rhinoceroses, which is now confined to Assam and some parts of the Himalayan terai in India. Chitalwala (1990: 79–82) suggests that the marshy habitat is not at all a precondition to the existence of rhinos nor should it be treated as an indicator of any particular climate as suggested by Zeuner earlier (1963). In Africa, for instance, rhinoceroses live in areas that are semi-arid with xerophytic vegetation. Besides, the repeated presence of drought-resistant summer crops— 'millets' in the Harappan phase of Gujarat— perhaps suggests that the climate might not have been very different from today. This argument finds additional support from the pollen analysis from Nalsarovar in north Gujarat. On the basis of the pollen profile at the Nal lake Vishnu-Mittre and Sharma (1979: 79) have indicated predominantly Savannah grasslands type vegetation in the region of north Gujarat during the 2nd millennium BC till the present (Table 1a and b). For Gujarat, the available localized data is

Table 1a. Pollen profile of Nalsarovar: north Gujarat

Zones	Dates	Assemblages
NS-1	Before 7000 yr. BP	Grasses, Chenopods
NS-2	7000–3500 yr. BP	Grasses, Chenopodium/Amaranthus
NS-3	3500–160 yr. BP	Dominance of grasses

Table 1b. Nalsarovar radiocarbon dates

Depth in cm	^{14}C BP	Lab #
0	160±200	TF-54
50	820±210	TF-55
100	1320±200	TF-56
150	1890±110	TF-57
200	2550±200	TF-58
250	3520±210	TF-59
300	4280±200	TF-60
350	4060±210	TF-61
400	4490±205	TF-62
450	6170±200	TF-63
500	7020±230	TF-64

generally interpreted as indicating no major difference between the 3rd/2nd millennium BC and today (Meadow, 1989; Rissman and Chitalwala, 1990). Therefore, it appears that there is some sort of consensus among scholars that there were no major climatic fluctuations in Gujarat during the Harappan period and the climate was perhaps not very different from today. To appreciate the palaeoenvironment and the influence that climate had on human affairs it is useful to briefly outline below the present environment and some current adaptations to it.

Present-day Geographical and Environmental Setting

The region of Gujarat is made up of two major geological zones. The first is the Deccan trap, which was formed as a result of huge flows of lava during Mesozoic. The trap formation dominates the lithology of Peninsular Gujarat, known as Saurashtra, the main parts of the southern mainland and Kutch. The second major feature is the recent alluvium brought from the adjacent highlands by the rivers of the Gujarat plain. The Mahi river conventionally divides this plain into northern and southern segments, although this is an arbitrary boundary.

The state of Gujarat can be divided into four geographical regions for analytical purposes. The

244

first two are the northern and southern halves, a well-defined corridor of alluvium bordered by salt waste and sea on the west, and hills that extend up to the Western Ghats. The third region is Kutch, a rocky and arid collection of outcrops which are surrounded on three sides by seasonally flooded salt flats called Ranns. The last, to the south of Kutch is the peninsula of Saurashtra or Kathiawad, which is surrounded by the sea on all but its northern border (Fig.1).

South Gujarat extends south towards Bombay and is bounded in the east by granite hills of the Sahyadri and Satpura ranges reaching up to 1000 m a.m.s.l. (Patel, 1977: 17). From these hills the Mahi, Dhadrar, Narmada, Kim and Tapti flow west to the Gulf of Khambhat. On the west the Gulf of Khambhat and the Arabian Sea bound it. The boundaries of the Gulf of Khambhat are

locally known as *bhal* and *bhalbaru* areas. Here the soils are sandy and salty and the ground water is brackish. Modern cultivators avoid this area due to heavy flooding. However, in some parts of the *bhal* tract wheat can be grown, when water recedes around October. The residual humidity is sufficient to permit the maturation of wheat without any further moisture input. In the central zone soils are of high quality but are saline to the west and thin on the western slopes. Sorghum is the most popular grain of south Gujarat. It is grown mainly as *kharif* crop but can be grown during *rabi* season as well. It is also the most popular fodder crop for cattle (Patel, 1977: 44).

The rainfall in south Gujarat ranges from 60 to 150 cm (Fig. 2) and increases sharply on the inner eastern hills, which has tropical dry deciduous forest vegetation. *Anogeissus latifolia, Terminalia*

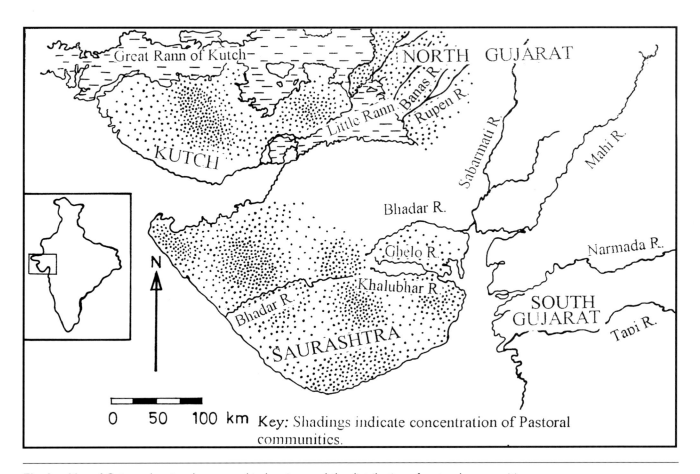

Fig. 1. Map of Gujarat showing the geographical regions and the distribution of pastoral communities.

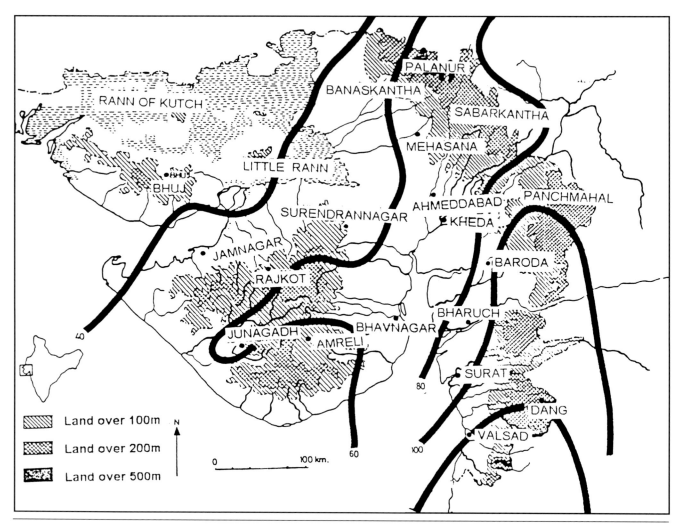

Fig. 2. Map of Gujarat showing the elevation and rainfall.

tomentora and *Tactone granlis* (teak) are the dominant trees (Gaussen *et al.*, 1968: 38). Grasses may occur but the canopy is more open. *Andropogan pumilus, Ischoemum angustifolium,* (Champion, 1935: 85) *Iseilma laxum, Apluda aristata, Heteropogon contortus* and *Cymbopogon martini* (Whyte, 1964: 99) are characteristic grasses of this area.

North Gujarat is a sandier and more arid plain. It is again bounded in the east by the Satpura hills and on the northeast by the Aravallis, which rise up to 900 m. To the northwest the discontinuous hills reach up to 100 m a.m.s.l. (Patel, 1977: 17–18). North Gujarat is bordered on the west by salt desert and the low-lying neck of the

Saurashtra Peninsula. The rivers flowing through the region include the Banas, Rupen, Sabarmati and Mahi. They are fed by monsoon rains and are ill suited for irrigation (Leshnik, 1968: 298). The northern plain is drier, receiving 80 cm of rain annually in the east grading down to 40 cm in the west. The soils are alluvium and have been derived from wind and water erosion of the Aravalli ranges. The percentage of silt and sand is much higher, reaching up to 90% in the west. Much of this area is a saline and alkaline treeless steppe that merges with the desert and thus is not very encouraging for agricultural activities. The landscape of north Gujarat is relieved by frequent occurrence of relict sand dunes, giving rise to numerous small lakes, which retain water for two to six months after the monsoon.

The extensive alkaline areas in this zone are known as *usar* lands and have characteristic grass vegetation. This area is dominated by *Acacia, Capparis, Euphorbia, Zizyphus* (Gaussen *et al.,*1968: 33) and *Prosopis spieigera* (Whyte, 1964: 98). The grasslands occur both as undergrowth and open patches. The dominant grasses are *Eriochloa ramosa, Chloris virgata, Eragrostis ciliaris, Dactyloctenium aegyptium, Sporobolus marginatus, Dichanthium annulatum, Cenchrus ciliari, Digitaria adscendens, Sporobolus helvolus, Sporobolus coromandelianus, Echinochloa colonum, Eleusine compressa, Eleusine brevifolia, Cynodon dactylon,* and *Aristida* sp. Legumes found growing in this area are *Heylandia latibrosa* and *Indigofera* sp. In the same area but adjacent to small salt water *nullahs*, riverbanks and the marshy areas, the dominant vegetation is *Dichanthium annulatum, Sporobolus marginatus* and *Eriochloa ramosa. Sporobolus marginatus, Cressa cretica* etc. are observed growing on pure salt crust. This tract is locally known as *Nani Banni* and has some of the best grassland types in India, that have very high protein content (refer Mann, 1916 and Burns *et al.*, 1916 for the analysis of grasses of north Gujarat). This area is associated with one of the world's best tropical breed of cattle known as *Kankrej* or *Wadhial* (George, 1985) and *Murrah* and *Mehsana* buffaloes (Whyte, 1964: 98). Besides, the area also supports a large population of sheep and goat and the Asiatic wild ass.

The soils in north Gujarat are rich in chlorides, sulphates and carbonates of sodium. This is especially true with the areas which are frequently flooded and have poor drainage. These conditions lead to a high rate of evaporation and hence deposition of dissolved salts. The Rann of Kutch is one of these situations (Possehl, 1980)

The third region is Kutch. Here the flow of the Deccan trap is patchy. The older facies dates to Jurassic times and comprises uplifted marine sediments, sandstone, limestone and shales (Wadia, 1957: 264; Spate and Learmonth, 1976: 23). The soils of Kutch are mixed as a result of the parent rock. The outcrops of Kutch face the sea on the southwest, while on all other sides desolate salt flats, the Ranns, bound them. The Ranns of Kutch are an alluvial deposit of clay and sand brought down by the minor rivers of Rajasthan as well as the eastern Nara of Sindh and were originally shallow bays connected by sea (Bombay Presidency, 1884: 3; Patel, 1977: 20; Gupta, 1977: 205). During the monsoons the Ranns get flooded with salt water to a depth of one to three feet (Bombay Presidency, 1880: 12). When water recedes with the change of season a thick deposit of salt remains. This region supports a large population of flamingoes and wild ass.

The climate of Kutch is very arid and rainfall ranges from 30 to 40 cm per year. The major forest regime is of *Acacia, Capparis, Zizyphus, Salvadora oleoides* and *Prosopis spicigera.* Like the western section of north Gujarat the southern section of the Greater Rann of Kutch is a vast grassland locally known as *Moti Banni,* and is favoured by breeders (Bose, 1975: 6). The dominant grasses are *Andropogan schoeuanthus, Andropogan sqnarrossus, Andropogan annulatas, Andropogan contortus, Andropogan filiformis, Andropogan helepensis, Anthistiria imberdis, Aristida adesenscionis, Aristida hystricula, Sporobolus indicus, Cynodon dactylon, Chloris barbata, Eleusine flagellifera, Eleusine aristata, Phrcigmites karke, Eragrostis ciliaris, Eragrostis cynosuroides, Eragrostis amabilis, Aeuropus vilosus, Halopyrum mucronatum, Dendroclamas strictus* and *Adiantum lunulatum.*

The fourth region is the peninsula of Saurashtra. It is connected to the mainland of the districts of Ahmedabad and Surendranagar. The central part of Saurashtra is made of irregular plateaus, bisected by peaks near Chotila in Surendarnagar district, Girnar in Junagadh district and Alech and Barda (Fig. 2) in the

Jamnagar district. Saurashtra has clayey black cotton soil derived from weathered bedrock. The vegetation is dry deciduous type forest and the rainfall ranges from 60 to 80 cm.

A distinctive feature of Gujarat today and perhaps in Harappan times as well, is the inherent seasonality in the ecological aspects of the region. It lies in a zone of the monsoon marked by an annual variation in rainfall of more than 30% of the mean (Rissman, 1985: 351). A great deal of fluctuation from year to year is evident; often an extremely dry year will follow a very rainy one. These shifts do not occur in a predictable sequence and thus make rainfall a risky resource.

Generally speaking, the agriculture of Gujarat is characterized by dry farming. Eighty-five percent of agriculture is rain fed. Crop production occurs chiefly in the monsoon season and the crops are harvested in autumn (*kharif*). The black cotton soil is labour-reducing. During the monsoon it swells with water and later develops cracks in the dry season (Rissman, 1985: 164). In much of Gujarat, hoeing with a blade harrow is practised as a substitute for ploughing. Annual ploughing to destroy weeds is not necessary because only one crop is grown in a year (Patel, 1977: 39). Subsistence crops of Gujarat are dominated by 'millets'—*bajra* (*Pennisetum typhoides*) and *jowar* (*Sorghum bicolour*). The village-based agriculturists maintain up to five animals as an incidental adjustment or an internal part of their farming enterprise, with the production of food and cash crops as their preliminary activity.

However, rainfall is the major source of moisture in Gujarat, and subsistence depends upon it. The level of subsistence depends upon the years of good monsoon. The dependence of subsistence production upon rainfall is reflected in the statistics on the relative frequency of crop failure in the state. Among the local administrative units, called *talukas*, of Gujarat, 22% experience total crop failure at least once every six years on an average (Rissman, 1985:

351). Another aspect of the precipitation regime is the localized nature of the cloudbursts. While one area may experience plentiful rainfall, an adjoining one may be passing through drought.

Another important means of livelihood in rural Gujarat is animal husbandry by pastoral transhumane communities like *Rabaris* and *Bharvads*, who do not traditionally own agricultural land and derive income and subsistence from breeding and/or herding cattle, sheep, camel and buffaloes. The profound nature of seasonality with dry and wet periods in Gujarat results in the movement of these people and their animals. Though these groups are distributed throughout Gujarat, a large number are present in the bordering villages of Nani and Moti Banni, Barda, Gir hills and in the northern and southern fringes of coastal Saurashtra (Fig. 1). It is against this environmental backdrop that we will view the archaeological and ethnographic data.

Archaeological Data

During the Late Harappan phase, the previously integrated regions of the Indus Valley Civilization broke into three localized cultures defined by Shaffer (1991) and Kenoyer (1991a) as the Localization Era. These cultures were named after the important sites where specific pottery styles were first discovered or the geographical regions in which such settlements were found (Fig. 3).

■ Jhukar Phase

In Sindh, sites such as Amri, Chanudaro and Jhukar—a site near Mohenjodaro, Mature Harappan phase assemblages give way to a complex named Jhukar. It has distinct forms of ceramics, metal objects and seals, but lacks Indus script, weights, and terracotta cakes (Piggott, 1962: 224–28; Wheeler, 1968: 58–9; Possehl, 1980: 16). The stratigraphic continuity between

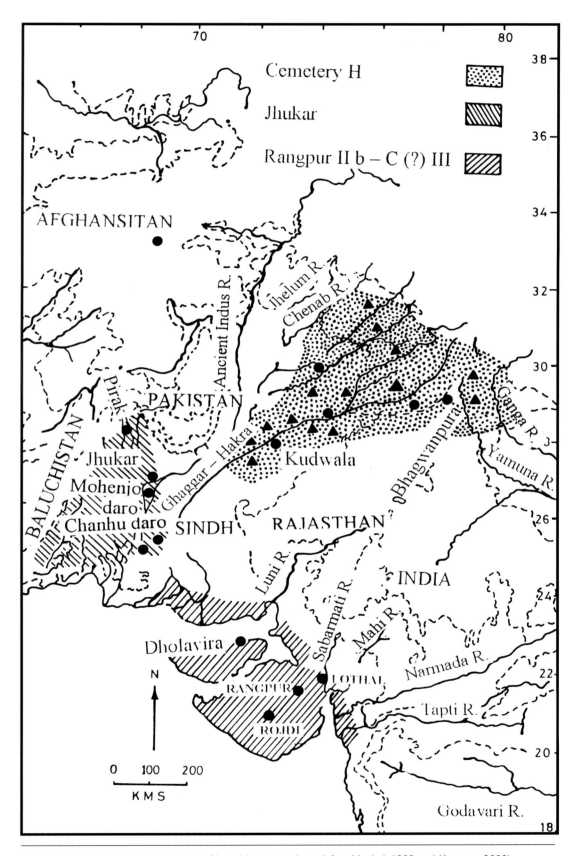

Fig. 3. Map showing cultural areas of Late Harappan phase (after Mughal, 1989 and Kenoyer, 2000).

Mature Harappan and Jhukar is clear at Amri and Chanhudaro, though Mackay (1943) stresses the divergence in the two material complexes, while Casal (1964: 63–69) and Possehl (1980: 15–17) have emphasized the similarity. At Chanhudaro this assemblage is associated with hut floors, paved with broken bricks and rough fire screens. In short, the Late Harappan phase in Sindh appears to have witnessed a decline of urban life and localization of artefact styles. The Jhukar phase incorporates sites in Sindh as well as part of Baluchistan.

■ Cemetery H

The Cemetery H phase refers to the northern regional culture that includes the large site of Harappa and sites further to the east in north India. The abandonment of Harappa was followed by an occupation named the Cemetery H Culture that was located in the portion of the site grid referred to as Area H (Vats, 1940). It is mainly known from burial deposits represented by elaborately painted ceramics. Piggott (1962: 235) and Fairservis (1975: 303) were of the opinion that the Cemetery H ceramics are unrelated to the Mature Harappan, but cultural continuity is indicated by Sankalia (1974: 397) and Allchin and Allchin (1982: 247). However, recent research at Harappa has shown that the transition from Harappan to Cemetery H culture (Punjab phase) was gradual (Kenoyer, 2000). Recent excavations conducted by the Harappa Archaeological Project in collaboration with the Department of Archaeology and Museums, Government of Pakistan at Harappa has indicated that baked brick architecture was constructed using both newly made bricks as well as reused bricks from earlier structures. There is evidence for over-crowding and encroachment rather than abandonment and decline (Kenoyer, 1991b).

At the impressive Cemetery H deposits along with Ghaggar-Hakra, where Mughal located fifty sites, some of the mounds are the largest of any

yet known in the Late Harappan Tradition, covering an area of 20 and even 38 hectares (Mughal, 1982: 93). Ongoing research at Harappa and excavations at one of these settlements would improve our knowledge of the Late Harappan way of life.

Continuity between the Harappan and Late Harappan phase in Sindh, Bahawalpur and the western Punjab still needs clarification, but is clearer in the case of eastern Punjab, Harayana and western Uttar Pradesh. In the former area a profusion of labels obscures the regional relationship, though these assemblages fit well into the overall scheme of the Early to Late Harappan phase. There is a stylistic coherence uniting degenerated Siswal (Surajbhan, 1975) and Bara (Sharma, 1982: 143), Ochre-Colour pottery (Sharma, 1971–72) and Late Harappan ceramics (Dikshit, 1982), which connects them distantly to cemetery H (Allchin and Allchin, 1982: 251).

■ Rangpur Phase (Late Harappan Phase)

The Rangpur phase refers to the entire region of Gujarat (Fig. 3 and 4). Here also, a high degree of continuity from the Mature Harappan to the Late Harappan phase has been indicated, though it is most evident in ceramics. Both the regional pottery types that are specific to Gujarat and the classical Indus forms undergo a period of change as the Mature Harappan ends. Alternation of certain ceramic forms and the introduction of the Lustrous Red Ware—a novel method of surface treatment, is thought to represent the Late Harappan phase in Gujarat. Nevertheless, at this juncture, a clarification about the Rangpur sequence that was used for more than three decades to define site periodization of Harappan Gujarat is needed, since the interplay between the subsistence system and the settlement pattern is a crucial variable in the interpretation of past economy and culture change.

In terms of the Rangpur sequence, period II A

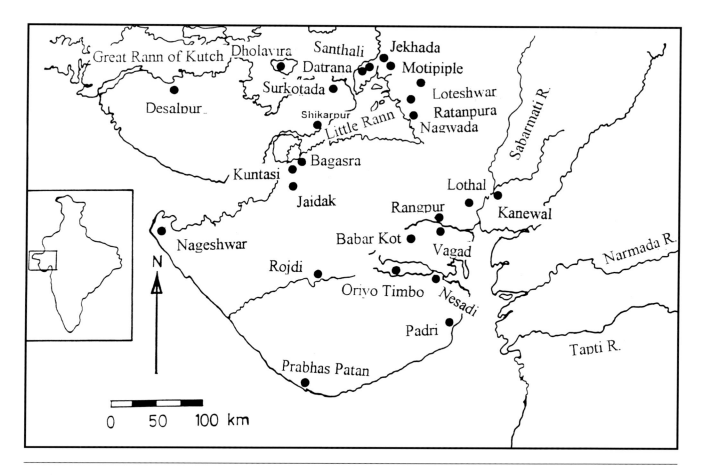

Fig. 4. Map of Gujarat showing major archaeological sites mentioned in the text.

was thought to be contemporary with Mature Harappan while period II signalled the onset of the Late Harappan phase. While certain forms like the 'Indus goblet' and beaker disappear, others seems to have simply altered in shape and decoration (Rao, 1963: 22, 59–65). The alteration of form continued in period IIC along with the introduction of a novel method of surface treatment, the technique of burnishing to a high lustre (Rao, 1963: 23). Pottery to which this treatment was applied is named Lustrous Red Ware. Both these phases were defined as the initial phase of Late Harappan in Gujarat (Bhan, 1989). The increase in frequency of Lustrous Red Ware (Rao, 1963: 24, 31), White Painted Black-and-Red Ware and Coarse Wares (Bhan, 1989) signals the beginning of Rangpur III. Rangpur III is the final period at Rangpur and the final period

of the Harappan Tradition in Gujarat as well. This period was defined as the final phase of Late Harappan in Gujarat (Bhan, 1989). However, the radiocarbon determinations from the excavations at Rojdi completely altered the earlier interpretations.

On the basis of radiocarbon determinations, Possehl and Rawal (1989) have indicated that all the settlements with pottery from Rojdi A and B phases should be dated to the urban phase and not to the Late Harappan phase. The emerging picture from their research is that there are two categories of settlements during the Mature Harappan phase in Gujarat. The first category of settlements, which Possehl and Rawal (1989) would like to designate as 'Sindhi' Harappan are supposed to share the material inventory of Mature Harappan sites of the Indus valley

251

(Possehl and Rawal, 1989) along with the regional ceramic types (Bhan, 1994). The excavated sites of Desalpur, Surkotada and perhaps Dhalovira in Kutch, Bagasra, Lothal and Nageshwar in Saurashtra and Nagwada, Jekhada and many other small settlements in north Gujarat represent this group (Fig. 4). Most of the settlements are found associated with some sort of craft activity, like shell bangle and ladle, stone bead or faience production. The other category designated as 'Sorath' Harappan, is represented by 162 small rural agricultural or pastoral camp settlements and more or less completely lacks evidence of substantial craft activity. These settlements were earlier thought to represent IIB-C of the Rangpur sequence or initial phase of the Late Harappan or Post-Urban Harappan phase in Gujarat. Herman (1997: 84, 88) questions the Rangpur sequence on the basis of its stratigraphy, relying heavily on the remarks made by Misra (1965) on the Rangpur sequence. Herman puts forward a core/Periphery model to explain this phenomenon.

Herman (1997: 84) suggests that during the Harappan period (second half of the 3rd millennium BC), the Kutch island and the Ranns of Kutch banks (with the Lothal in Bhal region) belonged to the urbanized Harappan core area of the Indus/Ghaggar-Hakra-Nara valley while the Sorath Harappan, Prabhas Patan and 'Padri' tradition in Saurashtra and the Anarta tradition in the north Gujarat plains formed an agro-pastoralists periphery, with the Ranns of Kutch banks and the coastal borders of Saurashtra as a narrow zone of interaction. However, on the basis of ceramic and stratigraphic analysis of a Harappan site at Bagasra in Saurashtra, Ajithprasad and his colleagues have indicated that though the Rangpur sequence is not completely vertical, the pottery sequence and the stratigraphy of Bagasara is fully in agreement with the Rangpur ceramic sequence. Like Rangpur, at Bagasra also the Mature Harappan ceramic assemblage is followed by Rangpur II A, II B-C

pottery forms (Ajithprasad et al., 1999). It is interesting to note here that excavations at Mature Harappan ('Sindhi' Harappan) sites in Gujarat have revealed the IIB-C ceramic types above the Mature Harappan deposits. However, it is not clear as yet if it is because the people of II B-C phase ('Sorath' Harappan) occupied these settlements after the Mature Harappan or if it has some chronological significance. The argument seems to have come a full circle and what we need at present are a set of consistent radiocarbon dates from Bagasra itself and many more sites so as to fix its proper chronological position in the Harappan tradition of Gujarat.

A majority of the settlements of this phase are located in Saurashtra. Out of the total 162 sites reported so far 152 (93.82%) settlements are in Saurashtra, while ten settlements (6.17%) are located on the eastern border of the Little Rann of Kutch in north Gujarat. The settlements are small with an average site size of 5.3 ha (Possehl, 1980: 65). Settlements at Laloi, Jaidak and Rojdi in Saurashtra approximately ranging between 4 and 7 ha have structures with stone foundations and stone walls circling them. Another category of the settlements of this phase has the remains of round hut floors as revealed at Jekhada[1] (Momin, 1983) and Kanewal (Mehta et al., 1980) in north Gujarat, Vagad (Sonawane and Mehta, 1985: 38–44) and Nesadi in Saurashtra (Mehta, 1984: 27–30). The recently located settlements of this phase in north Gujarat are extremely small in size and have very thin and patchy habitational deposit. They fall into the overall similar settlement pattern as well as the functional category, as has been discussed for other prehistoric settlements in north Gujarat (Bhan and Shah, 1990).

In the Greater Indus valley a subsistence pattern based on domestic cattle and the winter crops of wheat and barley is established to the 6th millennium BC (Meadow, 1989), though subsequent diversity in the Harappan subsistence is a much debated topic (Reddy, 1994: 25). Little information is available on the subsistence pattern of the Mature Harappan phase ('Sindhi') settlements of Gujarat, and interestingly neither wheat nor barley have been recovered so far. Rice grains (*Oryza sativa*) have been identified from the impressions in potsherds at Lothal and Rangpur II A (Ghosh and Lal, 1963: 161–175; Rao and Lal, 1985: 667–684). Rice grows wild today in the marshes of Rajasthan, in Madhya Pradesh (Vishnu-Mittre, 1961: 18) and Gujarat (Vishnu-Mittre and Savithri, 1982: 207), thus it cannot be concluded that it was cultivated for subsistence. Besides, the study of charcoal from Lothal (Rao and Lal, 1985: 667–684) and Rangpur (Ghosh and Lal, 1963: 161–175) indicates the presence of trees like *Acacia*, *Albizbia*, *Andina cordifolia* and *Tectone gradis*. The presence of *Tectone gradis* (teak) at Lothal is intriguing. Today it grows in eastern Gujarat in areas with rainfall ranging from 75 to 175 cm and comes under tropical dry deciduous type vegetation (Rao and Lal, 1985: 680). However, it is of interest to note here that the strategy and method used at both sites were not intensive enough to take them as representative of Harappan sites of Gujarat.

The first appearance of horse and donkey in the faunal data and the apparent 'millet' cultivation prompted Meadow (1989) to propose the second 'revolution' during the 2nd millennium BC. However, Weber (1991) has refuted this on the basis of evidence of 'millet' cultivation and early radiocarbon dates from Rojdi in Saurashtra, which as mentioned above are proposed to be contemporary to the Mature Harappan phase in Gujarat. The subsistence information of this phase ('Sorath' Harappan) comes mostly from the American and the Gujarat State Archaeological Department Excavation Programme carried out at Rojdi. The archaeobotanical samples available from Rojdi A and B include 'millets'—*Eleusine coracana* (ragi), *Panicum milliare* (little millet) and *Setaria tomentosa*, *S. glauca*, (yellow foxtail millet) *S.*

italica (foxtail millet). Weber (1989) argues that the *S. italica* did not become a prominent cultigen until 2000 BC. This has been disputed by Reddy (1994: 28) who is of the opinion that Weber's argument is based on a single site and this could be the result of changing demands by the community and the subsistence system. There are various theories regarding the centres of origin of the different millets which are not considered native to Gujarat and appear to have been brought in from elsewhere. However, the timing and dynamics of this process is much debated (Weber, 1989). Although the lowest levels at Rojdi revealed a few barley grains, Weber (1989) argues that barley could have been traded instead of being grown locally there.

In Gujarat, both the categories of Mature Harappan phase settlements indicate the predominance of cattle husbandry over the raising of sheep, goat, buffalo and pig. However, excavations at Nagwada in north Gujarat (Patel, 1985) and Nageshwar in Saurashtra (Hegde *et al.*, 1991) have yielded large amounts of wild animals like sambar, blackbuck, chital, gazelle and nilgai and other rather rare forms like hare, camel and wild ass (Patel, 1985: 47). Most of these animals inhabit grasslands interspersed with scrub jungle.

■ Late Harappan Phase

As already mentioned above certain changes were taking place around the turn of the 2nd millennium BC. Researchers have seen these changes as a result of failing or at least changes in international exchange networks (Jarriage, 1995: 5–30) and/or a change in river system (Misra, 1984). In Gujarat also such changes have been observed and a three-tiered settlement pattern seems to have been established.

At Desalpur I B and Surkotada I C in Kutch and settlements such as Lothal B, Rojdi C, Prabhas Patan II, Jaidak (Fig. 5), Rangpur II C and perhaps Babar Kot in Saurashtra appear to

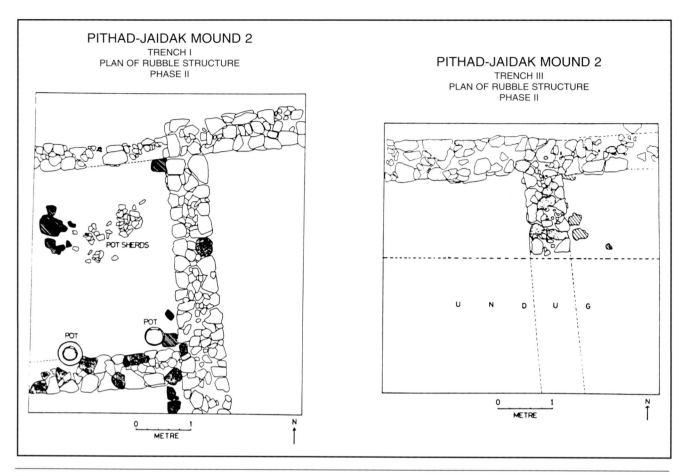

Fig. 5. Late Harappan structures at Jaidak, Pithad (Saurashtra).

have witnessed a continuation of a short period of urban organizations of the Mature Harappan phase into the Late Harappan phase. At Lothal and Surkotada also there is evidence for over-crowding and encroachment rather than abandonment and decline. Most of the 'Sindhi' settlements except Rangpur were abandoned after this stage. As a matter of fact some of the settlements (Rojdi, C) show expansion, though the Lustrous Red Ware have not been reported from any of the above-mentioned sites except at Rangpur II C (Rao, 1963). Herman (1995: 196; 1997: 100) is of the opinion that the presence of Lustrous Red Ware in this phase at Rangpur is debatable, because massive pits disturbed the layers. Now the question arises whether Lustrous Red Ware does really belong to II C or to III. This perhaps represents the transitional phase

from Mature Harappan to Late Harappan. Traditionally, the appearance of Lustrous Red Ware and changes in certain ceramic forms have been used to distinguish between Mature Harappan and the Late Harappan phases.

Another category of settlements, perhaps slightly later than the above-mentioned sites, are represented by round hut floors with potsholes along the periphery at Ratanpura (Fig. 7a) (Bhan, 1989), Kanewal (Fig. 6) (Mehta *et al.*, 1980) in north Gujarat, Vagad (Sonawane and Mehta, 1985: 38–44) and Nesadi (Mehta, 1984: 27–30) in Saurashtra. Except for Kanewal, none of the excavations are fully published yet. However, reexaminations of the pottery from Nesadi and Jekhada have indicated that some of the hut floors represented at these sites might belong to this phase. A preliminary reexamination of the

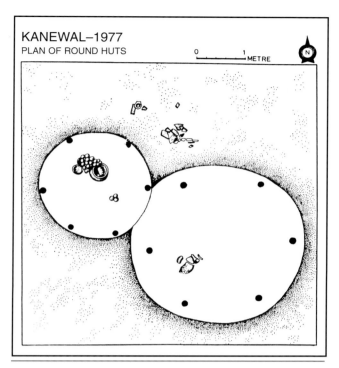

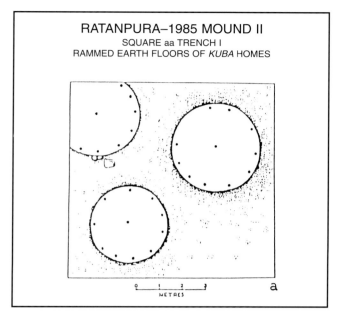

Fig. 6. Late Harappan Circular Hut Floors at Kanewal, north Gujarat (courtesy Department of Archaeology, M.S. University of Baroda, Vadodara).

Kanewal excavation report has indicated that at least the Trench V laid on Kesarisingh's Khetar representing two phases of circular hut floors positively belong to this phase. A successive occupation from 'Sorath' Harappan phase to Late Harappan phase is represented at these sites. At Oriya Timbo in Saurashtra, Ratanpura and Datrana in north Gujarat settlements of this phase are without any features (Oriyo Timbo and Datrana) or only have irregular patches of rammed floors (Fig. 7b).

In Gujarat radiocarbon dates are inconsistent. However, on the basis of the radiocarbon determinations (Table 2) available from Surkotada I C, Vagad I C, Rojdi C, Prabhas Patan III and Ratanpura in Gujarat, Ahar in Rajasthan, Chandali in Maharashtra and Navadatoli in Madhya Pradesh, a time period of c. 2000–1500 can be proposed for this phase in Gujarat, if the late date of Malvan phase I (TF-1084) and Rojdi C (PRL-1081) are not taken into consideration. The Oriyo Timbo (PRL-1427) and Prabhas Patan

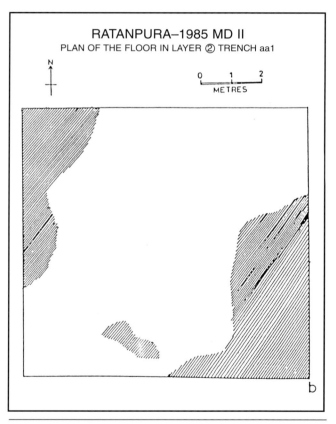

Fig. 7. (a) Late Harappan circular hut floors (b) Irregular rammed floor at Ratanpura, north Gujarat.

(PRL-91) dates calibrated to the beginning of the 3rd millennium BC are too early to be taken into consideration. For more precise dates, we will

Table 2. Radiocarbon dates for Late Harappan phase in Gujarat

Site	Lab #	Date (5730 Life, BC)	Mid Range for Corrected Dates (after Ralph et. al. 1973)	Reference
Surkotada, Period IC	TF-1294	1780±100	2170, 2110, 2050	Agrawal and Kusumgar, 1973.
	TF-1297	1795±100	2170, 2110–2130, 2050	Agrawal and Kusumgar, 1973.
Prabhas Patan, Period III	PRL-91	2020±170	2620, 2340–2460, 2190–2290	Kusumgar, Lal and Sarna, 1963.
	PRL-19	1245±165	1600–1640, 1460, 1220–1260	Kusumgar, Lal and Sarna, 1963.
	PRL-20	1485± 110	1780–1910, 1690, 1660–1650	Kusumgar, Lal and Sarna, 1963.
Rojdi C	BETA- 61767	1730±60	2110–2130, 2080, 2040	Herman, 1997.
	BETA-61768	1675±60	2080, 2040, 1920–1950	Herman, 1997.
	PRL-1084	1860±150	2340–2410, 2160, 2070	Herman, 1997.
	PRL-1081	1081± 15	800, 400–570, 210–360	Herman, 1997 Possehl, 1992.
Vagad	BM-2612	1810±50	2160, 2140, 090	M. S. Uni.
Ratanpura	BM-2615	1800±50	2160, 2120–2140, 2110	M. S. Uni.
Malvan	TF-1081	800±95	940–900, 800–900, 800	Herman, 1997 Possehl, 1992.
Oriyo Timbo	PRL-1427	1385±115	1690–1730, 1570–1600	Possehl, 1992.
	PRL-1424	1910±135	2420–2480, 2180, 2110	Possehl, 1992.
Ahar	TF-31	1275±110	1570–1600, 1490, 1310–1380	Kusumgar, Lal and Sarna, 1963.
	TF-32	1550±100	2020–2040, 1770–1870, 1660	Kusumgar, Lal and Sarna, 1963.
Navdatoli Chandoli	P-204	1600±130	2080, 1800, 1690	Ralph, 1959.
	P-472	1300±70	1570–1600, 1510, 1400–1450	Stuckenrath, 1963.
	P-473	1330±70	1600–1640, 1510–1550, 1480	Stuckenrath, 1963.
	P-474	1240±190	1650, 1460, 1170–1210	Stuckenrath, 1963.
	TF-42	1175± 120	1600–1630, 1320–1370, 1220–1240	Kusumgar, Lal and Sarna, 1963.
	TF-43	1040±105	1300, 1170, 1020	Kusumgar, Lal and Sarna, 1963.

have to wait for further excavations and more ^{14}C determinations.

Saurashtra witnessed a dramatic decrease in the settlement count and average site size. The majority of the settlements are situated near riverbanks on alluvial, residual black cotton or black cotton soil famous for its moisture-retaining capacity, the prime agricultural soil of Saurashtra. Until now, approximately 136 settlements of this phase have been reported from Gujarat, out of which nearly 37% (51) are located in Saurashtra. Conversely a large number of settlements 63% (87)—have been located in north Gujarat (Fig. 8).

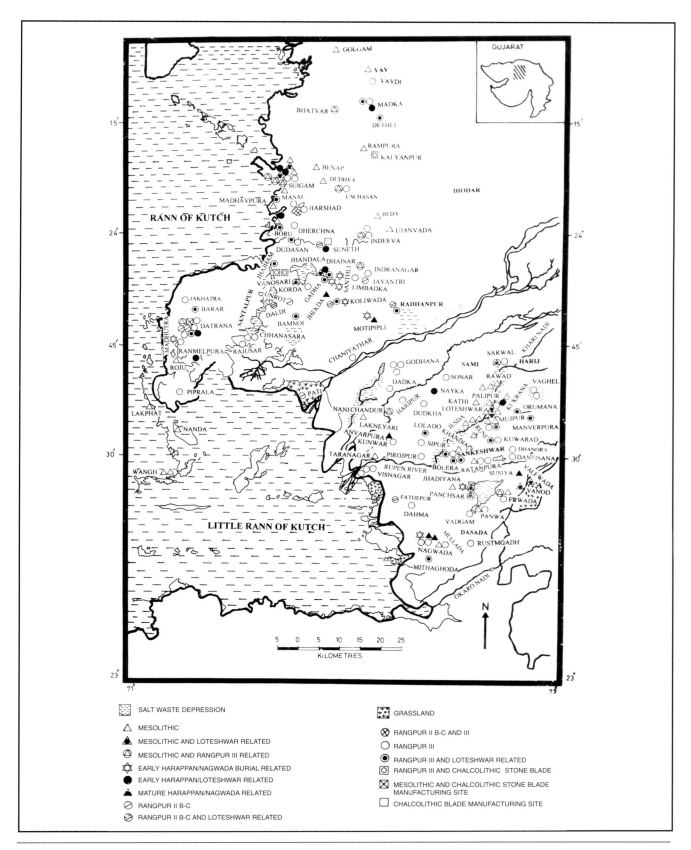

GOLGAM

VAV
VAVDI

BHATVAR
MADKA
DETHLI

RAMPURA
KALYANPUR

BENAP
SUIGAM
DUDHVA
MASAI
MADHĀVPURA
HARSHAD
USCHASAN
BEDA

RĀNN OF KUTCH
BORU
DHERCHNA
UJANVADA
INDERVA

DUDASAN
SUNETH

JHANDALA
DHAISAR
INDRANAGAR
VĀNOSARI
KORDA
JAVANTRI
SUNROT
GĀBHA
LIMBADKA
SANTHLI

JAKHATRA
BARAR
JHEKDA
KOLIWADA
RADHANPUR
DALDI
DATRANA
BAMNOI
MOTIPIPLI
CHHANASARA
RANMELPURA
RAJUSAR
CHANIYATHAR
ROJU

PIPRALA
LAKHPAT

GODHANA
SAMI
HARIJ
DADKA
SONAR
SARWAL
RAWAD
VAGHEL
NAYKA
KUKRANA
PALIPUR
NANDA
HARIPUR
KATHI
ORUMANA
NANICHANDUR
DUDKHA
LOTESHWAR
MUJPUR
LAKNEYARI
LOLADO
JESDA
MANVERPURA
ANVARPURA
KHANDIA
RUNI
KUNWAR
SIPUR
KUWARAD
DHANORA
TARANAGAR
PIROJPUR
SANKESHWAR
DANTISANA
BOLERA
RATANPURA
WANGH
RUPEN RIVER
JHADIYANA
SUSIYA
VALIWADA
VISNAGAR
VANOD
FATHEPUR
PANCHSAR
ERWADA
DAHMA
PANWA
VADGAM
DASADA
MULLADA
RUSTMGADH
NAGWADA
LITTLE RANN OF KUTCH
MITHAGHODA

DIODAR

GUJARAT

5 0 5 10 15 20 25
KILOMETRES

N

Fig. 8. Map showing the distribution of prehistoric settlements in the western section of north Gujarat.

SALT WASTE DEPRESSION

△ MESOLITHIC

MESOLITHIC AND LOTESHWAR RELATED

MESOLITHIC AND RANGPUR III RELATED

EARLY HARAPPAN/NAGWADA BURIAL RELATED

● EARLY HARAPPAN/LOTESHWAR RELATED

MATURE HARAPPAN/NAGWADA RELATED

RANGPUR II B-C

RANGPUR II B-C AND LOTESHWAR RELATED

GRASSLAND

⊗ RANGPUR II B-C AND III

○ RANGPUR III

◉ RANGPUR III AND LOTESHWAR RELATED

RANGPUR III AND CHALCOLITHIC STONE BLADE

MESOLITHIC AND CHALCOLITHIC STONE BLADE MANUFACTURING SITE

CHALCOLITHIC BLADE MANUFACTURING SITE

257

In north Gujarat, settlements of this phase are situated on the border of the Little Rann of Kutch and in the wastelands close to the Rupen river and its tributaries. The majority of the settlements are associated with relict sand dunes associated with inter-dune depressions, which retain water for three to four months after the monsoon. The settlements are small and have thin and patchy habitational deposits, with a high density of animal bones which at times outnumber the other material relics of the site. As a matter of fact, all other Chalcolithic settlements reported from this region show close similarities in their location as well as in the nature of settlements (Bhan, 1994, 2001). Most of the settlements of this phase are less than 0.5 ha. Some of the settlements like Chagda, Gudel, Khaksar and Popatpura in the

bhalbaru region of north Gujarat and Dhama, Mujpur, Nayaka, Vanod and Valewada in north Gujarat are so small that only a few potsherds and bones were recovered from these sites. Locating such sites is extremely difficult in any type of exploration and we suspect that a number of settlements may have been missed in our explorations. This perhaps explains why Misra and Leshnik (Leshnik, 1968) could not locate Sujinipur only a decade after Rao's survey (Rao, 1963: 177).

Information on agriculture of this phase comes mostly from Saurashtra and to some extent from Kutch, and is indicative of summer crop cultivation, mainly millets (Table 3). The archaeobotanical remains from Rangpur (Ghosh and Lal, 1963), and Surkotada (Vishnu–Mittre

Table 3. Archaeobotanical remains from Late Harappan sites of Gujarat

Surkotada, Period I C

Millet	*Eleusine coracanna, Setaria italica, Setaria virids L,* (7% of the total assemblage)

Others	Grasses and Sedges:	
	Grasses:	*Andropogon* sp., *Arundinella metzii, Brachiaria reptans, Dichanthium* sp., *Digitaria* sp., *Echinochloa stangina*
	Sedges:	*Cyperus micheleanus, Eriophorum* sp., *Fimbristylis ovata, Amaranthus* sp., (Vishnu-Mittre and Savithri, 1982: 214)

Rojdi C

Millet	*Eleusine coracona, Steria italica, Panicum miliare, Sorghum bicolor* (High % of *Steria* sp.)
Pulses	*Phaseolus, Pisum, lens*
Others	*Abelmoschus,, Amarthanthus, Chenopodium album, Dactyloctenium, Digitatia, Euphorbia, postrata* (Weber 1991; Weber and Vishnu-Mittre, 1989: 179)

Lothal

Millet	Present
Others	*Acacia* sps. (Rao and Lal, 1985: 679)

Rangpur

Millet	*Pennistetum typhoides*
Others	*Acacia* sp., *Albizia* sp., (Ghosh and Lal, 1963: 163, 169, 171)

Oriya Timbo and Babar Kot

Millets	*Eleusine Coracana, Setaria* spp. and *Panicum* sp.
Others	*Aizoon* sp., *Carex* sp., *Chenopodium/Amranthus* sp., *Cyperus* sp., *Digitaria* sp., *Zizyphus* sp. (Reddy, 1994, 1991: 80–81; Rissman, 1985)

and Savithri, 1982) are only from one or two samples and therefore results are limited. However, rigorous methods of recovery were employed at Rojdi (Weber, 1989), Oriya Timbo (Wagner, 1982; Reddy, 1994) and Babar Kot (Reddy, 1994). The millet crops recovered from this phase include *Panicum miliare, Eleucine coracana, Steria italica, Sorgum* sps. and *Echinochloa* sps. Besides, the examination of charcoal samples from Rangpur have indicated tree types like *Acacia* sps., *Albizia* sps., *Soymida febrifuga, Pterocarpus santalinus* (Ghosh and Lal, 1963: 161–75).

The analysis of archaeozoological remains from the majority of excavated sites of this phase shows a predominance of cattle over sheep, goat and buffalo. The faunal analysis carried out at Ratanpura revealed 67.68% of cattle, 13% sheep/goat and 3.06% buffalo, while the percentage of wild animals is as high as 16.04. This includes chital (7.58%), sambar (2.97%), nilgai (1.44%), pig, blackbuck and gazelle (1.35%). Other rare forms include hare, dog,

fowl, and wild ass (Bhan and Shah, 1990). The faunal analysis from north Gujarat is indicative of a higher exploitation of wild animals as compared to that of the settlements from Saurashtra.

The western section of north Gujarat, in which settlements are clustered together, has low rainfall (44-55 cm). Alluvial soil has high salinity and a very high percentage of silt and sand, reaching up to 90% and the rate of crop failure is also comparatively very high. Therefore, this area has a low cropping intensity, smaller population potential and a low level of infrastructure. Conversely, the eastern section of the region with higher soil fertility has a higher cropping intensity and thus large settlements with greater population potential and complex infrastructure (Desai, 1985: 103). These intensive explorations carried out in the eastern section of north Gujarat by Foote (1912), Sankalia and Karve (1949) and Misra and Leshnik (Leshnik, 1968: 259–309) have revealed a large number of settlements of microlith using communities (Fig. 9) and an

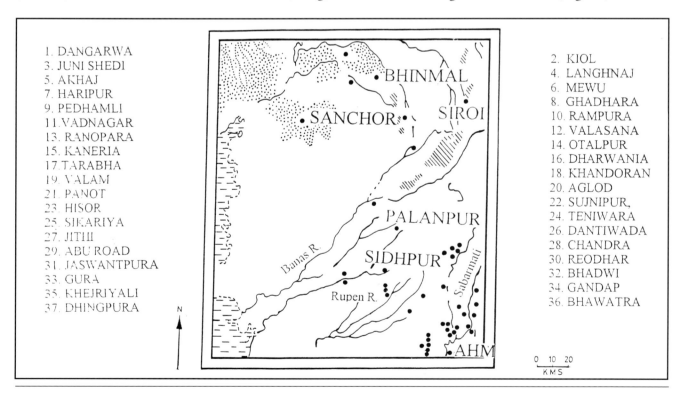

Fig. 9. Distribution of microlithic sites in the eastern section of north Gujarat.

absence of Bronze Age settlements, except one isolated settlement at Sujinipur. Some interaction with the Chalcolithic communities of the western section of the region is however reflected by the presence of pottery and copper tools recovered from the excavations at Langhnaj (Sankalia, 1974). The absence of Bronze Age settlements in the eastern section of north Gujarat and the nature of the settlements in the western section therefore indicate that the settlements were not located for harnessing the agricultural potential of the region as thought earlier (Hegde and Sonawane, 1986), but for some other reasons. The presence of excellent nutritious grassland in the western section, believed to be the best in India, and the availability of water in the inter-dune depressions, that retain water on an average for five to six months after the monsoon season, perhaps were the primary determinants for the development of settlements in this region. Thus on the basis of the location of these settlements and their ephemeral nature, we can suggest that the settlements were temporary camps of the people involved in pastoral activities. The overall increase in the number of the settlements during this phase in north Gujarat indicates that pastoralism perhaps increased significantly during this phase, although the area seems to have also attracted the early and Mature Harappans for similar reasons (Bhan, 1994; 2001). A similar situation has been noticed by Mughal in the Cholistan desert where a 26% increase in pastoral camp settlements has been recorded in the Late Harappan phase (1997: 56).

Recent palaeoethnobotanical studies carried out on the Late Harappan settlements at Oriya Timbo and Babar Kot in Saurashtra by Reddy (1994) has further substantiated and strengthened this view. Reddy's investigations have suggested that at Babar Kot cultivation of millet crops primarily as food grains by the occupants was important. The practice of sedentary agriculture was supplemented by sedentary animal husbandry. The greater emphasis on the cultivation of *Eleucine coracona* (which has a limited use as human food) compared to the other two millets—*Seteria italica* and *Panicum miliare* at Babar Kot was perhaps because it was cultivated as green fodder for animals. Reddy (1994) further suggests that the cultivation of millets at Babar Kot was for use both as animal fodder as well as for human consumption. It is suggestive of an economy more akin to sedentary agriculturists, with a minor pastoral component, that of sedentary animal husbandry (Reddy, 1994: 381–82).

Reddy's palaeobotanical analysis at Oriyo Timbo has indicated that millet was not cultivated at the site but instead was brought into the site as highly processed grain. She further suggests that these grains were either brought to this settlement through trade or exchange with farming groups in the area, or cultivated elsewhere (outside the range of the settlement) by the occupants and brought along with them on their seasonal migration to the site (Reddy, 1994: 382). Reddy's observations find additional support from Rissman's (1985) seasonality data from the annuli growth ring of cattle teeth that places the occupation of Oriya Timbo from March through July. Reddy (1994: 383) also stresses that the subsistence economy of Oriyo Timbo was not exclusively based on animal husbandry, since wild plants and animal exploitation were significant components of their subsistence strategies.

The natural incentives provided by the environs of north Gujarat seem to have attracted various pastoral/hunter-gatherers to north Gujarat as early as the 3rd millennium BC (Bhan, 2001). Pastoral activities also seem to have increased substantially during the Late Harappan phase (Bhan, 1989; 1994), which is reflected in the increase in the number of settlements of this phase in the arid locale of north Gujarat. This argument, though forwarded as early as 1986 (Bhan, 1989), has been unable to gain confirmation either for or against

(Possehl, 1997: 456; Mughal, 1989: 217; refer Sonawane and Ajithprasad, 1994: 136 for divergent view). This has highlighted the problem of accurately differentiating pastoral sites from hunter-gatherer and/or agriculturists. Therefore an ethnoarchaeological study of pastoralism in Gujarat was undertaken and the preliminary results of this study are briefly presented below. Ethnographic observations of the contemporary pastoral communities of Gujarat provide an insight into the pastoral economy and help to further strengthen the proposed argument.

Ethnoarchaeological Studies

Livestock raising in Gujarat is almost exclusively carried out by professional breeders. A distinctive feature of pastoral communities like the Rabari, Bharvad, Jat and to some extent Charan is that they make a living primarily by breeding and/or herding animals. They are collectively called *Maldharis* (Fig. 10*)*. At present there is no restriction of principle which governs the type of livestock they raise. Their livestock are of the type best suited to the local ecological situations. However, the meaning of *Maldhari* perhaps denotes those who concentrate on cattle-breeding rather than cattle-tending (Hellbusch, 1975: 123). The environment imposes conditions upon the animal breeder that dictate the type of animal which can be bred successfully, tradition not withstanding. An area, once suitable only for cattle may be transformed by extended overuse to one suitable for smaller animals like sheep and goat. Environmental factors influence the choice of the animal but cultural tradition connects these communities with specific animals. In general

Fig. 10. *Maldharis* on their migration to south Gujarat.

Fig. 11. World famous tropical breed of cattle—*Kankrej* or *Wadhial.*

Rabaris are camel-breeders (Fig. 11), the Charan and Jats breed cattle along with buffalo, and the Bharvads are sheep-and-goat breeders. It seems the tradition was more strictly observed in the past than at the present (Hellbusch, 1975: 124). Jats are the only Muslim breeders of Gujarat and inhabit the Moti-Banni area of Kutch. The Hindu breeder groups are distributed all over Gujarat, but a large number inhabit the bordering villages of the Banni tract in Kutch, the eastern border of the Little Rann of Kutch in north Gujarat (Nani-Banni), Barda and Gir hills, and the northern and southern fringes of coastal Gujarat. Certain traits of Jats in Kutch are shared with other animal-breeding communities of Gujarat (i.e. Bharvad, Rabari and Charan). Among those of ethnohistorical interest are caste-panchayat,

caste-assemblies, and the veneration of goddesses (Hellbusch, 1975: 122). Unfortunately, statistical data regarding these pastoral communities does not exist.

These professional breeders pursue their business with considerable skill and knowledge. They are most careful about mating, practise early castration and herd their animals separately, take them to the best grazing grounds at the best season and produce excellent animals with the lowest possible expenditure (Keatinge, 1917: 17, 31). These breeding communities of north Gujarat are responsible for the development of world famous tropical breeds of cattle like *Kankrej* or *Wadhial*—which are similar to zebu cattle with the characteristic hump and dewlap (Fig. 11), that is proudly displayed as a symbol on

Harappan seals. In the very recent past they were exported to North and South America for grading up indigenous cattle (Imperial Gazetteer, 1909: 11). Besides, breeders of this region are also responsible for the development of *Murrah* and *Mehsana* buffaloes and are equally adept at raising a local breed of sheep, which yields a fine quality of wool, most of which is exported. It must have taken thousands of years for these breeders to evolve the best breed of animals suited to tropics—which would withstand fly and mosquito nuisance and tropical diseases and could live on monsoonic grass or roughage.

Such breeds can only be produced by a long and uninterrupted period of breeding on sound principles, which are thoroughly understood by these professional breeders. Animals are not allowed to cover their own progeny. They select their stud animal with great care, choosing the second or third calves of exceptionally well-shaped, big, young bulls of good pedigree.

There is scanty, though useful, mention of these breeders in ancient literature. The *Periplus of Erythraean Sea* mentions the great herds of cattle in Saurashtra and export of *ghee* along with other produce (Huntingford, 1980: 81). Abul Fazal describes Kutch in the *Ain-e-Akbari* (1583-1590 AD) as a territory with an excellent breed of horses, good camels and goats. The reference to Rabaris appears in *Rasmala* and other chronicles of Gujarat, which mention Lakho Fulani, who was a son of a Rabari woman and ruled Kutch during the 10th century AD (William, 1958: 76). Later during the early 20th century we find frequent flattering mention of these breeders in British policy discussions.

Ethnoarchaeological studies have been carried out in Gujarat over a period of six years (1989–1995). A number of variables regarded as possible factors in the choice of site location were recorded during a survey. These include site maps, location, distance and gradient, and a known source of water. A variety of cultural remains left in these camps were also recorded

and mapped. Besides, a special attempt was made to record the type of vessels these pastoralists carry with them

The studies revealed that these breeders congregate during the monsoon season all along the river banks of north Gujarat specially Rupen, the bordering area of the Ranns, and also around the salt marshy wastelands known as *Padthar* or *Kharphat*. In Saurashtra they occupy the foothills of forested areas. In north Gujarat and Kutch the presence of nutritious grasses, water in inter-dune depressions and the availability of various salts and minerals in this area attract even those pastoralist groups who may not undertake long summer migrations to south or east Gujarat later. Pastoralists here construct huts (Fig. 12) of thorn bushes over hastily made circular plaster floors which are often linked together to form animal enclosures. Usually the camps are set on a dune top or its slopes to avoid accumulation of rainwater. One of the obvious restrictions is on the location of camps near or on the black soil and other soils, which retain moisture content. In such soil types the domestic stock gets infected by muddy hooves and is also ideal for transmitting a variety of diseases. The water source is at least 100 m or more from the camp area perhaps to avoid being the victim of mosquitoes.

In contrast to the expectations of many archaeologists who think of pastoral camps as ephemeral sites without substantial traces that would enter the archaeological record, the monsoon campsites (Figs. 13 and 14) we studied in the Barda hills of Saurashtra boast of a wealth of architectural features. As mentioned above, slopes of foothills are usually preferred for locating the camp settlements. Here the complexes are provided with one or two circular huts along with animal enclosures. Usually stone is used to construct the animal pens as well as the living areas which are provided with hearths. At regular intervals the walls of the animal enclosures have been provided with outlets to

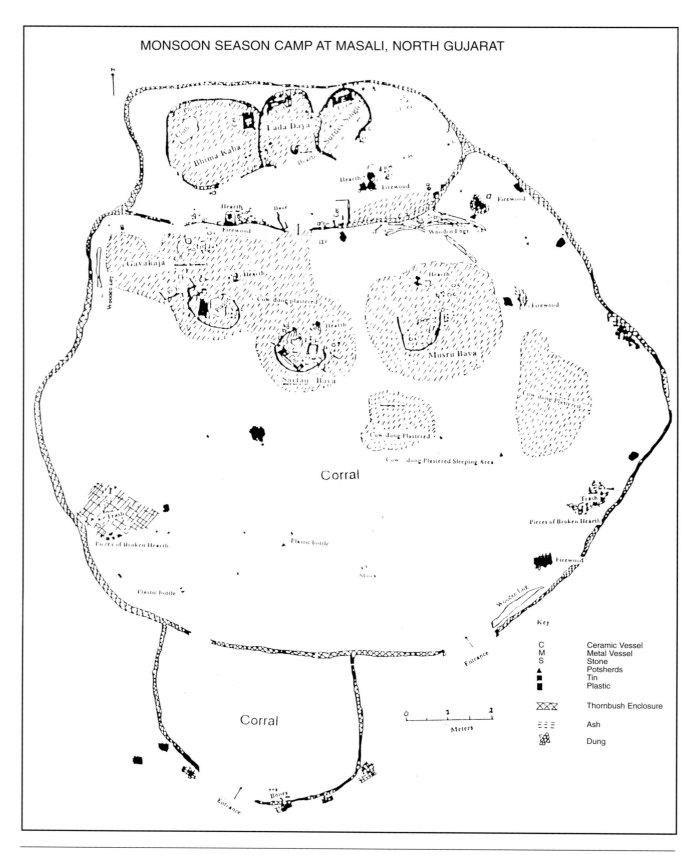

Fig. 12. Monsoon season camp at Masali, north Gujarat.

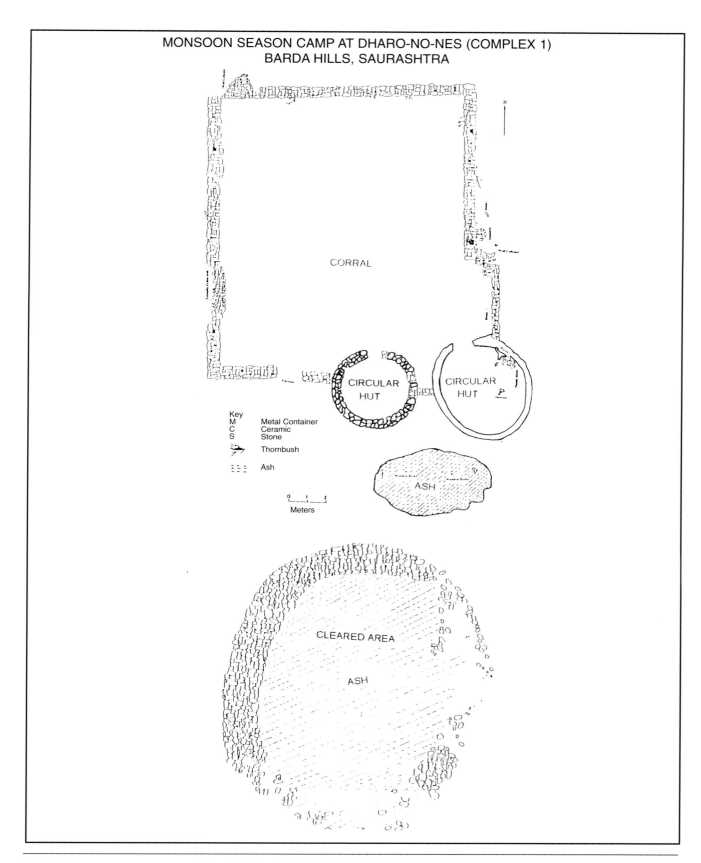

Fig. 13. Monsoon season camp at Dharo-No-Nes (Complex 1), Barda Hills, Saurashtra.

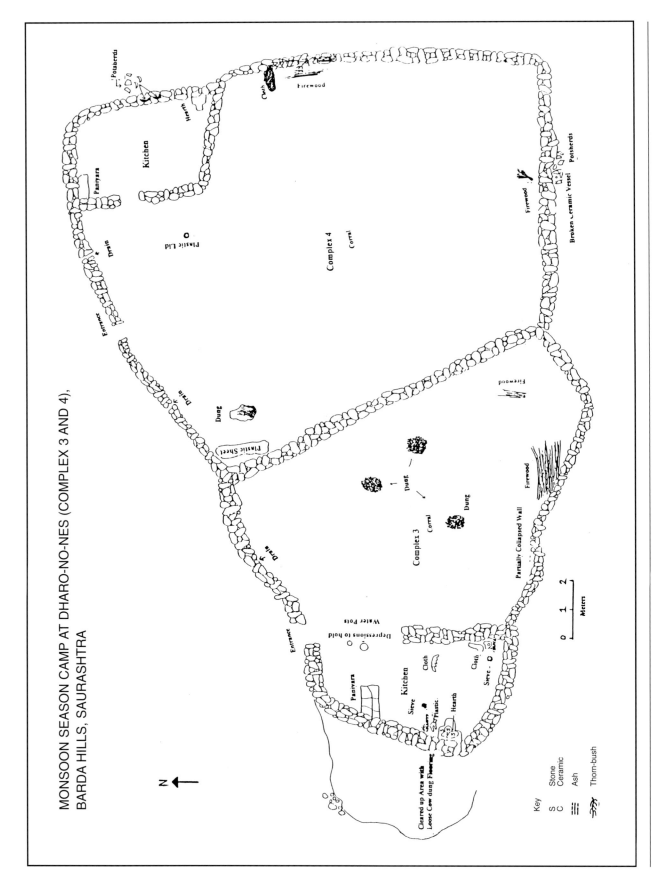

MONSOON SEASON CAMP AT DHARO-NO-NES (COMPLEX 3 AND 4),
BARDA HILLS, SAURASHTRA

Fig. 14. Monsoon season camp at Dharo-No-Nes (Complex 3 and 4), Barda Hills, Saurashtra.

266

drain out rainwater. Some of the complexes have raised stone platforms to hold water vessels.

The year of these breeders usually commences with Diwali, which falls in the later part of October or early November. At this time each breeder and his family leave their monsoon camps and at the onset of summer they disperse into south and eastern Gujarat. Of specific interest to archaeologists are the material consequences and the location of such camps. During the summer season they move with their herds, halting 1–2 days on the harvested fields or urban vacant land, leaving behind ephemeral hearth scars and little debris. At times they may also spend 4–20 days in summer camp after negotiating with landholders. Occupational debris is ploughed back into the fields leaving little evidence of the summer encampments, while the rainy season camps usually have substantial stone or thorn bush animal enclosures and plastered platforms as living space. These features are of special interest because same camps are often repeatedly used. While evidence for individual monsoon season occupation in certain cases perhaps may not preserve very well, the fact that these features are at times refurbished and reused make them enduring from the archaeological point of view.

Morphological studies of the vessels of five pastoral camps have indicated that the vessels with constricted long necks and narrow orifices are more commonly used than those with short necks and wide orifices—given their importance in transporting water and dairy products during migrations in pastoral economies. This has been further substantiated by the ethnographic studies carried out in two pottery manufacturing centres in Kutch (Choksi, 1995: 107). This is one of the important indicators that can be used to delineate pastoral assemblages. It is interesting to note here that the ceramic analysis of one of the Late Harappan settlements of north Gujarat has revealed interesting results. The analysis carried out by one of our graduate student from the

'Kiln' area at Datrana has indicated that more than 60% of vessels recovered from the area have comparatively long necks and narrow orifices (Devi, 2000: 51–65). Similar studies from other sites of north Gujarat perhaps will be rewarding.

Conclusion

The Harappans of Gujarat were perhaps sure of two things: first, in not too distant future the monsoon would be bad enough to ruin their subsistence production; second, things would be better somewhere else. This type of monsoon places a heavy premium on subsistence production. It is in this environmental background that the proliferation of pastoral encampments in the arid locale of north Gujarat should be viewed. In summary, it can be proposed that the proliferation of settlements in the arid locale of north Gujarat, that provided nutritious grasslands rich in various minerals and salts, water in inter-dune depressions and the presence of sand dunes provided an excellent breeding ground for the animals. The decentralization of the Harappan State in the early 2nd millennium BC as suggested by Reddy (1994) perhaps accompanied by a degradation in environment allowed for a wider range of choices of economic adaptations and particularly facilitated the integration of two distinct but complementary economies—agriculture and pastoralism in the semi-arid environment of Gujarat.

Taking the environmental and locational criteria of the present-day pastoral camps into consideration it seems that the area has played a very crucial role in the development of settlements in north Gujarat from very early times. The increase in the number of pastoral camp settlements during the Late Harappan phase was perhaps the result of increased environmental degradation combined with a breakdown in the Harappan integration sphere around c. 2000 BC. It can be further proposed

Pottery in India is still made in the same way as in the Indus civilization. (Photo Takeshi Takeda)

that the settlements close to the Ranns or *Kharphats* with greater archaeological debris were monsoon season camps, while settlements like Dhama, Mujpur, Nayaka, Vagel, Vanod, Valewada, Popatpura, etc., with comparatively very little habitational deposit were summer encampments. Dispersion and aggregation of the prehistoric population in north Gujarat is seen as a result of seasonal availability of animal food resources. However, an interdisciplinary approach is warranted to study the seasonality of these sites. Of particular importance in such studies would be the use of ethnoarchaeo-botanical, (like the one developed by Reddy, 1994 for Saurashtra), zooarchaeological data, ehnographic models and improved excavation methods like 'open area excavation' and 'single layer plan' as discussed by Harris (1979). Whether this hypothesis presented here is proven correct or not, further investigation in north Gujarat based on the model presented here will contribute to better understanding of the Harappan Tradition in South Asia.

Acknowledgements

This paper comes from the archaeological research work of more than two decades and ehnoarchaeological research undertaken by the author during 1989-1995 in Gujarat. I would like to acknowledge the Ford Foundation research grant for carrying out the ethnographic work in Gujarat. I would like to thank Dr. Seetha Narahari Reddy, for accompanying me in the initial stages of the ethnographic fieldwork. I was highly benefited by her comments and advice. I would like to especially acknowledge the work of Dr. Paul Rissman and Dr. Seetha Narhari Reddy without which this paper could not have been written. However, they are not responsible for any error or misinterpretation that might have unintentionally occurred during the presentation of their data. Professor J.M. Kenoyer and Dr. Massimo Vidale also share my acknowledgements for very many discussions and constructive criticism on this research.

Notes

1. Jekhada is also spelt as Zekhada. We prefer to maintain Jekhada in our text because it is the name referred to in the maps.
2. A strong argument for a late date of Surkotada IC has been put forward recently on the basis of recalibrated radiocarbon dates and the ceramics of this phase. Refer Herman (1997: 99)

References

Agrawal, D.P and R. Kusumgar (1973): Tata Institute Radiocarbon Date list X, *Radiocarbon*, 15(3): 547–585.

Agrawal, D.P and R.K. Sood (1982): Ecological Factors and the Harappan Civilization. In Possehl G.L. (ed.): *Harappan Civilization*, American Institute of Indian Studies, Delhi, pp. 223–252.

Ajithprasad, P., V.H. Sonawane, A. Majumdar and K. Dimri (1999): The Harappan Cultural Sequence at Rangpur and New Data form Bagasra, Gujarat. *Paper presented at the annual meeting of ISPQS*, Pune.

Allchin, B and F.R. Allchin (1982): *The Rise of Civilization in India and Pakistan,* Cambridge Press, Cambridge, 379 pp.

Allchin, R. and B. Allchin (1997): *Origins of Civilization.* Viking-Penguin India, New Delhi, 287 pp.

Bhan, K.K (2001): In the Sand Dunes of North Gujarat, *Marg.* Forthcoming Publication on Harappan Civilization.

Bhan, K.K. (1994): Cultural Development of the Prehistoric Period in North Gujarat with Reference to Western India, *South Asian Studies* 10: 71–90.

Bhan, K.K. (1989): Late Harappan Settlements of Western India, with Specific Reference to Gujarat. In Kenoyer J.M. (ed.): *Old Problems and New Perspective in the Archaeology of South Asia*, Wisconsin Archaeological Report, Vol. 2: 61–74.

Bhan K.K. and D.R. Shah (1990): Pastoral Adaptation in the Late Harappan Tradition of western India with Specific reference to Gujarat. Paper presented in the seminar on *Rising Trends in Palaeoanthropology: Environmental Changes and Human Responses (Last Two Million Years)*, Pune.

Bombay Presidency (1880): *Gazetteer of the Bombay Presidency,* Vol. 5, Cutch, Palanpur and Mahi Kanta Bombay.

Bombay Presidency (1884): *Gazetteer of the Bombay Presidency,* Vol. 7. Kathiawad.

Bose, A.S. (1975): Pastoralism nomadism in India. In L.S. Leshnik and G.D. Sonhteimer (eds.): *Pastoralists and Nomads in South Asia*, O. Harrasowitz, Wiesbaden, pp. 1–15.

Bryson R.A and A.M. Swain., (1981): Holocene Variations of Monsoon Rainfall in Rajasthan. *Quaternary Research* 16: 135–45.

Burns W., R.K. Bhide, L.B. Kulkarni and N.M. Hannmanti (1916): Some wild Fodder Plants of the Bombay Presidency. *Department of Agriculture, Bulletin # 78*: 1–17.

Casal, J.M. (1964): *Fouilles d'Amri*, Commission des Fouilles Archaeologiques, Paris, pp. 63–69.

Champion, H.G. (1935): A Preliminary Survey of the Forest Types of India, *Indian Forest Records*, 1: 1–204.

Chitalwala, Y.M. (1990): The Disappearance of Rhino from Saurashtra: A Study of Palaeoecology, *Bulletin of Deccan College*, Postgraduate and Research Institute, 49: 79–82.

Choksi, A. (1995): Ceramic Vessels: Their Role in Illuminating Past and Present Social and Economic Relationship. *Man and Environment*, XX (1): 87–108.

Desai, A. (1985): *Spatial Aspects of Settlement Pattern, A Study of the Narmada Commond Area of Mehsana District, Gujarat*, New Delhi, 103 pp.

Devi, Y. Krishnabarnati (2000): *Metrical Analysis of Sorath Harappan Pottery from Datrana*. Unpublished M.A. Dissertation submitted to the Department of Archaeology, M.S. University, Baroda.

Dikshit, K.N. (1982): Hulas and the Late Harappan Complex in the western Uttar Pradesh. In Possehl, G.L. (ed.): *Harappan Civilization*. American Institute of Indian Studies, Delhi, pp. 339–52.

Dikshit, R.B. (1938): *Prehistoric Civilization of the Indus Valley*, Madras, University of Madras Press.

Enzel, Y., L.L. Ely, S. Misra, R. Ramesh, R. Amit, B. Lazar, S.N. Rajaguru, V.R. Baker and A. Sandler (1999): High–Resolution Holocene Environmental Change in the Thar Desert, Northwestern India. *Science*, 248: 125–128.

Fairservis, W.A. (1975): *The Roots of Ancient India*. (2nd ed.). Chicago University Press, Chicago, pp. 303–480.

Fairservis, W.A. (1956): Excavation in the Quetta Valley, West Pakistan. *Anthropology Papers of the American Museum of Natural History* 45 (Part 2), pp. 169–402

Foote, R.B. (1912): *Indian Prehistoric and Proto-historic Antiquities in the Indian Museum—Note on Age and Distribution*, Government of Madras, Madras.

Gaussen, H.P. Legris, F. Blasco, V.M. Meher-Homji and J.P.Troy. (1968): Notice de la feuille Kathiawar. In *Extrait des Travaux de la Section Scientifique et Technique de L' Institute Francais de Pondicherry*. Hors Serie # 9.

George, S. (1985): *Operation Flood: an appraisal of current Indian dairy policy*, Oxford University Press, Delhi, 320 pp.

Ghosh, S.S. and K.R. Lal (1963): Plant Remains from Rangpur, *Ancient India*, 18–19: 161–77.

Gupta, S. K. (1977): Holocene silting in the Little Rann of Kutch. In Agrawal, D. P. and B. M. Pande (eds.): *Ecology and Archaeology of Western India*. Concept Publishers, Delhi, pp. 181–193.

Harris, E.C. (1979): *Principles of Archaeological Stratigraphy*, Academic Press Inc., New York, pp. 136.

Hegde, K.T.M., K.K. Bhan, V.H. Sonawane, K. Krishnan and D.R. Shah (1991): *Excavation at Nageswar: A Shell Working Site on the Gulf of Kutch*, Archaeological Series # 18, M.S. University, Baroda,

Hegde, K.T.M. and V.H. Sonawane (1986): Landscape and Settlement Pattern in Rupen Estuary. *Man and Environment*, X: pp. 23–30.

Hegde, K.T.M. (1977): Late Quaternary Environment in Gujarat and Rajasthan. In Agrawal, D.P. and B.M. Pande (eds.): *Ecology and Archaeology of Western India*. Delhi: Concept Pub. Co., Delhi, pp. 169–180.

Hellbusch, S.W.(1975): Changes in Meaning of Ethnic Names as Exemplified by the Jat, Rabari, Bharvad, and Charan in Northwestern India. In Leshnik L.S. and G.D. Sontheimer (eds.): *Pastoralists Nomads in South Asia*. Otto Harrassowitz, Wiesbeden, pp. 117–136

Herman, C.F. (1995): The Rangpur-sequence of 'Harappan' Gujarat, (India): a reassessment. In Allchin, B. and R. Allchin (eds.): *South Asian Archaeology (1995)*, Oxford and IHB Publishing Co. Pvt. Ltd., New Delhi, In press MSS.

Herman, C.F. (1997): 'Harappan' Gujarat: The Archaeological–Chronological Connections. *Paléorient*, 22 (2): 77–112.

Huntingford, G.W.B (ed.) (1980): *The Periplus of the Erythraen Sea*: London, 81 pp.

IAR, 1984–85; 1990–91: *Indian Archaeology: A Review*, New Delhi.

Imperial Gazetteer of India 1909, Vol. II.

Jarriage, J.F. (1995): Du néolithique à la civilisation de l'Inde ancienne: contribution des recherches archéologiques dans le nord–ouest du souscontinent indo-pakistanais. *Art Asiatiques*, L: 5–30

Keatinge, G.F. (1917): *Note on Cattle in Bombay Presidency*, Department of Agriculture, Bombay, pp. 11–12.

Kenoyer, J.M. (2000): Culture Change during Late Harappan Period at Harappa: New Insights on Vedic Aryan Issue. Paper submitted for publication in *Indo–Aryan* Volume (MSS).

Kenoyer, J.M., (1991a): The Indus Valley Tradition of Pakistan and Western India. *Journal of World Prehistory*, 5 (4): 331–385.

Kenoyer, J.M. (1991b): Urban Process in the Indus Tradition: A Preliminary Model from Harappa. In Meadow, R. (ed.): *Harappa Excavations 1986–1990*, Prehistoric Press, Madison, pp. 29–60.

Kusumgar, R., D. Lal and R.P Saran (1963): Tata Institute Radiocarbon Date List I. *Radiocarbon*, 5: 273–282

Leshnik, L.S., (1968): Prehistoric Explorations in north Gujarat and Parts of Rajasthan, *East and West*, 18 (3–4): 295–310.

Mackay, E.J.H. (1943): *Chanhu-daro Excavation 1935–36*, Boston Museum of Fine Arts, American Oriental Series.Vol. 20 American Oriental Society, New Haven, pp. 132–137.

Mann, H. (1916): *Fodder Crop of Western India*, Department of Agriculture #77 Bombay, pp. 1–53.

Marshall, J.H. (1931): *Mohenjo daro and the Indus Civilization*, A. Porbsthian, London.

Meadow, R. (1989): Continuity and Change in the Agriculture of the Indus Valley: The Palaeobotanical and Zooarchaeological Evidence. In Kenoyer, J.M. (ed.): *Old Problems and the New Perspectives in the Archaeology of South Asia. Wisconsin Archaeological Report*, Vol. 2: 61–74.

Mehta, R.N. (1984): Valabhi: A Station of Harappan Cattle Breeders. In Lal, B.B. and S.P. Gupta (ed.): *Frontiers of the Indus Civilization*, New Delhi.

Mehta, R.N., K.N. Momin and D.R. Shah (1980): *Excavations at Kanewal*, Archaeological Series # 17, M.S. University, Baroda.

Misra, V.N., (1984): Climate, a factor in the rise and fall of the Indus Civilization: Evidence from Rajasthan and Beyond. In Lal, B.B. and S.P. Gupta (eds.): *Frontiers of the Indus Civilization*. Books & Books, New Delhi, 546 pp.

Misra, V.N. (1965): Book Review: Ancient India, Bulletin of Archaeological Survey of India Nos. 18–19 (1962–63). *The Eastern Anthropologist*, 18 (I) 44 –52.

Momin, K.N. (1983): Excavations at Zekhda, *Puratattva*, 12, (1980–81): 120–125.

Mughal, M.R. (1997): *Ancient Cholistan: Archaeology and Architecture*. Ferozsons (Pvt.) Ltd., Karachi, 170 pp.

Mughal, M.R. (1989): Jhukhar and the Late Harappan Cultural Mosaic of the Greater Indus Valley. In Jarriage, C. (ed.): *South Asian Archaeology*. Prehistoric Press, Wisconsin, Madison, pp. 213–22.

Mughal, M.R. (1982): Recent Archaeological Research in Cholistan Desert. In Possehl, G.L. (ed.): *Harappan Civilization*, American Institute of Indian Studies, New Delhi, pp. 85–96.

Patel, A. (1985): Vertebrate Archaeofauna From Nagwada: A Preliminary Study. M. A. Dissertation for the Department of Archaeology, M.S. University, Baroda.

Patel, G. (1977): *Gujarat's Agriculture*, Ahmedabad.

Piggott, S. (1962): *Prehistoric India to 1000 BC* (2nd edition), Cassell, London, 294 pp.

Possehl, G.L (1997): The Transformation of the Indus Civilization, *Journal of the World Prehistory* II (4): 425–472.

Possehl, G.L. (1992): The Harappan Civilization in Gujarat: The Sorath and Sindhi Harappans, *The Eastern* Anthropologist, 45: 117–154.

Possehl, G.L. (1980): *Indus Civilization in Saurashtra*, B.R., New Delhi, 264 pp.

Possehl, G.L. and M.H. Rawal (1989): *Harappan Civilization and Rojdi*, Oxford and IHB Publishing Pvt. Ltd., New Delhi, 197 pp.

Raikes, R.L and R.H. Dyson (1961): The Prehistoric Climate of Baluchistan and the Indus Valley, *American Anthropologist*, 63: 265–281.

Ralph, E.K., (1959): University of Pennsylvanian Radiocarbon Date III, *American Journal of Science Radiocarbon Supplement* I: 45–58.

Ralph, E.K., H.N. Michael and M.Han (1973): Radiocarbon Dates and Reality. *MASCA News Letter* 9 (1): 1–20.

Rao, R.K. and K. Lal (1985): Plant Remains from Lothal. In *Lothal: A Port Town (1955–62)* Memoirs of the Archaeological Survey of India # 78 Vol. II, New Delhi.

Rao, S.R. (1963): Excavations at Rangpur and other Explorations in Gujarat, *Ancient India*, 18: 5–207.

Reddy, S.N. (1994): *Plant Usage and Subsistence Modelling: An Ethnoarchaeological Approach to The Late Harappan of Northwest India*. Ph.D. thesis submitted to the University of Wisconsin, Madsion.

Reddy, S.N. (1991): Archaeological Investigation at Oriya Timbo (1989–1990): A Post Harappan Site In Gujarat, *Man and Environment*, XVI (1): 79–81.

Rissman, P.C., (1985): *Migratory Pastoralism in the Western India in the Second millennium B.C: the evidence from Oriya Timbo (Chiroda)*. Ph.D. thesis submitted to the University of Pennsylvania, Philadelphia.

Rissman, P.C and Y.M. Chitalwala (1990): *Harappan Civilization and Oriyo Timbo*, Oxford and IHB Publishing Company Pvt. Ltd., New Delhi and American Institute of Indian Studies, New Dehli, 155 pp.

Sankalia, H.D. (1974): *Prehistory and Protohistory of India and Pakistan*, Deccan College, Pune, 397 pp.

Sankalia, H.D. and I. Karve.(1949): Primitive Microlithic Culture and People of Gujarat, *South Western Journal of Anthropology* LI (1).

Shaffer, J.G. (1991): The Indus Valley, Baluchistan and Helmand Tradition: Neolithic through Bronze Age. In Enrich, R. (ed.): *Chronologies in Old World Archaeology*, 3rd edition, Vol. 1, University of Chicago Press, Chicago, pp. 441–464.

Shaffer, J.G and D.A. Lichtenstein. (1989): Ethnicity and change in Indus Valley Cultural Tradition. In Kenoyer, J.M. (ed.): *Old Problems and New Perspective in Archaeology of South Asia*. Vol. 2. 117–126. Wisconsin University Archaeological Report.

Sharma, Y.D. (1982): Harappan Complex on the Sutlej India. In Possehl, G.L. (ed.): *Harappan Civilization*. American Institute of Indian Studies, Delhi, pp. 141–166.

Sharma, Y.D (ed.) (1971–72): OCP and NBP: 1971 *Puratattva*, 5: 1–100.

Singh, G. (1977): Stratigraphic and Palynological Evidence for Desertification in the Great Indian Desert, *Annals of the arid Zone*, 16 (3): 310–20.

Singh, G. (1971): The Indus Valley Culture, seen in the Context of Post Glacial Climatic and Ecological Studies in northwest India, *Archaeology and Physical Anthropology of Oceania*, 6 (2): 177–89.

Sonawane, V.H. and P Ajithprasad (1994): Harappa-Culture and Gujarat, *Man and Environment* XIX (1–2): 129–130.

Sonawane, V.H. and R.N. Mehta (1985): Vagad—a rural Harappan Settlement in Gujarat. *Man and Environment, IX*: 38–44.

Spate, O.H., and A.T.A. Learmonth (1976): India and Pakistan: A Regional Geography 3rd (ed.) Methuen, London.

Stuckenrath, Robert Jr., (1967): University of Pennsylvania Radiocarbon Dates X. *Radiocarbon,* 5: 82–103

Surajbhan (1975): *Excavation at Mitathal (1968) and other Explorations in the Sutlej–Yamuna Divide.* Kurukshetra University, Kurukshetra.

Vats, M.S., (1940): *Excavations at Harappa.* Government of India Press, Delhi, 488 pp.

Vishnu-Mittre (1974): Plant Remains and Climate From the Late Harappan and Other Chalcolithic Cultures of India—A Study in inter-relationship, *Geophytology,* V (4) 1: 46–53.

Vishnu-Mittre (1961): Plant economy in Ancient Navdatoli-Maheshwar. In Clutton-Brock, J., Vishnu-Mitre and A.N. Gulati (eds.): *Technical Report on Archaeological Remains.* Deccan College, Poona, pp. 13–52.

Vishnu-Mittre and R. Savithri (1982): Food Economy of the Harappans. In Possehl, G. L. (ed.): *Harappan Civilization,* American Institute of Indian Studies, Delhi: 205–222.

Vishnu-Mittre and C. Sharma (1979): Pollen Analytical Studies at Nal Lake (Nalsarovar), *Palaeobotanist* 26 (1): 95–104

Wadia, D.N. (1957): *Geology of India.* Macmillan, London, 536 pp.

Weber, S. (1991): *Plants and Harappan Subsistence,* American Institute of Indian Studies, Oxford and IHB Publishing Company Pvt. Ltd., New Delhi, 200 pp.

Weber, S. (1989): *Plants and Harappan Subsistence: An Example of Stability and change from Rojdi.* Ph. D. thesis for the Department of Anthropology, University of Pennsylvania, Philadelphia.

Weber, S. and Vishnu-Mittre (1989): Palaeobotanical Research at Rojdi in Possehl, G.L. and M.H. Rawal (eds.): *Harappan Civilization and Rojdi,* Oxford and IHB Publishing Pvt. Ltd.: New Delhi, pp. 177–181.

Wagner, G.E. (1982): *Flotation for Small—Artefact Recovery at Oriyo Timbo.* Philadelphia: University Museum, University of Pennsylvania.

William L.F.R. (1958): *The Black Hills: Kutch in History and Legend,* Weidenfeld and Nicolson, London, 276 pp.

Wheeler, R.M. (1968): *The Indus civilization,* Cambridge History of India, Cambridge University Press, Cambridge, 144 pp.

Whyte, R.O. (1964): *The Grassland and Fodder Resources of India,* Indian, Council of Agricultural Research, New Delhi.

Zeuner, F.E. (1963): *Environment of Early Man with Special Reference to the Tropical Regions,* Baroda.

Food production in the Indus civilization was based on wheat cultivation and livestock grazing. A wheat field in Rajasthan, India. (Photo by Takeshi Takeda)

Chapter 17

Solar Forcing of Climate Change and a Monsoon-related Cultural Shift in Western India around 800 cal. yrs. BC

BAS VAN GEEL, VASANT SHINDE AND YOSHINORI YASUDA

Introduction

Natural variations in atmospheric ^{14}C, which are expressed as wiggles in the radiocarbon calibration curve (Stuiver and van der Plicht, 1998) limit the possibilities for fine-resolution dating of changes in vegetation and climate. To avoid this limitation van Geel and Mook (1989) stressed the importance of ^{14}C wiggle-match dating of stratigraphic sequences. This dating strategy not only offers improved dating precision, but it can also reveal relationships between ^{14}C variations and short-term climatic fluctuations caused by solar variations.

In the temperate climatic zones the Subboreal-Subatlantic transition was a sudden and strong shift from a relatively dry and warm period to a humid and, especially at the onset, a cold episode (van Geel *et al.*, 1996, 1998, 2000). This climate change, as reflected in the Bryophyte species composition of European raised bogs, was dated with the ^{14}C wiggle-match dating strategy (Kilian *et al.*, 1995, 2000; van Geel *et al.*, 1996; Speranza *et al.*, 2000), and it appeared to be characterized by a temporary sharp rise of the atmospheric ^{14}C content. This rise occurred between 850 and 760 cal. yrs. BC, and it was caused by a sudden decline of solar activity (decline of solar wind, permitting more cosmic rays to penetrate into the atmosphere, and therefore a higher production of the cosmogenic isotope ^{14}C). Van Geel *et al.*, (1996, 1998, 2000; see also Dergachev *et al.*, 2000) combined palaeoecological evidence with archaeological information for the impact of that climate change on human populations.

We focus on evidence for climate change in India (a dryness crisis), starting around 850 cal. yrs. BC, and a contemporaneous cultural shift. We refer to a similar climate shift in Central Africa, and we give a possible palaeoclimatological explanation for the dry-wet transition in the temperate zones, and the wet-dry transition in the tropics.

Archaeological Evidence for a Dryness Crisis and the End of the Chalcolithic Culture Around 800 cal. yrs. BC in Western India

The Chalcolithic culture of western India is one of the best studied archaeological cultures in India. Large-scale excavations have revealed various aspects, including the socio-economic organization of settlements. Based on palynological data from Rajasthan lake deposits, the following climatic sequence for the Holocene period was reconstructed by Krishnamurty *et al.*, (1981):

Before 8000 BC	Severe aridity
8000–7500 BC	Relatively wet
7500–3000 BC	Relatively dry
3000–1700 BC	Sudden increase in wetness

1700–1500 BC	Relatively dry
1500–1000 BC	Relatively wet
1000–500 BC	Arid

This climatic sequence is probably applicable to the larger semi-arid region of western and Central India.

The Chalcolithic people of western India established their settlements in the proximity of fertile black cotton soils with a perennial supply of water. Most of the rivers in western India were semi-perennial, and farmers established their settlements near the river meanders. The second author surveyed more than a hundred Chalcolithic sites (Shinde, 1998). The area of the tributaries of the Tapi river were densely populated during the Chalcolithic period. Phase V (Late Jorwe) is the last phase (c. 1000–700 BC) of the Chalcolithic culture (Shinde, 1989). A deterioration in the standard of living (economic decline; impoverishment in the farming system) during the early part of the 1st millennium BC is strongly believed to be due to a drop in rainfall (Dhavalikar, 1973, 1988; Shinde, 1989; Naik and Mishra, 1997). Aridity forced people to a shift from sedentism to sheep/goat pastoralism (semi-nomadic existence). At Inamgaon the faunal assemblage changed during the late Jorwe phase from one dominated by domestic cattle to wild deer (Naik and Mishra, 1997, referring to the unpublished thesis by S. Pawankar, 1996). A sterile layer, formed under dry conditions (Mujumdar and Rajaguru, 1965), was found between the Protohistoric (Chalcolithic) and Early Historic period at Nevasa. Mishra *et al.,* (1999) studied palaeoenvironments of the Thar desert and dated an aeolian sand deposit on top of a Chalcolithic layer by GLSL (Green Light Simulated Luminescence) to c. 850 BC.

It must have been difficult for the Chalcolithic farmers to cope with the changed climatic conditions and it was probably for that reason that they deserted their settlements in Gujarat, Central India and the northern Deccan. However, they continued to live in the Bhima basin until 700 BC. Considering the fact that at Inamgaon, people continued to cultivate crops such as barley, peas, and oil seeds till around 700 BC, it seems that this area until then received at least the bare minimum rainfall required for growing these crops. The Megalithic people from south India, equipped with effective iron implements and fast-moving horses, began to arrive in the Deccan around 800 BC and they probably were responsible for the end of the Chalcolithic culture in this region.

Summarizing from the archaeological record it is evident that from the beginning of the 3rd millennium to the end of the 2nd millennium BC, western India had favourable climatic conditions and a relatively dense population. Most probably a climatic shift (dryness, shortly after 1000 BC) and the arrival of Megalithic people in the Deccan around 800 BC, were responsible for the decline of the Chalcolithic culture.

Evidence from the Arabian Sea

Based on the study of an early to middle Holocene laminated core from the upper Pakistan continental margin, Staubwasser (1999) concluded that the Indian monsoon was strongly affected by solar irradiance changes. Lückge *et al.,* (2001) studied monsoonal variability based on laminated sediments from the northeastern Arabian sea and they recorded an onset of aridification at 3000 BP, which was probably caused by a weakening of the monsoon system. Recently Neff *et al.* (2001) and also Agnihotri *et al.* (2002) found more and strong evidence for solar forcing of the intensity of the Indian monsoon.

Evidence for a Dryness Crisis in Central Africa after c. 850 cal. yrs. BC

From palynological evidence in Cameroun, Reynaud-Farrera *et al.,* (1996; see also van Geel et al., 1998) recorded a drastic change in the vegetation cover as a consequence of dryness after c. 850 cal. yrs. BC. This change from high representation of rainforest taxa to increasing abundance of heliophilous taxa (dry conditions; extension of savanna) was also observed elsewhere in the Central African rain forest belt (Elenga et al., 1994; Giresse et al., 1994; Wirrmann et al., 2001). Shortly afterwards farmers migrated into the area, availing themselves of what was from the human standpoint a regional climatic improvement.

The Sharp Rise of Δ^{14}C Around 850 cal. yrs. BC and Evidence for an Abrupt, World-wide Climate Change: A Possible Mechanism

From radiocarbon measurements of dendrochronologically dated wood, changes in atmospheric ^{14}C-content during the Holocene have been calculated and published as Δ^{14}C-data. There were numerous minor fluctuations and several major changes of Δ^{14}C. One of the most pronounced short-lived increases in the ^{14}C-content of the atmosphere during the Holocene occurred between 850 and 760 cal. yrs. BC.

From archaeological and palaeoecological studies in the Netherlands and elsewhere, van Geel et al., (1996, 1998, 2000) concluded that the start of this rise in atmospheric radiocarbon occurred when there was a climate change to cooler, wetter conditions in the temperate zones (Subboreal-Subatlantic transition) and a shift to dryness in the tropics. The rise in atmospheric radiocarbon points to a sudden decline in solar activity (Stuiver and Braziunas, 1989). But could a relatively small reduction in solar activity induce a relatively large change in global climate? van Geel and Renssen (1998) outlined the possible mechanisms for such a phenomenon. A reduction of solar ultraviolet radiation may also lead to a decline in ozone production in the lower

stratosphere (Harvey, 1980). The latter process could be the trigger mechanism responsible for the inferred climate changes, as may be deduced from climate modelling studies by Haigh (1994, 1996). She performed simulations with climate models to study the relation between the 11-year solar activity cycles, ozone production and climate change. First, Haigh (1994) used a chemical model of the atmosphere and found that a 1% increase in UV radiation at the peak of a solar activity cycle generated 1–2% more ozone in the stratosphere. Subsequently, Haigh (1996) used this increase as an input in a January climate model experiment. In the simulation, this produced a warming of the lower stratosphere by the absorption of more sunlight. In addition, the stratospheric winds were strengthened and the tropospheric westerly jet streams displaced poleward. Since the position of these jets determines the latitudinal extent of the Hadley Cells, this poleward shift resulted in a similar displacement of the descending parts of the Hadley Cells and a strengthening of the monsoon. The changes led to a poleward relocation of the middle latitude storm tracks.

Although solar variations during the Holocene have a different time scale, an effect opposite to the one simulated by Haigh (1996) may have played a role in the climate change around 2750 BP (c. 850 cal. yrs. BC). The observed strong increase of atmospheric ^{14}C during the period around 2650 BP was caused by reduced solar activity. Such a reduction in solar activity (less solar UV) would also have resulted in a decrease in the stratospheric ozone content. If one assumes that this decrease in stratospheric ozone content leads to an opposite effect to the one simulated by Haigh (1996), a contraction of the latitudinal extent of the Hadley Cell circulation and a weakening of monsoon would follow. A contraction of the Hadley Cell circulation and an associated weakening of the monsoons around 850 cal. yrs. BC would be consistent with drier conditions in the tropics.

Conclusion

A sudden and sharp rise in the atmospheric ^{14}C content around 2750 BP was contemporaneous with an abrupt and global climate change. At middle latitudes of the Northern and Southern Hemisphere, this change was to a cooler and wetter climate; and in the tropics (Central Africa, India), it was to a drier climate, as evidenced by both archaeological and palaeoecological data.

The variations in atmospheric ^{14}C content and climate may be tentatively explained by a reduction in solar activity. The possible mechanism is based on the idea that a reduced solar input could have reduced the stratospheric ozone content. The latter process may have been the trigger mechanism responsible for a decreased latitudinal extent of the Hadley Cells and a weakening of the monsoon.

Our idea about the climate development with its effects on cultures is based on a limited amount of data. This is certainly not sufficient for making a final statement, and more studies and a detailed ^{14}C dating programme of the recorded archaeological evidence is necessary. Although there are these uncertainties we postulate that a decline of solar activity around 850 cal. yrs. BC forced a world-wide climate change, and caused the decline and disappearance of the Chalcolithic culture in western India.

References

Agnihotri, R., K. Dutta, R, Bhushan and B.L.K. Somayajulu (2002): Evidence for solar forcing on the Indian monsoon during the last millennium. *Earth and Planetary Science Letters,* 198: 521-527.

Dergachev, V., B. van Geel, G. Zaitseva, A. Alekseev, K. Chugunov, J. van der Plicht, G. Possnert and O. Raspopov (2000): The earliest records of Scythians in Eurasia and sharp climatic changes around 2700 BP. In Paraskevopoulos, K.M. (ed.): *Physics in culture, the solid state physics in the study of cultural heritage.* Aristotle University of Thessaloniki, pp. 208-216.

Dhavalikar, M.K. (1973): Development and decline of the Deccan Chalcolithic. In Ghosh, A. and D.P. Agrawal (eds.): *Radiocarbon and Indian Archaeology,* Tata Institute of Fundamental Research, Bombay.

Dhavalikar, M.K. (1988): *The First Farmers of the Deccan* Ravish Publishers, Pune.

Elenga, H., D. Schwartz and A. Vincens (1994): Pollen evidence of late Quaternary vegetation and inferred climate changes in Congo, *Palaeogeography, Palaeoclimatology, Palaeoecology,* 109: 345-356.

Giresse, P., J. Maley and P. Brenac (1994): Late Quaternary palaeoenvironments in the Lake Barombi Mbo (West Cameroon) deduced from pollen and carbon isotopes of organic matter, *Palaeogeography, Palaeoclimatology, Palaeoecology,* 107: 65-78.

Haigh, J.D. (1994): The role of stratospheric ozone in modulating the solar radiative forcing of climate, *Nature,* 370: 544-546.

Haigh, J.D. (1996): The impact of solar variability on climate, *Science,* 272: 981-984.

Harvey, L.D.D. (1980): Solar variability as a contributing factor to Holocene climatic change. *Progress in Physical Geography,* 4: 487-530.

Kilian, M.R., J. van der Plicht and B. van Geel (1995): Dating raised bogs: new aspects of AMS ^{14}C wiggle matching, a reservoir effect and climatic change. *Quaternary Science Reviews,* 14: 959-966.

Kilian, M.R., B. van Geel and J. van der Plicht (2000): ^{14}C AMS wiggle matching of raised bog deposits and models of peat accumulation. *Quaternary Science Reviews,* 19: 1011-1033.

Krishnamurty, R.V., D.P. Agrawal, V.N. Misra and S.N. Rajaguru (1981): Palaeo-climatic influences from the behaviour of radio-carbon dates of carbonates from sand dunes of Rajasthan. *Proceedings of the Indian Academy of Sciences* (Earth Planet Science), 90: 155-160.

Lückge, A., H. Doose-Rolinski, A.A. Khan, H. Schulz and U. von Rad (2001): Monsoonal variability in the northeastern Arabian Sea during the past 5000 years: geochemical evidence from laminated sediments. *Palaeogeography, Palaeoclimatology, Palaeoecology,* 167: 273-286.

Mishra, S., M. Jain, S.K. Tandon, A.K. Singhvi, P.P. Joglekar, S.C. Bhatt, A.A. Kshirsagar, S. Naik and A. Deshpande-Muhkerjee (1999): Prehistoric cultures and Late Quaternary environments in the Luni Basin around Balotra. *Man and Environment,* 24: 39-49.

Mujumdar, G.G. and S.N. Rajaguru (1965): Comments on 'Soils as Environmental and Chronological Tool'. In Misra, V.N. and M.S. Mate, (eds): *Indian Prehistory,* Deccan College, Research Institute, Poona, pp. 248-253.

Naik, S. and S. Mishra (1997): The Chalcolithic phase in the Bhima Basin, Maharashtra: a review. *Man and Environment,* 22: 45-58.

Neff, U., S.J. Burns, A. Mangini, M. Mudelsee, D. Fleitmann and A. Matter (2001): Strong coherence between solar variability and the monsoon in Oman between 9 and 6 kyr ago. *Nature,* 411: 290-293.

Reynaud-Farrera, I., J. Maley and D. Wirrmann (1996): *Végétation et climat dans les forêts du Sud-Ouest Cameroun depuis 4770 ans BP: analyse pollinique des sédiments du Lac Ossa.* Comptes Rendus de l'Académie des Sciences Paris, 322, Série II a: 749–755.

Shinde, V. (1989): New light on the origin, settlement system and decline of the Jorwe Culture of the Deccan, India. *South Asian Studies,* 5: 60–72.

Shinde, V. (1998): *Early settlements in the Central Tapi Basin.* Munshiram Manoharlal Publishers, New Delhi, 140 pp.

Shinde, V. (2002): Chalcolithic phase in Western India (including Central India and the Deccan Region). In Paddayya K. (ed.): *Recent Trends in Indian Archaeology,* ICHR Publication, New Delhi.

Speranza, A., J. van der Plicht and B. van Geel (2000): Improving the time control of the Subboreal/Subatlantic transition in a Czech peat sequence by ^{14}C wiggle-matching. *Quaternary Science Reviews,* 19: 1589–1604.

Staubwasser, M. (1999): Early Holocene variability of the Indian monsoon and Arabian Sea thermocline ventilation. Dissertation Thesis Mathematisch-Naturwissenschaftliche Fakultät der Christian-Albrechts-Universität zu Kiel.

Stuiver, M. and J. van der Plicht (1998): INTCAL 98: Calibration issue. *Radiocarbon,* 40: 1041-1159.

Stuiver, M., and T. F. Braziunas (1989): Atmospheric ^{14}C and century-scale solar oscillations. *Nature,* 338: 405–408.

van Geel, B. and W. G. Mook (1989): High-resolution ^{14}C dating of organic deposits using natural atmospheric ^{14}C variations. *Radiocarbon,* 31: 151–156.

van Geel, B., J. Buurman and H. T. Waterbolk (1996): Archaeological and palaeoecological indications of an abrupt climate change in The Netherlands, and evidence for climatological teleconnections around 2650 BP. *Journal of Quaternary Science,* 11: 451–460.

van Geel, B., J. van der Plicht, M.R. Kilian, E.R. Klaver, J.H.M. Kouwenberg, H, Renssen, I. Reynaud-Farrera and H.T. Waterbolk (1998): The sharp rise of $\Delta^{14}C$ c. 800 cal. yrs. BC: possible causes, related climatic teleconnections and the impact on human environments. *Radiocarbon,* 40: 535–550.

van Geel, B. and H. Renssen (1998): Abrupt climate change around 2,650 BP in North-West Europe: evidence for climatic teleconnections and a tentative explanation. In: Issar, A.S. and N. Brown (eds.): *Water, Environment and Society in Times of Climatic Change,* Kluwer, Dordrecht, pp. 21–41.

van Geel, B., C.J. Heusser, H. Renssen and C.J.E. Schuurmans (2000): Climatic change in Chile at around 2700 BP and global evidence for solar forcing: a hypothesis. *The Holocene,* 10: 659–664.

Wirrmann, D., J. Bertaux and J. Kossoni (2001): Late Holocene paleoclimatic changes in Western Central Africa inferred from mineral abundance in dated sediments from Lake Ossa (Southwest Cameroon). *Quaternary Research* 56: 275–287.

A farmer and his cattle cultivating paddy in the rain in Guizhou province, China. (Photo by Takeshi Takeda)

Part IV

Monsoon and Civilizations

In the Asian monsoon region, boats and ships are important means of transportation. Jiangsu province, China. (Photo by Takeshi Takeda)

Chapter 18

Rains, Mother Earth and the Bull

SHUBHANGANA ATRE

Introduction: Monsoon, the Lifeline of Indian People

It is a matter of purpose that directs human reactions to environment. There is a folk tale that illustrates this, which narrates two diametrical views regarding the desirability of rains. The story is about a villager who had two daughters. One was married to a farmer and the other to a potter. One day the villager thought of visiting his daughters to see if they and their husbands were doing well. He first visited his elder daughter and her husband who was a farmer. The farmer was worried because the rains were delayed that year and without rains the sowing operations could not take place. The villager assured his son-in-law that the merciful god would send the rains soon and went away to see his second daughter and her potter husband. To his surprise, he found that the potter was in a joyous mood because it was still dry. He said that he could do with the dry weather for some more weeks because it gave him time to bake more pots. The delayed rains spelt prosperity to him. We get to witness two opposite views regarding the arrival of monsoons in modern times as well. A city dweller may find the rainy season very inconvenient while the farmers in rural areas would pray for it.

However bothersome the rains might be for a few people, nobody denies the utter necessity of water and hence of rains in our life. The anxiety regarding timely and adequate rainfall becomes more pronounced in a country where its unreliability is a common occurrence. Quite often, in India, the course of the monsoon causes either severe drought or untimely and excessive rains. In both circumstances the crops and livestock are at stake and people suffer. This situation is not of recent origin but has always existed. It is intended here to bring out the inevitability of the monsoon's influence on the ideology and material culture of the human race.

Furthermore, it is not only the actual occurrence of a bad monsoon or its fear that is reflected in our ideology and culture. When the monsoon is good and on time, it is time to celebrate. The eternal and recurrent drama of vegetative regeneration, witnessed in every rainy season is a marvel to be rejoiced over, and an experience of emotional exhilaration.

In a pastoral-agricultural society either the desperation or exhilaration caused by bad or good rainfall prompts people to devise various rituals. These rituals, their symbolism, and accordingly the mythological links reflect either gratitude or anxiety. It will become clear in the course of this paper that the viewer's perspective of symbols used in such rituals is not limited to a defined range of meanings. It includes a wider array of ideational subtleties, which is continuously subjected to addition and omission depending on the spatial and temporal context.

Nevertheless, some of these subtleties, and the motifs derived therefrom seem to remain embedded at the core of any given symbol. They maintain a kind of perpetuity despite any change of context. They seem to be intuitively understood and instinctively used by people at all times and at all places. The 'rain-generative fluids-bull' theme and its symbolism seems to have gained widespread perpetuity amongst pastoral-agricultural communities. It has formed an integral part of mythologies of many such communities from protohistoric to modern times.

It is not possible within the purview of this paper, and also not intended, to present an exhaustive review of literary occurrences of rain symbolism. This paper mainly cites from modern works that have studied Indian symbolism with the help of Vedic, post-Vedic, and Puranic literature. Its amplitude and variety of applications generally corresponds with mythologies of pastoral agricultural cultures from other parts of the world. However, by no means is it presumed that only Vedic symbolism encompasses the entire gamut of symbolic expressions of the ancient people of India and other civilizations. The treatment of these symbolic expressions in this paper may appear to be 'Frazerian', (Ingold, 1986: 244) meaning that the material may seem to have been drawn from divergent sources. However, considering the perpetuity and universality of certain archetypes reflected in symbolic expressions it may be considered as of natural course. Besides, the limitations of this presentation are due to the express purpose of looking only at some of the frequently used citations of such symbolism, in order to understand their manifestations within material assemblages.

The 'Sky Father' and the 'Mother Earth' occur frequently in mythologies of the ancient world. The working premise for this paper is that the faith system of various ancient peoples looked upon the primeval pair of the 'Sky' and the 'Earth' as the progenitors of all beings. These belief systems indeed manifest in the material assemblage of the respective cultures. However, it must be stated that this is not a systematic study of symbolism of individual cultures. The examples drawn here serve only to illustrate the universality of certain motifs, with full awareness of the time-related subtleties, as manifested by a variation in detail.

The earth is perceived verily as the primeval womb, a receptacle, open to receive the seed from the sky; and the seed is poured into her in the form of rains. In keeping with this logic it can be observed that the 'Bull of Heavens' is the prime symbol associated with 'rain-belief' among human communities—especially pastoral-agricultural communities—scattered through space and time. At the core of this symbol is the presumption of the unrivalled virility of the animal, and the anticipation of his role as a fecundating agent.

In order to understand 'rain symbolism' in the Indian context, various gods need to be identified with the 'Bull' symbol based on their role as the primeval male. The sacrificial rites prescribed in order to promote vegetative fertility are discussed with the help of a few examples. They are organized under following titles:

The Earth Mother and the 'Cow';
'Waters-Rains' in mythological and theogonic symbolism;
'Rains' and theology;
Occurrence of the 'Bull Motif' in mythology;
Theogonic myths and symbols from ancient civilizations outside India;
Rain/Fertility rituals and bull sacrifice.

The Earth Mother and the 'Cow'

The authors of ancient civilizations as pastoralists and agriculturists were naturally preoccupied with the fertility of their land and herds. In the land where the first monsoon showers spell relief from the scorching heat as well as from the drudgery of fetching water from long distances, it is not surprising that the season also heralds a period of leisure and merriment. It is borne in mind that this would be only possible if the land yields well. The land needs to be cared for, just like an expectant mother. In addition, the land does not bring forth only vegetation but also the animate and inanimate world.

Ultimately, everything under the sun is made of her substance. With immediate reference, the 'Field', and in a broader sense the 'Earth', is the Mother of all animate and inanimate beings. As the primeval mother she has nothing to hide, nothing to be ashamed of. She is ever lying bare beneath the sky, awaiting its fructifying showers, conceived as *retas*/semen of the sky bull. She is known as *Uttānā Mahi* (Dhere, 1978: 56, 194)— the earth lying flat with her face toward the sky.

It may be mentioned here that in the sphere of symbolism human minds often keep striking similar paths. The concept of Earth Mother lying horizontal either to receive the seed or to give birth is recurrent. 'In the Navajo language the earth is called Naestsan, literally, the "horizontal" or "recumbent" woman (Eliade, 1963: 157).'

It is significant in this instance that Indrāṇi, Indra's consort, is identified with the earth or the arable land (Dange, 1970: 63, 1971: 105 fn. 65) and, Indra is called *urvarāpati*. (Ṛgveda 8.21.3) The dictionary meaning of *urvarā* is 'land in general; fertile soil that brings forth every kind of crop'. However, Dange points out that it is 'the land under the plough' and since, 'Indrāṇi is...identified with the field...it is only natural that Indra be the husband of this 'ploughed field'. (Dange, 1971: 104, 105 fn.) A verse in Atharva Veda (10.6,33) helps to establish that *urvarā* is the land made fertile by ploughing (Chitrav, 1972). In the light of *urvarā* as the ploughed field, other dictionary meanings of the term seem to be of immediate significance. The term is denoted as, 'mixed mass of fibre/wool or a humorous term for curled hair' (Apte, 1957). These are apparently instances of extended meanings of the term based on the knotty appearance of the ploughed field. Hence, if Indrāṇi is identified with *urvarā,* then both can be said to be representing the ploughed field lying barren and bare under the sky before the onset of rains. It is for Indra, the bull, to shower it with the life-generating rains. Once the land is moist enough she is not to be ploughed (Dhere, 1978: 92–102).

The earth is revered as *Mahānagni,* the great uncovered field lying open to sky awaiting to receive the seed (Dange, 1979: 216–222); *Aditi Uttānpāda,* the 'Infinity' is usually interpreted as the vast expanse of the earth. Kramrisch identifies it in the role of a parturient mother (1956). In the same vein, her consort is called *Mahānagna.* Dhere citing from V.S. Agrawala, suggests that the *Mahānagna* can be identified with Śiva in the form of *Nandi,* the bull. This association naturally anticipates the earth to be conceived as the 'cow'. Vedic Aditi is described as the cow. The commentators used the term *go-gau* to mean *Pṛthvi* and the composers of Purāṇa texts perfected the concept of the earth-cow that could be milked of desired wealth (Dhere, 1978: 169–170).

Hindus venerate the cow as a mother, a phenomenon that generally baffles or amuses outsiders. The motif of a nourishing mother cow is epitomized as the wish-fulfilling cow Surabhi, but she is not looked upon as the consort of the Bull of Heavens. Rather she stays in *rāsatala,* the netherworld. However her daughter Sarvakāmdughā Nandini is Indra's wishing-cow (Fausboll, 1981: 92–93). O' Flaherty asserts that she is the good mother, the white goddess comparable to the concept of the magic wishing-cow found in Norse, Iranian, and Irish traditions. She cites from the first chapter of the Mahabharata that, 'the magic wishing-cow, is the earth milked of good and evil substances by gods and demons. She is churned out of the ocean of milk; from which all else is churned forth, in turn flows from the udder of the wishing-cow' (O' Flaherty, 1981: 241).

'Aditi, personified as a female who gives birth by crouching with legs spread, is the cosmic origin of space itself and of earth, as well as being mother of the gods of the sky' (O' Flaherty, 1981: 79). Dandekar, (1979), in the context of the Harappan Mother Goddess, has noted Przyluski's opinion that 'the Vedic concept of Aditi is the result of the Aryan borrowing from some non-Aryan populations who adored a Great Mother'. He has also quoted Johansson that 'the feminine counterpart of Varuṇa is Mother Earth—Aditi or Pṛthvi...Probably Aditi was primarily the wife of Varuṇa...' (1979: 39, 23 f.n.).

Varuṇa is one of the main gods to be associated with the concept of fertility as he is the controller of 'water'. He is 'moisture' personified. The concept of Aditi or Pṛthvi..., the earth in way of her vast expanse as Varuṇa's wife can be translated as the moistened earth after the rains. The moistened earth is looked upon as menstruating, now mature to hold and nourish the seed in her womb (Dhere, 1978: 90–104). It is a sin to plough the land when she is menstruating. At other times also it was a sin to dig the earth because it was as good as inflicting wounds on the Mother Earth's body. The belief was widespread in the ancient world. Eliade quotes one American-Indian prophet saying, 'Am I to take a knife and plunge it into the breast of my mother... You tell me to dig up and take away the stones? Must I mutilate her flesh so as to get her bones?' (1975: 155).

'Waters-Rains' in the Mythological and Theogonic Symbolism

'In order to arrive at a proper understanding of the fact that in India we meet with a distinct deity for the rainfall, we must first consider what is recorded not only from former times but up to the present day, both of the rain's power and violence and its utility and blessing for the soil, when the land has been parched by the burning heat of the sun for three whole months... Parjanya denotes originally only rain-cloud... But later the rain-cloud was personified and Parjanya used in the sense of the Rain God... But the Parjanya is originally identical with Indra' (Fausboll, 1981: 93–98).

In the mythological context the rains/waters and associated elements like fire (Sun in heaven and Agni on earth), winds and earth, are very clearly woven around the concept of 'fertility'. *Prima facie* these elements do appear as contradicting forces but in the mythological universe they are complementary to each other, and also quite often they appear as variants of one and the same natural element. For instance, the 'three bulls, that sprinkle their *retas* over the pastures and the worlds' are mentioned as 'being recognized by the tradition as the Sun in heaven, Vāyū in the middle region and the earthly fire on the terrestrial plane' (Dange, 1970: 18).

The flexibility of symbolic implications and applications is of course extended to individual gods as well; who are generally personifications of natural elements. 'The identification of the bull in the pasture,' according to Dange, 'is complete when some of the gods are said to have horns the context being the gain of rain. Thus, the Maruts are said to wear the horns and are connected with the welfare of the cows and the release of rain...and further...Indra is himself the bull and is also compared with the sharp-horned bull, which is said of Agni and also Soma...Parjanya is the four-horned bull...and Soma sharpens his horns. Agni flames are horns and he is the bull...Even the sacrificial horse is said to be *hiraṇya-sṛṅga*...The culmination of the bull-imagery is in the belief that the highest principle in the Vedic concept, i.e., ṛta has horns... Closely associated with the bull and the cow in the pasture, we have the symbolic cow Sabardughā and the Pṛṣṇī, in her triple aspect. As her counterpart we have the male Pṛṣṇī who is the Aditya or the Sun... who "seeds" the earth and also the male sabardughā which comes as the unique epithet of Soma...' (1970: 22).

The quest to understand the creative and operative processes of the cosmic existence has always provided impetus to human imagination; a vital force behind myth-making. Vedic ideology believed in these processes as inherent components of a cosmic ongoing sacrifice. The Ṛgveda (4.58.3) describes the cosmic sacrifice as a bellowing bull with four horns, three feet, two heads, and seven hands. Buddha Prakash (1966: 28) has identified this bull with Tvaṣṭṛ, but Dange points out that, 'nowhere in the Ṛgveda is Tvaṣṭṛ, associated with the word Vṛṣabha or Vṛṣan, which is common with certain other gods...The real nature and implication of this bull...has been a point of much guesswork, suffice it to say here that he is the seeding bull, whose retas is rain. The idea of rain being the retas is clear in the expression apām retānsi, which the sacrifice to Agni is said to impel' (1970: 18).

Water has always been thought to symbolize potent procreative fluid, retas. It is there in the form of primordial waters. Rainwater is considered as the 'heavenly seed of the clouds' (Ṛgveda 9.74.1; 1.100.3), that fecundates the earth. When stable, it nourishes the germinated seed. Dange (1970: 12) explains the word Vāstū occurring in the Ṛgveda as meaning pastureland and the god of pasture, Vāstoṣpati is supposed to be Rudra. He finds the episode of Vāstoṣpati's birth as 'closely associated with rain'.

Indian mythology is replete with allegories explaining the associative contexts that allot a unique place to each element in the cosmic theme. These allegoric descriptions also indicate the course of the formative process of the Vedic and Puranic pantheon.

'Rains' and Theology

Indian theology considered many divinities as rain-giving. It includes Parjanya, Indra, Sūrya, Maruts, Dyaus Pitar, Varuṇa, and so on. These gods representing natural elements were also perceived as virile beings, who poured their vital seed/semen on the earth. Their seed/semen was equated with rain showers. In that capacity they were described as the 'Bull—Vṛṣa', the one who showers his potent seed. Bull, the powerful animal was the closest and the strongest motif

apparent to the mind of the pastoralist-agriculturist. This motif was so ingrained in the collective Indian psyche that an outstanding man is still honoured by the epithet 'Narapuṅgava', meaning the bull among men.

At this juncture, an interesting observation by a friend may be stated here. In Maharashtra, members of a community known as Naridiwale, still move from place to place with their trained bull that is supposed to represent the Naridi—Śiva's vehicle. It is still customary to ask the bull whether the rains will be good this year. This, in a way, is the testimony to popular memory that connects rains with the bull so closely.

There prevails a general consensus that Śiva is not a Vedic deity. In all probability it belongs to a pre-Vedic stratum in the formation of the Hindu pantheon. 'Rudra, the Vedic equivalent of Śiva, begins to take his place clearly as the deity of transcendent darkness...only in the Upaniṣads... The complex Saivite cosmology differs in its form and expression from the Vedic, although it has had an obvious influence on some of the later Vedic texts and Upaniṣads. Most of the terms of the two systems can and have been equated with one another in later Hinduism, so that Hindus today can hardly believe that there may have been originally two distinct systems.... Śiva is everything. According to the aspect of his divinity envisaged he.... (is) represented by his various names....' (Danielou, 1985: 188ff).

As Rudra, Śiva also wields the thunderbolt like Indra (Ṛgveda 2.33.3), and indeed at places he is identified with Indra as the Lord of Heaven. One of the explanations of his name describes him as 'the howler', who gives rain and lightning to men. He is also identified with Agni (Danielou, 1985: 194, 195). Śiva is just one of his epithets to prevail over his Vedic name Rudra. In his aspect as Śarva, he is the 'support of all beings, animate or inanimate, the deity whose substance is earth...' (Danielou, 1985: 204). As Paśupati he is equated with Prajāpati, the Lord of Progeny (Danielou, 1964: 209). As Sadyojāta, the

If the horse was a symbol of arid Asia, the cow is the symbol of the wet Asian monsoon region. In India, the cow is regarded as a sacred animal. Two zebus in Rajasthan, India. (Photo by Takeshi Takeda)

suddenly born, he is likened to Soma (Danielou, 1985: 212).

The reason to summon the similarities of his attributes in his various aspects with other deities like Indra, Agni, Prajāpati, and Soma, is just to draw our attention to the fact that at some or the other time all of them are envisioned as the 'bull'. Śiva himself rides the bull, Nandi, who represents the procreative energies. By the same logic, deities like Parjanya, Indra, Agni, etc., who also represent different aspects of the procreative energies have been regarded as bull *Nandi*. However, it should be noted here that Rudra's identification as rain-giving or storm-god is debated by some scholars (Dandekar, 1979: 215–221).

Parjanya, though originally meaning only a rain cloud, later attained an independent status as a god identical with Indra. '…The identification is two-fold. Parjanya or the sun is the divine bull, and the bull in the pasture is divinity…believed to be Indra…the pastoral bull' (Dange, 1970: 19). The Bṛhad Āraṇyaka Upaniṣad views the cosmic creation as a sacrifice and we are told that the gods made Parjanya, the rain-god, into a fire altar, enkindled the wind-fire into it and offered Soma-oblations. As a result rains were born. Further the earth was made the sacrificial altar and the rains were offered as oblations and food-grains were born (6.2.10, 11).

Indra, is held responsible either for sending or holding back the rains. He is *Vṛṣa-Karman,* the rain-maker and *Vṛṣamanah,* the sprinkler (Dange, 1970: 99). There are various speculations regarding the origin of Indra's name. One of it relates his name to the word *Indu,* meaning 'a drop' because he gives rains. As mentioned earlier one of Indra's epithets is *urvarā-pati* (Ṛgveda 8.21.3).

Generally, as the release of waters caused by Indra by killing the demon Vṛtra, is interpreted as the release of celestial or cosmic waters. However Dange argues, 'In a considerable number of instances the fight between Indra and Vṛtra gets closely associated with the release of waters, which cannot be flatly taken to indicate the cosmic waters…The fight between Vṛtra and Indra…, gets not only associated with rain, but with the flooding of the earthly rivers….' (1970: 14–15).

Dange (1970: 15) also draws our attention to the commonly occurring motif of 'fight-causing rains' in the mythology of various countries. On a mundane level we may associate this particular motif to the thunderstorms that precede the monsoon showers and on a symbolic level it is obviously indicative of the unbreakable link between valour and virility.

Indra also sent torrential rains to Vraj when he was angry because people there stopped worshipping him on Kṛṣṇa's advice (Bhāgavata Purāṇa 10.24–23). 'In the Vedas, Indra appears as the deity of the sphere of space, the dispenser of rain who dwells in the clouds. Feared as the ruler of the storm, the thrower of the thunderbolt, he is also the cause of fertility…Agni, Indra and Sūrya then represent the three forms of fire: the fire of the earthly world, the thunderbolt or fire of the sphere of space, and the sun, the fire of the sky' (Danielou, 1985: 106).

In this regard we may recall that the 'three bulls, that sprinkle their *retas* over the pastures and the world', are 'being recognized by the tradition as the Sun in heaven, Vāyū in the mid region and the earthly fire on the terrestrial plane' (Dange, 1970: 18).

Many scholars have discussed the concept of the three forms of fire. Dandekar, in the context of two other Ṛgvedic divinities, viz., Āpām Napāt and Trita Aptya, observes, 'It is, indeed, asserted that no cosmological fact is more frequently alluded to in the Ṛgveda than this three-fold division of fire…First Agni was born in the (sic) heaven; the second time from us; and the third time, he was born among the waters… Terrestrial and solar fire, as being permanent and appearing every day, were naturally regarded as the first and the second forms, while its fleeting and rarer manifestation was looked upon as the third...Agni is characterized as *Āpām garbha*...(thus) Āpām Napāt represents the third or the lightning form of Agnies. Macdonell, who strongly advocates this view, further says that the Vedic god Trita (sic) also is no other than the third or the lightning form of Agni...Lightning is the chief agent in the thunderstorm, and its manifestation precedes the release of heavenly waters..., it is thus a matter of course that Trita should be associated with Indra in the conflict with the drought fiend....' (1979: 300: 301).

Eliade, summing up the symbolic significance of Indra's various appellations says, 'Whether we are reading of his thunderbolts that strike Vṛtra and release the waters, of the storm that precedes the rain, or of the absorption of fabulous quantities of *Soma*, of his fertilization of the fields or of his gigantic sexual potencies, we are being continually presented with an epiphany of the forces of life' (1963: 139). The steadfast association of Indra with rains is expressed beautifully in the couplet composed by Muktesvara, a Marathi poet of the 16–17th century. He describes Lord Śiva, the magnificent, as manifest in nature (Dhere, 1969: 32). It reads:

Pṛthvīkhaṅd yā śalunkā
Anant parvat Śivapiṅdika
Indre māndile abhiṣeka
Gaganapātri jaladhārā

(*The Earth itself is the base in the form of yoni— Śaluṅkā, the limitless ranges of mountains are the Śivaliṅga the phallus, upon which Indra has begun to shower through the pitcher that is the sky.'*)

Another god who is, to date, associated with rains in popular memory, is Varuṇa, often invoked with Indra in the Ṛgveda. However, in the earlier literature he is linked, not with rains but various water bodies on the earth. *Vārunya* means stagnant water. Nevertheless, he is described as of dark blue colour like a rain-cloud (e.g.,

Jaladharshyamo Varuṇo–Mahābhārata, 3.42.25). As mentioned earlier, Aditi, in the role of Varuṇa's wife could be the moistened earth. It is very significant in this context that *Kardama*, mud, is the father of Varuṇa (Monier-Williams, 1956). *Kardama* also means *ardra mṛttika* (Visnusmrti, 23.41). Aditi, as mentioned earlier is the moistened earth and *ardra mṛttika*, also means the moistened soil. That, by implication makes Varuṇa, Aditi's son and the concept of 'son becoming consort' has been very common in ancient mythologies. *Kardama* is obviously a masculine term and its identification with Aditi as the moistened earth need not baffle us because ambiguity of gender in allegoric descriptions or the assumption of an androgynous nature of primeval progenitors is by no means an uncommon phenomenon. Varuṇa's role as a god of fertility is further confirmed as he is often alluded to as a *Deva-gandharva* and *Nāga* (serpent) or a king of *Nāgas* (Monnier-Williams, 1956), for 'the Varuṇa-pāśas' are identified with snakes (Dange, 1969: 59 fn. 145).

Varuṇa is the one who maintains *ṛta*, the cosmic order. Significantly, regularity or irregularity of rains and subsequently that of the cycle of regeneration on earth are inevitable sequels of the time-related status of the cosmic order. If the cosmic order is disturbed, that is if *anṛta* prevails because of overwhelming sins committed by people, then as the popular belief goes, one of the immediate consequences to become apparent is the setting of a drought as punishment.

Varuṇa, in later Vedic literature, along with Agni, Yama, Viṣṇu, is often paired with Mitra, the Sun God. Ahur and Mithras are the corresponding deities in the Avesta. Either the association or the equation of Indra, and Sūrya, the Sun God notably accentuates the symbolism of 'rains-semen-fertility'. The association of Varuṇa and Sūrya also projects similar symbolism.

Sūrya, the sun, in his turn, is equated with Yajña-Prajāpati and in that capacity both

Prajāpati and the sun represent the progenitor. Dyaus Pitar, the 'Sky' also is the primeval father. 'Dyaus covers the Earth and fertilizes her with his seed, that is with rain' (Danielou, 1985: 92). Along with the earth he completes the dyad that represents the primeval parents.

Dyaus Pitar corresponds with the Greek supreme god Zeus Pater. Zeus Pater shares with Indra and Varuṇa, the attribute of a virile, fierce sovereign who punishes his subjects with either floods or droughts. Kramer (1961: 250) refers to a story 'told in Hesiod's *Catalogue of Women*, that Zeus sent the great flood to destroy mankind and free the gods of sorrow.' On the other hand, the story of Demeter and her daughter (Kore) Persephone who was kidnapped to the netherworld portrays Zeus as the benefactor who helps Demeter in bringing her daughter to Earth for six months. This story credits Zeus with the act of bringing a prolonged drought to an end.

It is also significant that Greek Zeus is the equivalent of Roman Jupiter and 'Strabo describes the Indians as worshipping Jupiter Pluvius, probably meaning Indra.' (Danielou, 1964: 111).

Occurrence of the 'Bull' Motif in Mythology

Rain and bull are parallel motifs because of their capacity of pouring (semen) (Dange, 1970). The Sanskrit terms *Vṛṣṭi* and *Vṛṣan*, meaning rain and bull respectively, are derived from the root *vrs* that means to rain or pour. It is significant that the Sanskrit term *megha*, meaning rain-cloud is derived from the root *mih*, which means to emit, pour down. So a *megha* is the one that pours down. Although it is not very often that a *megha* is compared to a bull, in one instance it is quite significant where an indirect allusion can be found. It is a couplet in the praise of the Āśvins, who are the gods of agriculture:

Vṛṣa vām megho vṛṣṇā pipayā
Gor na seke manuṣo dasayān

Dange contends, 'The word *megha* in the present context is understood as the cloud; but even so the name is due, not only to the simple act of sprinkling, but has to be understood in the image of the bull that "sprays" into the cow—actual or the earth—which is so clear in the verse (*Gor na seke*). The action becomes beneficial to men, as it results in the rain' (1970: 23).

The other term for rain-cloud '*abhra*', on the other hand means, the one that covers waters, implying that it does not let the water fall. It holds the water. The rain-cloud thus has both the attributes of Indra, inasmuch as it pours and also holds rains. Significantly, Vṛtra, the demon that was killed by Indra, also means 'covering' (from the root 'vṛ') and the weapon with which he was killed is *vajra*, the thunderbolt. The gods who sometimes helped Indra in his battle are Rudra and Maruts, the storm gods. It may be mentioned here that, in the Sumerian myth of Gilgamesh, the Bull of Heaven represents a seven-year drought and famine that god Anu sent at Goddess Istar's behest. The bull is killed by Gilgamesh and Enkidu, who himself is a bull-man (1974: 12). Thus the bull-motif here appears in the capacity of the one who holds waters and also the one who releases it.

Indra is called the Bull for his valour and virility (Bhāgavata Purāṇa 9.6). One of his three sons is also named Ṛṣabha (Vṛṣabha), the Bull (Bhāgavata Purāṇa 6.18.7). Besides, Soma, Indra's favourite drink is also likened to a bull (Ṛgveda 9.19.5; 9.108.8; Atharva Veda 2.5.7).

Thus, it is not only Indra to be described as the bull. The sun is referred to as Yajna-Prajāpati because he embodies the cosmic sacrifice by the virtue of his self-consuming fire. The sun as Yajna-Prajāpati is the progenitor of the animate and inanimate world and so he is also visualized as the bull (Motilal Sarma Gaud cited in Danielou, 1985: 93). Referring to various versions of the legend of Prajāpati committing incest with his daughter, Dange examines 'the expression, which compares the "father" with the bull who unites with his *prajā* (offspring).' He further contends that it 'clearly indicates that the bull is the *pati* of his own *prajā*. Thus he is the *Prajāpati*...It is to be noted that we have a male (said to be the god Prajāpati... in the later traditions) and he is compared with the bull...that approaches his progeny' (1971: 182–83).

It is obvious that the bull motif signifies the male principle, the prime player in the drama of creation, the provider of the vital seed. In that capacity, the identities of gods keep mixing and merging into each other as we have seen so far. There may appear even ambiguity of gender as noticed earlier in the instance of *Kardama,* the masculine, which is also *ārdra mṛttikā,* the feminine. The God of Rain, Parjanya is visualized both as the bull and the cow. 'A more explicit androgyny occurs in the figure of the rain god Parjanya, some of whose complexities may be related to the androgynous figure of Heaven/Earth....' (O'Flaherty, 1981: 25).

When it comes to the formative process of symbols representing either celestial or cosmogonic phenomena, it may be observed that the ancients often chose the medium of theriomorphs. Theriomorphs as symbols are both intelligible and unintelligible at any given time. They are intelligible on the wider horizon of tacit, intuitive perception and so one can relate to it collectively and individually. However, they may be left unintelligible, because of their indeterminate cognitive matrix. Nevertheless, the awe created because of this very unintelligibility endows symbols with divine disposition instituting the belief of actually participating in the cosmogenic phenomena among individuals. Hence the complexities, ambiguities with regard to gender and species, represented by such symbols ought not to be viewed as impermeable barriers.

Thus, multiplicity of symbols, signifying a single phenomenon and also a single symbol signifying multiple and diverse, yet linked phenomena could easily be of common occurrence. In other words, making use of diverse symbols on the basis of some common attributes may convey a single concept. Likewise, a single symbol may be used to convey diverse concepts, again on the basis of some common factors. It so happens, especially in the instance of theriomorphs because rather than the form itself the idea communicated through it is always overriding.

As long as the theriomorph is a socially and culturally potent key to the epitomized meaning it continues to be used. 'Theriomorphic symbols may lose their valence when the animate themselves lose charisma' (O'Flaherty, 1981: 260). For example, the bull as the most powerful animal remains the prime motif in Indian mythology. For some time it seems to have become concurrent with the horse in Vedic mythology, but did not continue in later periods as can be seen from the ubiquity of the bull and the sparseness of horse figurines in archaeological assemblages of the Protohistoric period.

Both animals symbolize male power but as far as the fertility of the earth is concerned the horse carries minimal significance. Besides, horses left without care in the Indian climate hardly develop into the powerful animals that could evoke a feeling of veneration. The strength of the bull motif is of course natural in the pastoral-agricultural context because of its indispensability for fecundating the cows and tilling of the land. Both acts are interpreted on the same symbolic platform in mythological perspective.

Theogonic Myths and Symbols from Ancient Civilizations Outside India

The complexity of theriomorphs is well reflected in a depiction of Nut, the ancient Egyptain heavenly cow. Kramer observes, 'Four different Egyptian concepts of the sky are attested here: a cow, an ocean, the woman Nut, and a roof. All of these concepts were accepted as correct...There is no question that at the very beginning of their

In the Asian monsoon region, a water buffalo may be offered as a sacrifice in rituals or bullfights may be held as an augury. Water buffaloes frolicking in water in Guizhou, China. (Photo by Takeshi Takeda)

history, about 3000 BC, the Egyptians were aware that the concept of the sky could not be understood directly by means of reason and sensual experience. They were conscious of the fact that they were employing symbols to make it understandable in human terms. As no symbol can possibly encompass the whole essence of what it stands for, an increase in the number of symbols might well have appeared enlightening rather than confusing' (Kramer, 1961: 20–22).

Another deity in the Egyptian pantheon also appears as the cow, the Mother Goddess, known as Hathor with whom Nut was fused later. Everyday she gave birth to a calf who grew into a bull, 'the Bull of Heaven' that symbolizes the sun. This bull was also identified with the king. 'In addition to the concept of the fertile bull, the idea that the king was the strong bull, the victor over his enemies, is attested as early as the very beginning of writing in Egypt' (Kramer, 1961: 31).

Sumerian mythology is also not devoid of this bull-cow imagery that relates to waters. 'One of the more detailed and revealing of the Sumerian myths concerns the organization of the universe by Enki, the Sumerian water-god who was also the God of Wisdom. The myth begins with a hymn of praise addressed to Enki which exalts Enki as the god who watches over the universe and is responsible for the fertility of field and farm, of flock and herd' (Kramer, 1961: 98).

Enki is the god who proceeds from one place to another, from rivers to the ocean, blessing the waters and the fields, flocks and herds with fertility and prosperity and setting shrines of various deities wherever he sets his word of wisdom. Among the lands visited by him are Magan, Dilmun and Meluhha[1]. 'Enki now turns from the fate and destiny of the various lands which made up the Sumerian inhabited world, and performs a whole series of acts vital to the earth's fertility and productiveness...He begins by filling the Tigris with fresh, sparkling, life-giving water—in the concrete metaphorical imagery...Enki is a rampant bull who mates with the river imagined as a wild cow...Finally Enki "called" the life-giving rain, made it come down on earth, and put the storm-god Ishkur in charge' (Kramer, 1961: 99–100).

The Anatolian myths narrate a story about a vanished god, which Kramer feels, is the storm god, who is the provider of the rain. He has to fight with the dragon of drought (1961: 173–74). There is no direct reference to the storm god as a bull, but in another narrative there is a mention of the storm god by the name Teshub, and one of his sacred bulls (Kramer, 1961: 158–59). At this point we may recall the Indra-Vṛtra battle and also the earlier mention of the commonly occurring motif of 'fight-causing rains' that can, as a visible phenomenon, be associated with thunderstorms preceding the monsoon showers.

In this regard the Canaanite tale of Baal, the Lord of Earth, the God of Fertility and Life, is quite illustrative. '...Baal grants both rain and dew, he functions as a water-giving god during all twelve months of the year' (Kramer 961: 184). He is the 'Rider of Clouds' (Kramer, 1961: 193, 199, 200, 210). The rain-girl Tallai is one of Baal's three daughters/consorts (Kramer, 1961: 196). The other two are Pidrai, the Girl of Light and Arsai, the Earth. Although, this narrative centres on Baal, El the head of the pantheon, is the main actor, the decision maker, in this tale. He is the bull (Kramer, 1961: 192, 204), the father of Anath[2] as well as of the seven gods who are entrusted with the responsibility of establishing seven-year cycles of abundance. The myth is the precedent to be invoked for re-establishing in time the primeval event, (Kramer, 1961: 185–90).

Baal becomes the king of gods by dethroning Yamm, the sea-god. But El accepts to surrender Baal to Yamm and then the narrative follows in sequence wherein Baal slays Yamm, eventually to perish himself at the hands of Mot[3]. It is said that

Mot entered Baal's palace through a window. Kramer points out that, 'All this is connected with the functioning of Baal as the storm-god, because a rain and thunderstorm ensue. Perhaps it is somehow connected with the "windows" of heaven mentioned in Genesis... as the source of rain.' It is significant that before falling dead on the earth Baal sired a tauromorphic son on a heifer (Lüders, 1961: 209).

In the entire episode the 'fight-causing rain' motif, though not in a high relief, is certainly to be perceived as an underlying but necessary principle. As Kramer has explained, 'Canaan is characterized by a succession of seasons that normally produce a fertile year. With some luck a number of such fertile years follow one after the other to form a fertile cycle. But unfortunately, rain does not always materialize in the rainy season; nor is there always sufficient dew in the summer. Moreover, locusts may plague the land and devour the crops. A series of bad years is the major natural catastrophe against which the fertility cult was directed. The meteorological history of Canaan, where Baal was pitted against Mot in the minds of the people, required the concept that the conflict between the two gods took place repeatedly. In the frame of reference of Canaanite religious psychology, each of the two gods was both vanquished and triumphant many a time in the course of any century' (1961: 195).

Rain/Fertility Rituals and Bull Sacrifice

Animal sacrifice and even human sacrifice has not been uncommon phenomena in human communities. The occasions could be many and the reasons several. The commonly known among them are thanksgiving or pacificatory rites, celebrations of a good harvest or a victory, commemoration of ancestors, and so on. However, anthropologists have recorded ample examples of sacrifices among various agrarian communities which are designed to re-enact the drama of the yearly regeneration of vegetative world (Briffault, 1927; Frazer, 1936).

One explanation, offered for animal and even human sacrifices based on the interpretation of mythologies of various communities is that '...the history of religions knows gods who disappear from the surface of the Earth, but disappear because they were put to death by men...The violent death of these divinities is *creative*. Something of great importance for human life appears as the result of their death. Nor is this all: the new thing thus shares in the substance of the slain divinity and hence in some sort continues his existence. Murdered *in illo tempore*, the divinity survives in the rites by which the murder is periodically re-enacted;...' (Eliade, 1963: 99). As Eliade avers further these violent deaths are creative because they are related to vegetation (1963: 110).

In many of these myths the 'concept of the corn-spirit leads to the sacrifice of animals whose sap is believed to produce good crops or promote rain... In ancient Egypt Osiris himself was believed to be the corn spirit...and a bull was sacrificed as representing him...and after his death he was believed to be born in the form of corn, being watered by a cow-headed deity—his mother...In one of the Mithraic sculptures of Assyria, Mithra is shown as killing a bull from whose wound, not blood but corn-stalks are shown to be sprouting' (Dange, 1970: 44).

The Indian mythology may or may not carry exactly comparable strains of the corn spirit symbolism associated with the violent death of a god that was represented either as an animal or a human victim. However, the focal concern of any sacrifice is no doubt the arrival of rains and the consequent fertility of the soil.

Dange asserts that the animal victims in the horse-sacrifice and the Śulgava—bull sacrifice— were indeed identified with gods like Rudra, Indra and Soma. Thus the Vedic sacrifice did concur with the symbolism of the death of gods representing the corn spirit and the ritual of animal sacrifice enacting the drama of a god's death. He reminds, that 'In the animal sacrifices at the Soma sacrifice, the animals were a link between the plants and the gods...We have to remember that the pressing of Soma was mourned even as he was getting ritually killed...There is, hence, no doubt that the Vedic people believed in the sacrifice as imbibing the killing of the Soma and the animals as the god. The bull had already become a favourite with the Vedic people as a pastoral god whose sap was believed to nourish the cattle and vegetation...He is Puṣan. His association with Soma identifies his sap with the juice of Soma. He is the very foetus of waters, or we may render it as—the bull is the source of water' (1970: 44–45).

Much of the killing outside the Vedic sacrificial system is in order to avert the anger of minor deities and spirits as well as to ward off calamities such as droughts. Till today offering of animal blood seems to have remained a major part of the folk religion pertaining to the pastoral-agricultural way of life. The sacrificial symbolism developed by Vedic and later Vedic texts, no doubt reflects the essence of the archetypes that build up folk belief systems. The archetypes, of course, cannot be exclusively claimed by any single belief system.

The threads of the Protohistoric folk belief system are not yet completely lost to us as they still run through festivals like *Polā*, celebrated in parts of rural Maharashtra. The word Pola is derived from the term *pol* meaning a stud-bull, though today it does not refer to worship of the stud-bull alone. The tale that is told on the evening of this festival indicates the killing (ritual sacrifice?) of a brother of the wife by a farmer. The victim was revived by his sister, the farmer's wife, by throwing cucumber seeds in all directions and calling him back. The tale, according to Dange 'is clearly a remnant of an agricultural ritual-killing of the bull who is, symbolically, said to be the brother of the woman, who probably denotes the field' (1970: 52).

As mentioned earlier, the horse lost its significance very quickly in Indian pastoral-agricultural symbolism but the bull never did so. The bull symbol keeps appearing in various media constantly through the passage of time and there is no dearth of it in archaeological assemblages. The bull motif among pottery designs and bull figurines are a common occurrence in the deposits of Protohistoric and historic sites. The decorative role and the stylistic development of the bull motif on pottery are of course well acknowledged in archaeological reports and studies. However, the role of the bull in the faith system has not been paid adequate attention.

Occurrence of either clay or terracotta figurines is well associated with the Neolithic-Chalcolithic cultures. Among them female and bull figurines are predominant; a fact that underlines their close association with the pastoral-agricultural way of life. The religious connotation of these figurines is generally accepted but sometimes there lies a faint streak of hesitation in doing so. It is generally thought that, 'the function of these bull figurines cannot be determined with certainty...These bull figurines were probably votive offerings as this practice is very common among tribal societies all over the country...' (Misra *et al.*, 1993: 152). The doubt especially is raised while analysing the stylistic differences, especially so, if the figurines are well-made and carefully decorated. The presumption that crudely made figurines denote ritual purpose while carefully made, well-decorated specimens represent evolved popular art, generally passes without contention.

The ritual connotation, though often presumed for want of a better explanation is only occasionally asserted very firmly. Ethnographic parallels or inferences derived from modern cult practices are also often resorted to but without attempting to restate the ritual context of the figurines. Terms like 'bull cult' or 'animal worship' may also be often used to interpret them that may or may not lead us anywhere. Nonetheless, from the foregoing discussion we may assert that the clay or terracotta figurines recovered from occupational deposits of Neolithic-Chalcolithic and even from some historical sites are part of the fertility and rain-making rituals. More so, in all probability the animal and even human figurines were offered as token sacrifice.

In the context of token sacrifice one observation by Jayaswal and Krishna is very pertinent. They write, 'Our study of the animal figures from various archaeological contexts reveals that many of the cultures appear to have preferred forms of one animal type against the others. For example, the bull at all the Chalcolithic and Neolithic sites—Kulli and Jhob group of sites of Pre-Harappan Baluchistan, Harappan cities, Kayatha representing the Central Indian Chalcolithic, and Piklihal and Tekkalkotta the Neolithic culture stages, and the ram at Bhagawanpura in the overlap period of Late Harappan and the Painted Grey Ware culture, seem to have a definite preference over the others. Does it imply that the Neolithic-Chalcolithic communities attached more preference to the bull figures, and the Bhagawanpurians to those of the ram than the figures of other animals?' (Jayaswal and Krishna, 1986: 131).

Bhagawanpura is viewed as an important site (District Kurukshetra, Haryana) where an overlap (Sub-period IB, dated to c. 1400–1000 BC) between the Late Harappan and Painted Grey Ware culture is very clearly observed. It is stated that the 'ram was a highly popular animal during this period. Many figures of rams are available.' On the other hand, it is also confirmed that bull figurines were rare (Joshi, 1993: 29).

The close association of the ram with the bull as an emblem of male virility is well known. In that case it may be inferred that this is an incidence of one animal taking precedence over another but not necessarily that of any change in the ideology basic to the formation of symbols.

This may also be attributed to regionality owing to the sources of two diverse cultures that had come together in this instance. Be that as it may, it is a known fact that the 'ram' has been as closely associated a symbol with Mother Goddess worship as is the bull.

So far, it has been attempted to substantiate the proposition that the role of the bull has been perceived by agricultural-pastoral people as a rain-making god/animal and subsequently, as a fecundating agent of the Earth Mother. In that capacity it was revered and sacrificed. Thus we may bear in mind that in all probability the bull-figurines do not mark just votive offerings but token sacrifice as well. If the sacrifice, either real or token, was offered by people desiring a good monsoon and good harvest, etc., then it may turn out that a bad season, and the ensuing desperation would prompt people to offer sacrifices in larger numbers than is usually customary. In that instance, a quantitative analysis of bull figurines occurring at Chalcolithic-Neolithic sites with respect to their frequency from various levels, supported by other environmental data, may prove to be an effective tool to understand the oscillations of the monsoon in the ancient period. This possibility is expressed here with a full understanding of the need for developing a scientific paradigm followed by in-depth research.

Acknowledgements

I wish to express my gratitude for the help rendered by friends like Dr. Pradnya Kulkarni and Dr. Prasad Joshi in checking original Sanskrit references.

Note

1 Magan is now identified with the coastal region of Makran; Dilmun with Bahrain; and Meluhha with the ancient region of Harappan Civilization.
2 The Virgin Goddess of Fertility.
3 God of Death.

Reference

Apte, V.S. (1957) Practical Sanskrit-English Dictionary, Prasad Prakashna, Pune.

Bhagavata Purana (1950): Niranaysagar Press, Mumbai.

Briffault R. (1927) *The Mother.* George Allen & Unwin Ltd., London.

Buddha P. (1966): *Rgveda and the Indus-Valley Civilization,* Vishvesvarananda Institute, Hoshiarpur.

Chitrav S. (1972): *Atharvavedache Marathi Bhashantara* (Marathi Translation of Atharva Veda), Amriteshvar Devasthan, Pune.

Dandekar, R.N. (1979): *Vedic Mythological Tracts,* Ajanta Publications, Delhi.

Dange, S.A. (1969): *Legends in The Mahabharata,* Motilal Banarasidas, Delhi.

Dange, S.A. (1970): *Pastoral Symbolism from the Rgveda* (The Bhau Vishnu Ashtekar Vedic Research Series), University of Poona, Pune.

Dange, S.A. (1971): *Vedic Concept of 'Field' and The Divine Fructification.* University of Bombay, Mumbai.

Dange, S.A. (1979): *Sexual Symbolism from the Vedic Ritual,* Ajanta Publications, Delhi.

Danielou, A. (1985): *The Gods of India,* Inner Traditions International Ltd., New York.

Dhere, R.C. (1969): *Vividha* (in Marathi), Nilakantha Prakashan, Pune.

Dhere, R.C. (1978): *Lajjagauri* (in Marathi), Shrividya Prakashan, Pune.

Eliade M. (1963): *Myth and Reality* (World Perspectives Series), George Allen & Unwin Ltd., London.

Eliade M. (1975): *Myths, Dreams, and Mysteries,* Harper & Row Publishers Incorporated, New York.

Fausboll, (1981): *Indian Mythology: According to Indian Epics,* Cosmo Publications, New Delhi.

Flaherty, O. and W. Doniger (1981): *Sexual Metaphors and Animal Symbols in Indian Mythology,* Motilal Banarasidass, Delhi.

Frazer, J. (1936): *Golden Bough,* Macmillan & Co., London.

Ingold, T. (1986): *The Appropriation of Nature,* Manchester University Press, U.K.

Ions, V. (1974): *The World's Mythology,* The Hamlyn Publishing Group Ltd. London.

Jayaswal, V. and K. Krishna (1986): *An Ethno-Archaeological View of Indian Terracottas,* Agam Kala Prakashan, New Delhi.

Joshi, J.P. (1993): *Excavations at Bhagawanpura 1975–76,* Archaeological Survey of India, New Delhi.

Kramer, S. N. (1961): Mythology of Ancient Greece, in *Mythologies of the Ancient World,* Anchor Books, New York, pp. 221–271.

Kramrisch, S. (1956): An image of Aditi Uttanpada, *Artibus Asiae* 19: 259–70.

Limaye, V.P. and R.P. Vadekar (eds.) (1958): Brhad Arnyaka Upanisad in *Astadasa-Upanisad.* Vaidik Samsodhana Mandal, Pune.

Misra, V.N., V.S. Shinde, R.K. Mohanty and L. Pandey (1993): Terracotta Bull Figurines from Marmi: A Chalcolitic Settlement in Chittorgarh District, Rajasthan. *Man and Environment,* Vol. XVIII (2), pp. 149–52.

Monnier-Williams (1956): *A Sanskrit-English Dictionary,* Clarendon Press, Oxford. O'Flaherty, Wendy Doniger (1981), *Sexual Metaphors and Animal Symbols in Indian Mythology,* Motilal Banarasidass, New Delhi.

Sonatakke, N.S. and C.G. Kashikar, (eds.) (1983): *Rgveda Samhita.* Vaidik Samsodhana Mandala, Pune.

Roth, R. and W. Whitney, (eds.) (1924): *Atharva Veda,* Berlin.

Chapter 19

Vassavasa—A Precursor to Complex Buddhism

MANJIRI BHALERAO

Monsoons are the main source of water for India. The monsoon cycle is characterized by the rainy season which spans the months of June to middle September, and especially heavy rains around the month of July. The rains coincide with the month of *Ashadha* according to the traditional Indian calendar. It is called *vassa-ritu* (Sanskrit *Varsha Ritu*). Although the rains are known as the life-giver, they are also known to create many obstacles in the normal functioning of the daily life of the people.

This was more so in the past especially in the case of the wandering ascetics. Many *sanyasi* (ascetic) communities existed in ancient India who believed that one had to renounce one's home and relatives in search for the Ultimate Truth. This act was technically called *Parivrajya* (i.e., going forth). It was practised in India for many centuries, even before the birth of the Buddha. The cannons of these wandering communities prescribed some regulations for the rainy season. The Buddhists called it the *vassa*, the Jainas the *Pajjusana* and the Brahmanical wanderers labelled it as the *Dhruvashila*, i.e., to have a fixed residence during the rainy season (Dutt, 1962:53).

The wanderers of sects other than the Buddhists could stay at any place they wished. They could even live alone if they liked; there was no obligation on them to live in an ascetic community. In contrast, Buddha had permitted the *Bhikkhus* to stay only in the company of fellow monks. The period of the *vassa* actually began with the full moon day of Ashadha or a month later and continued for the three following months, ending on the full-moon day of Kartika (*Mahavagga*, III.2.2). The keeping of the *vassa*, i.e., residing at a certain place during the rainy season, served two purposes. Firstly, the vegetation that grew on roads was not trodden upon by the feet of the wandering *Bhikkhus*, and secondly, the *Bhikkhus* would be saved from the dangers and troubles, which they encountered in their journey from one place to another (Bhagwat, 1939: 137). Initially, the *Bhikkhus* were not allowed to stay in homes. They were expected to wander around and not stay at one place for more than a night, except during the rainy season. They, therefore, used to take shelter in the woods, under the trees, on hillsides, in mountain caves, in cemeteries and open plains, etc. This proved to be very inconvenient in the rainy season and, eventually, a rich merchant from Rajagriha expressed his willingness to construct residences for the monks. For the first time, then, Buddha allowed the *Bhikkhus* to use five kinds of abodes i.e., *viharas*, Addhayogas, storied dwellings, attics and caves. It is believed that this merchant constructed sixty dwelling places in one day (Chullavagga, VI.1.3). Buddha asked him to donate those cells to the present and future *Sangha* of the four quarters. Buddha is believed to have said that he is a wise man who builds pleasant dwellings and lodges learned men there (Chullavagga, VI.2.3). This probably inspired many lay followers, who came forward on their own and offered many *viharas* for the use of the *Sangha* of the four quarters.

The permission to reside in these *viharas* brought many other liberties, which the Buddha himself had not expected when he initiated this lifestyle for a *Bhikkhu*, e.g., the begging for food. Once the *Bhikkhus* started staying at one place, donors came forward with offers of daily meals for the *Sangha*. Buddha himself had accepted such donations in his lifetime. The earlier ideal of a wandering ascetic, whose subsistence depended on the alms given by the laity gradually faded away.

The resting places in which the *vassa* was kept were fixed by natural boundaries like streams, lakes, hills, ridges, anthills, etc. These locations were neither too near nor too far removed from habitation, so that the *Bhikkhus* could easily get alms. *Mahavagga* mentions that the *vassa* should not be kept in a place where the majority of people were non-believers because then the *Bhikkhus* would not get enough support from the laity (Bhagwat, 1939: 137–138).

The early *viharas* were located in the countryside. Sometimes, the lay devotees donated their pleasure gardens, i.e., the *Aramas*. In these gardens residential cells were constructed for the monks. These structures were strictly temporary in nature, built only to last for the three rainy months (Chullavagga, VI.11.3). However, even this short spell of living together gave rise to a sense of communal life among the resident monks. Certain institutions, customs and practices were developed, which were of a congregational character e.g., the recital of the *patimokkha*, ceremonies like the *pavarana* (invitation) and *kathina* (distribution of robes). The earlier requisite of using robes made of rags was conveniently overlooked and a ceremony called *kathina* was promoted. In this, the *upasakas* or the lay devotees offered robes to the *Bhikkhus*, but in the name of the Sangha as the *Bhikkhus* were not allowed to keep any private and personal possessions (*Mahavagga*, VIII.5.2).

The pious lay devotees started giving long-term endowments for the subsistence of the resident monks of the monasteries. Even the kings patronized the monks by donating the revenues of some villages. All these developments brought about a change in the *Bhikkhus'* monastic lifestyle. The temporary residences were turned into more or less permanent abodes. The *Bhikkhus,* who kept *vassa* at a particular residence, made it a point to return to the same residence the next year. This made the monastic life of the *Bhikkhus* more peaceful, as the like-minded and those who had habitually lived together gradually settled down at one place. However, the ancient ideal of the wandering ascetic was never completely given up by the *Bhikkhus*. The purpose of constructing monasteries was always put forward as providing residence during the rainy season.

In practice, however, it can be observed that they hardly remained the *Bhikkhus* who went begging for alms from door to door for their existence. Their daily bread and butter was no longer a problem for them. Generous donations from the pious lay devotees fulfilled all their needs. Now they had to manage, maintain and organize all the donated money and property, in return for which they had to satisfy the spiritual and religious needs of the laity. Due to a close contact with them and an introduction to their popular and folk religious notions, the monastic community developed many rituals and religious institutions.

Numerous *viharas* were constructed and donated for the use of the *Sangha*. Along with the structural *viharas*, rock-cut caves for shelter during the rains were made as well. The earlier plain *viharas* fulfilled the basic needs of the *Sangha*, i.e., an assembly hall, and living quarters for monks. However, the later *viharas* consisted of storehouses, kitchens, dining halls, wells, bathrooms, *cankramana* (wandering) places, etc. Initially no need was felt for having a place for the object of worship. For the monks, the Buddha *vacanas* (i.e., the sayings of the Buddha) were the sacred words. *Yoga* and meditation formed the religious practices of the monks. As for the laity, the religious instructions given by the monks were sacred. The goal of developing a moral conduct was placed in front of the laity by the Buddha. For the monks, Buddha was a religious leader, a pathfinder. However, for the laity, he was the object of devout faith, a saviour and an almost superhuman being. The laity might have tried to combine popular religious practices with their reverence for the Buddha. From this attempt started the worship of the *stupas*. The already existing cult of the *stupas* was unconsciously applied by the lay Buddhist followers for the worship of Buddha. It is evident from literature like the *Milindapanno1* (Questions of King Milinda) that monks did not initially take this practice very seriously. They believed that it was meant only for the laity and the monks should rather practise understanding and contemplation (Dutt, 1957: 157).

As time went by, munificent donations from the lay followers started coming in for construction, decoration and the maintenance of the *stupa*. Gradually, the *stupas* became big establishments, which needed the complete attention of an organization. Even royal personages like King Ashoka joined hands with the laity in constructing *stupas* all over India. The monks could not neglect the importance of the *stupa* in the religious life of the majority of the followers of their religion. At a certain point in the history of Buddhism, the monks themselves started donating for the construction and decoration of the *stupas*. The organization and management of the donations were controlled by the *Sangha*. The earlier perception of the *stupa* cult as the expression of popular Buddhism, changed completely. It became a monastically dominated cult.

A statistical analysis of the early donative inscriptions from Bharhut and Sanchi indicates that a considerable portion of the donors comprised monks and nuns, as high as 40% at Bharhut. At Sanchi, out of the 437 inscriptions, 163 were monastic donors. Another interesting fact is that huge monastic residential complexes accompanied almost all the early *stupas*, both structural and rock-cut. These two units developed simultaneously on plan and elevation. But both of them were controlled by the local *Sangha*. It can be observed that the monastic community played a major role in the development of Buddhist religious architecture.

This brings about another shift from the rules of the Vinaya, i.e., the right over the donations received. It is clearly mentioned in the Chullavagga that the right of property was completely vested in the hands of the *Sangha*. No single individual was allowed to possess any private property, not even any lodging and furniture (Chullavagga, XI.I.14). Thus the problem arose that from which funds could the monks and nuns make their donations if they were not supposed to possess any private property. This led to further complexities in the religion, one of which was the beginning of image worship. In the initial stages of religious development, the Buddha was worshipped in symbolic forms and not in his anthromorphic form. However, around the 1st century AD, images of the Buddha started appearing in north India and quickly became very popular. A very interesting feature of this image cult are the donative inscriptions, which the sculptures carry with them. Out of the 18 Kharoshthi inscriptions edited by Konow (1929) which record the setting up of an image and in which the name of the donor is preserved, 13 were of monks. Lüders (1961) has also noted that similar inscriptions were found at Mathura. Such inscriptions account for almost exactly the same percentage, i.e., 18 out of 28 inscriptions were of monks and nuns. Even in case of the western Indian cave temples, studied by Burgess (1883) the inscriptions connected with images are almost always associated with monks. On the basis of such statistical data, scholars like Gregory Schopen have concluded that the image cult was a monastically initiated cult (Schopen, 1985: 27). The need of an image of the Buddha was probably felt more by the monastic community than by the laity. Or the monks had developed this cult to compete with the image worship in other religious systems. This further led to the development of the Mahayana ideals, and images.

Be that as it may, in the 3rd century BC the structure of the *stupa* was incorporated in monastic residential complexes, in case of the structural *viharas*. But in the case of the rock-cut caves, we see that the *stupa* was placed in a big hall called the *chaityagriha*. This had two separate structures for worship and residence. In *c.* 1st century AD, the *stupa* was placed in the *vihara* cave, parallel to the *chaityagriha*, as is evident at Nasik, Caves X, III (Nagaraju, 1981: 268). It became a regular feature of the late Hinayana period (the 1st – 3rd century AD) caves in western India (Dhavalikar, 1984: 30–37). This indicates that the monks probably wanted a separate *stupa* only for themselves rather than sharing it with the laity. It is quite possible that they had developed certain rituals, which they wanted to perform secretly in their own chambers.

As noted earlier, the monks were not supposed to keep any private or personal possessions for themselves. Whatever they received belonged to the *Sangha*. However, this obligation was overruled many times in the history of the religion. One such instance could be seen in Cave VIII at Kuda. At one end of the bench in this cave is a hollow (55 cm square, 40 cm deep) with a ledge provided at its mouth for a flat lid (Nagaraju, 1981: 243). A similar instance is also noted in one of the caves in the Jakhinwadi group of the Buddhist caves at Karad. Such hollows were probably used as vaults to keep certain valuable possessions of the monks living there.

It is important to note that these peculiarities in the nature of the possessions and property of the monks really began when they started accepting heavy donations from the laity. This in turn was the result of the cenobitic lifestyle of the earlier wandering ascetics. The participation of the laity in the construction of the *vassa*-residences had an important role to play in the creation of the resulting novel practices of the Buddhists. Although the ideal of *vassavasa* (the residence during the rainy season) was not categorically practised in later years, the theoretical ideal was always glorified by the monks. It is also possible that a part of the monastic community strictly observed this ideal, as we find donations made to the mendicant, who kept *vassa*, in a particular cave at Nasik, in the 2nd – 3rd centuries AD (Senart, 1905–06: 90). Some of these inscriptions were written in months other than the rainy season, indicating that some monks stayed there for the whole year (Senart, 1902-1903: 71,73 and 1905–06: 59, 60, 67, 73, 90, 94).

In this period, i.e., in the 2nd – 3rd centuries AD, *akshayanivis* or perpetual grants, were made to the *Sangha* (Senart, 1905–06: 88, 89). These became a steady source of income. Different hierarchical positions were created for monks in the *Sangha* according to their seniority and capability to handle various financial and organizational matters. We find separate residential single or double-celled quarters made in this period at the cave sites like Nasik, Junnar, etc. Along with other religio-economical reasons, this hierarchy among the *Sangha* could be one of the reasons for such single-celled units, as suggested by Dhavalikar (1984: 79).

This state of the religion was perhaps achieved after assimilating different innovative ideas in the simple plain ascetic sectarian movement. The incorporation of such novel concepts began only after the adoption of a semi-settled lifestyle by the mendicant community in the rainy season. This ultimately resulted in the emergence of an advanced complex religion; witnessing three distinct phases in its development, i.e., Hinayana, Mahayana and Vajrayana.

Note

This is a Pali text. Milind is the Indian name for the Greek ruler Menander. He had some doubts about the Buddhist philosophy and expressed them to the monk, Nagasena, who resolved them. Their dialogue is presented at length in this text that is dated to c. 1st century BC.

References

Bhagwat, D. (1939): *Early Buddhist Jurisprudence*, Oriental Book Agency, Pune.

Burgess, J. (1883): *Report on the Buddhist Cave Temples and their Inscriptions*, Archaeological Survey of Western India, Vol. IV, London.

Davids, R and H. Oldenberg, (1965): (Tr.) Cullavagga: *Vinaya Texts* Pt II, Motilal Banarsidas, Delhi.

Dhavalikar, M.K. (1984): *Late Hinayana Caves of Western India*, Deccan College, Pune.

Dutt, S. (1957): *The Buddha and Five After Centuries*, Luzac and Company Ltd., London.

Dutt. S. (1962): *Buddhist Monks and Monasteries in India*, George Allen & Unwin Ltd., London.

Davids, R. and H. Oldenberg (1965): (Tr.) *Mahavagga: Vinaya Texts* Pt. II, Motilal Banarsidas, Delhi.

Konow, S. (1929): *Kharashthi Inscriptions with the exceptions of those of Ashoka, Corpus Inscriptionum Indicarum*, Vol. II, Pt 1, Calcutta.

Lüders, H. (1961): *Mathura Inscriptions*. Unpublished Papers ed. by K.L. Janert, (*Abhandlungen der Akademic der Wissenschaften in Gottingen*), Philo.-Hist. Kl., Dritte Folge, Nr. 47, Gottingen.

Nagaraju, S. (1981): *Buddhist Architecture of Western India*, Agam Prakashan, Delhi.

Schopen, G. (1985): Two Problems in the History of Indian Buddhism: The Layman/Monk Distinction and the Doctrine of the Transference of Merit, in *Studien Zur Indologie und Inranistik*, Vol. X: 9–47

Senart E. (1902-1903): Karle Inscriptions, *Epigraphia Indica*, Vol. VII, Government of India, Calcutta.

Senart, E. (1905–06): Inscriptions in the caves at Nasik, in *Epigraphia Indica*, Vol. VII: 59-96.

The linga is a symbol of fertility. Rajasthan, India.
(Photo by Takeshi Takeda)

Chapter 20

Monsoon and Civilization: With Special Reference to Kanheri

SHOBHANA LAXMAN GOKHALE

In all the three main Indian religions, i.e., Brahmanism, Jainism and Buddhism, the four monsoon months (15th June to 15th October) are considered as an auspicious time. Even today, in orthodox families, people observe all manner of religious vows and practices. These practices have not only maintained the rhythm of Indian tradition but have also preserved the ethos of ancient Indian civilization and culture, its values and lessons, for the people.

Even today, in the monsoon period, several cultural activities are performed by rural and urban communities alike to welcome the monsoon. Cultural programmes are organized to greet a good monsoon. There is a beautiful synchronism between these religio-cultural activities, traditions and social psyche.

The western coast of the Indian Peninsula lies in the monsoon zone. It is studded with more than 1200 Buddhist caves. These caves have offered an excellent basis for understanding the real spirit and merits of the ancient Indian civilization. Amongst them all, Kanheri has a distinctive character.

Kanheri is 10 km to the southeast of Borivali, a suburb of metropolitan Mumbai, the commerical capital of India located on the western coast of Maharashtra. The name Kanheri is derived from the Sanskrit name 'Krṣṇagiri' (Gokhale, 1991) which means the black mountain. The Kanheri hills command the view from the Mumbai harbour to the Bassaein creek in the north. The site of the caves is lonely and picturesque. There are 104 Buddhist caves. According to the descriptions given in the *Vinaya text* (Davids and Oldenberg)[2], Kanheri was modelled like a *'Jetavana'* monastery. There are dwelling caves, benches, service halls, store-rooms, water places and graceful sculptures. The artisans have followed the text and exhibited their superb skill. Kanheri enjoyed a favourable monsoon and royal patronage since the Mauryan period. King Ashoka promoted a religious atmosphere by erecting a *stupa* and rock edicts at Sopara, a prosperous port near Mumbai. Ashoka's liberal attitude and goodwill towards all irrespective of caste and creed made a deep impact on the religious as well as political environs of the western coast. The ruins of the Kanheri monastery give us some idea of the magnificence of the place. The echoes of a flourishing civilization are still evident in these ruins.

The architectural splendours at Kanheri indicate that it was a paradise for cave cutters. The inscriptions at Kanheri span the 2nd to the 9th century AD. Kanheri enjoyed a flourishing trade and the patronage of the Sātavāhanas, Traikūṭakas and Rāshṭrakūṭas. It was named as *Mahārajā Mahāvihāra*—the sovereign of all the monasteries.

The 1st century BC to 2nd century AD was a period of remarkable progress in trade and industry. Hippalus made the great discovery in AD 45, when he noted the existence of monsoon winds blowing regularly across the Indian Ocean, enabling ships to sail faster. This was the time when the western coast of India witnessed the growth of brisk foreign trade with the Roman world. Sopara, Kalyan and Chaul were flourishing ports. Thus, the western coast was humming with trade activities and was also dominated by the Buddhist religion (*Aparānta*). The four monsoon months were regarded as the time of *vassāvāsa*. The monks of Kanheri observed the rules framed by the Blessed One. Thus, the spiritual thoughts of Buddhism percolated into the caves along with the monsoon rains.

Monastic institutions were the most remarkable contribution of Buddhism to ancient Indian civilization. Their original object was to give suitable accommodation to monks for carrying on their studies and meditation. These monasteries gradually developed into academic centres for producing the right type of men, well grounded in religion and philosophy to propagate the teaching of Buddhism.

Kanheri, which enjoyed both favourable monsoon and a favourable geographic position, was established as an important educational centre.

The monsoon rains were channelized at Kanheri. The caves facing west have channels on top and tanks have been excavated near the caves where the rain water was collected. All these tanks were donated by common people in order to accrue religious merit.

There is one interesting instance. An inscription in cave no. 50 records that the son of Nandanikā made a meritorius gift of a tank (*poḍhi*) and a cave. The merit was shared by his mother, wife, sons, grandsons, great grandsons, great granddaughters, daughters-in-law and all the relatives and unborn children. The inscription has drawn an ideal picture of a joint family. Conjugal fidelity and an ideal relationship between husband and wife were the distinguishing features of Indian society at the time.

The inscriptions at Kanheri have revealed a very simple picture of the social life. People of diverse occupations made donations of caves, cells, water-cisterns and paths to accrue religious merit, for themselves, their family members and for the welfare of all living beings. It indicates a great attachment of people to religion and for the attainment of spiritual blessings and happiness, donors made various types of grants.

The cave site of Kanheri enjoyed the three stages of Buddhism, i.e., Hinayāna, Mahāyāna and Vajrayāna. There are litany scenes in cave nos. 2, 41 and 90 of *Avalokiteśvara* rescuing his devotees. He is depicted as saving people from eight or ten dangers such as the fire, elephant, lion, cobra, crocodile, etc. These ten dangers have been described in '*Milindapaña*' (Vedekar, 1940: 195).

The introduction of female figures, the eleven-headed *Avalokiteśvara* and litany scenes suggest the popularity of Vajrayāna Buddhism. The litany scenes and *makara* motifs in sculptures indicate the traders' support to the establishment.

Cave no. 3 has a figure of Tārā, who enjoyed an exalted position similar to that occupied by the great goddess Durgā in Brahmanism. She is considered to be the personified energy of her consort (Gokhale, 1991), *Avalokiteśvara*. She is holding a full-blown lotus in her left hand and a torch in her right hand, a superb example of plastic art. The moon digit is conspiciously shown on her chignon. She was the guiding star of navigation. This form of *Dīptārā* is the only example of Tārā found so far in western India (Fig. 1).

The sculpture of Tārā is on the left wall of cave no. 3 and to the left of the colossal figure of the standing Buddha. Tārā occupies a very small place on the wall and is shown as a companion of *Avalokiteśvara*. To her left he is holding a lotus and a rosary. The sculpture of Tārā is certainly later than the colossal Buddha and *Avalokiteśvara*. This is clear as she occupies a distinctive position rather than blending in

Fig. 1. Dīptārā at Kanheri.

with the total plan. Her measurements are as follows:

Height	73 cm
Breadth	21 cm
Waist line	8 cm
From waist to feet	43 cm
Face	15.5 – 8 cm

There is a unique eleven-headed image of *Avalokiteśvara* in cave no. 41. It is in high relief and stands beside the Amitābha Buddha (Fig. 2). Instead of a single image of the Buddha, which is usually found in the tapering headdress of the *Avalokiteśvara* ten additional faces are piled in three tiers over the head, and the image has four hands. The feet are in a broken condition. *Avalokiteśvara* holds a lotus bud with a long stalk in one left hand; another left hand holds a bottle, and a right hand in *'abhayamudrā'* holds a rosary. Each of the eleven faces has a crown, and there are earrings, necklace, dhoti and ornaments around the waist. The image is usually dated to the 6th century AD. This sculpture is one of its kind, and is a superb illustration of the imagination and skill of the learned Buddhist philosophers and the artists of that period. It is a unique example of a cultural exchange between Brahmanism and Buddhism. By this time the Mahāyāna activities were resumed with full force at Kanheri. *Avalokiteśvara,* the bodhisattva of compassion, was the popular subject. As Kanheri is located on both the sea and inland trade routes, it served as a refuge for traders as well as an abode for Buddhist monks.

The measurements of *Avalokiteśvara* are as follows:

Height	145 cm
Breadth	37 cm
Height of eleven heads	32 cm
Breadth of eleven heads	21 cm
Each individual head	8 – 8 cm

1. To interpret this unique piece it is necessary to make a close study of the cultural environment of Kanheri.
2. One has to take into consideration the contemporary awe-inspiring Saivite excavation at Elephanta and the vigorous artistic developments at Jogesvari.
3. India's cultural contacts with Central Asia China.

Amongst the caves of western India, Kanheri is the only site that illustrates in good measure the various stages in the history of political power and trade. In the 2nd century AD, a flourishing trade and the patronage of the Sātavāhanas gave Buddhism a firm foundation. The English

Fig. 2. The Eleven-headed *Avalokiteśvara* and Amitābha.

translations of the Tibetan text of the verses of Nāgārjuna's letter to Gautamiputra Sātakarṇi (Lozang Jamspal Santina, 1983) has recently been published. From that letter it appears that Gautamiputra, son of Queen Balasrī, might have embraced Buddhism.

The inscriptions at Kanheri not only throw light on ancient Buddhist monastic institutions but they also give evidence for the first time of a teachers' tradition in western India. This tradition continued even in the early Rāshṭrakūṭa period. The Kanheri inscription of Pullaśakti (Mirashi, 1955), who was a feudatory of Amoghavarṣa I, records a donation to the Buddhist vihara at Kanheri, a part of which was utilized for purchasing books. Taking into consideration this educational tradition and the advent of the Mahāyāna sect, there must be Buddhist scholars who were authorities on the

Mahāyāna texts such as *Sukhāvativyūha* (Vaidya, 1961), *Amitāyurdhyānasutra* (Maxmuller, 1894) *Daśabhumisūtra* (Winternitz, 1991), etc. The whole of *Mahāyanasutra* is devoted to the Bodhisattva *Avalokiteśvara* who refuses to assume Buddhahood until all beings are rescued. *Avalokiteśvara's* sole task is to bring the doctrine of salvation to all beings and help sufferers. He is closely related to the Amitābha Buddha—Lord of Blessed Land of Bliss in the west. Surprisingly the whole sculptural representation of the Amitābha Buddha and the Eleven-headed *Avalokiteśvara* is in the western corner of the cave. The subject matter of '*Daśabhumisūtra*' is a discourse on the ten steps by which Buddhahood may be attained. *Guṇakāraṇḍavyūha* (Vaidya, 1961) describes *Avalokiteśvara* as *Ekādashshīrṣa*. It may be a conversion of the eleven violent gods of the Vedic age—the Ekādasha Rudra Brahmanism.

310

As suggested by Spink (1968) the Śaivite caves at Jogesvari and Elephanta might have been excavated by the craftsmen trained during the Vākāṭaka regime. The eight-headed Śiva image from Parel belongs to this period. It is possible that the learned Buddhist scholars might have invited those craftsmen to Kanheri.

There are the remains of a brick *stupa* in front of the main chaitya at Kanheri (Fig. 3). The copper plates of the Trāikuṭakas (K. 245–494–95 AD) found in the *stupa* record the construction of the chaitya which is made of stone and brick and is dedicated to Sāriputra, by a monk named Buddharuchi, son of Puṣyavarman, who hailed from Sindhudesh. The inscription records one of the epithets of the Buddha *'Daśabali'*.

The earliest evidence of Buddha's epithet *Daśabali* is from Nagarjunikonda (Sircar, 1942). The Devanimori (Sircar, 1965) stone casket inscription of the early 3rd century records the construction of a chaitya in the name of *Daśabali*. The third evidence is from the inscription of Toramana (AD 500) from Punjab, West Pakistan (Sircar, 1965).

It is interesting to note that the epithet *Daśabali* has occurred in the Ikṣvāku inscription which was the origin of Mahāyāna sect. The eleven-headed image of *Avalokiteśvara* at Kanheri with *Daśabali* in the head-dress is a masterpiece of Buddhist iconography. A highly intellectualized spiritual experience of *Daśabali* Buddha is beautifully transferred into sculpture.

The ten powers of the Buddha are well described in Majjhimanikāya (Bapat, 1958:

Fig. 3. Remains of the brick stupa of Sāriputra at Kanheri. Lower left in front of the main Chaitya.

121-136) in the *Simhanādasūtra*. They are as follows:

1. Sthānāsthānabalam—Knowledge of correct and faulty conclusion.
2. Karmavipākañanbalam—Result of one's actions.
3. Sabbatthagāminipatipadā—Knowledge of everything under all circumstances.
4. Nānādhātuñanabalam—Diversity of specific experience.
5. Nānādhimuktiñanabalam—Diversity of disposition.
6. Sarvādhyānavimokṣa Samādhiñanabalam—Knowledge of meditation with its different stages.
7. Purvanivāsānusmrtiñanabalam—Knowledge of one's former state of existence.
8. Indriyaparāparāñanabalam—Knowledge of what goes on in the senses and intentions of others.
9. Chityutpattiñanabalam—Knowledge of thoughts.
10. Āsavakhayañanabalam—Freedom from depravities.

Four qualities of the Buddha
1. Āsabhathāna—Eminent qualities.
2. Abhaya—Confidence.
3. Vesārajja—Highest knowledge.
4. Khema—Tranquility

The *Mahāyanasutrasangraha* lists ten powers of the Bodhisattva as follows:
1. Adhimuktibalam—Faith.
2. Pratisamkhayanabalam—Power of computation.
3. Bhāvabalam—Power of self-culture.
4. Patibhāṇabalam—Illumination
5. Patipattibalam—Proper behaviour
6. Samādhibalam—Power of concentration.
7. Puṇyabalam—Power of merit.
8. Kṣantibalam—Power of forbearance
9. Nāṇanbalam—Power of knowledge.
10. Prahāṇabalam—Power in abandoning.

Under the direction of learned monks of Kanheri, highly skilled artisans made sculptures of *Avalokiteśvara* which incorporated ideas from classical literature (Figs. 4 and 5). Taking into consideration inscriptional and sculptural supporting evidence it may logically be interpreted that the powers of the Bodhisattva and the four hands are his natural additional qualities. As *Avalokiteśvara* refused to assume Buddhahood until all beings were rescued, he is shown as an attendant of Amitābha Buddha.

The third point which is to be taken into consideration is communication between India and China which was nurtured by the visit of Hiuen-Tsang during this time.

Fig. 4. *Avalokiteśvara*, Kanheri.

312

Fig. 5. A scene in cave no. 90, Kanheri.

This absolute type of *Ekadaśamukha–Avalokiteśvara,* i.e., eleven-faced Kuan-yin, was popular in China in the late 7th and 8th Centuries AD. While discussing *A Colossal Eleven-faced Kuan-yin of the Tiang Dynasty,* Sherman Lee and Wai-Kam-Ho (1959) have shown that this deity was popular in China. He has remarked, 'The canonical works or copies of icons imported from India seem to have been the only logical means by which the Chinese were introduced to this deity.' He has further pointed out that the earliest and only existing representation of the Eleven-faced One in India is from Kanheri. The earliest complete iconographic description of the Eleven-faced Kuan-yin occurs in a Sanskrit sutra translated into Chinese by Yaśogupta in AD 561–577, i.e., *Avalokiteśvara Ekadaśamukha Dharani.* A second condensed version from the

same Sanskrit origin was translated by Atigupta in AD 653, i.e., *Ekadaśamukhahṛdimantra hṛdaya-Sūtra.* The third was done only two years later by the great Chinese traveller Hiuen-Tsang. The authors have made a comprehensive review of all the images of the eleven-headed *Avalokiteśvara* available in China along with various interpretations. Sherman Lee and Wai-kam-Ho have pointed out:

'In China the Sutras give full description of the image. The earliest translation of the *Avalokiteśvara Ekadashamukhadhāraṇi* by Yaśogupta (AD 561–577) gives the full iconographic requirements.' The body is one foot and three inches tall with eleven heads. The three faces that face forward are Bodhisattvas, the three on the left side are to be given angry faces and the three on the right should be given faces

313

resembling those of Bodhisattvas but with canine tusks projecting upwards. On the rear is another head which is shown as laughing out loud, and on the top of the main head is still another which must be a Buddha face. He holds a washing bottle (*kundikā*) in his left hand from the mouth of which project lotus flowers, and he extends his right hand grasping his pendant chains of jewels and makes the *Abhayamudrā*. Subsequent translations agree with this description of the deity. There are, however, different interpretations of the eleven faces of the deity. 'One source points out that ten of the faces are representations of the ten *bhumis* or grounds, of the fundamental causes which lead to Buddhahood, represented by the single face on the top. Another Dhāraṇi devoted to this Bodhisattva specifies that the three peaceful faces represent the virtue of the Great Void (*Śunya*), the three angry faces, the virtue of the Great Wisdom (*Prajña*) and the three faces with canine tusks, the virtue of Great Compassion (*Karuṇā*) and finally the top face signifies the non-differentiated unity of compassion and wisdom. With this perfect combination of the three virtues to answer the prayers of all living beings, the Eleven-faced Kuan-yin is able to grant his believers, who invoke his name, the ten victories over all the major causes for fears and suffering, as well as the four rewards which include rebirth in the Western paradise.'

Unfortunately, the *Dhārani Sutras* (Nanjo, 1883) which described the Eleven-headed *Avalokiteśvara* were taken to China by Buddhist monks and this form of *Avalokiteśvara* became popular in China so much so that the image is variously interpreted whereas in India because of the loss of the literary tradition it remained in isolation. The *Mahāyānasutrasangraha* does not provide any detailed description of the image of the Eleven-headed *Avalokiteśvara*. There are remains of painting on the faces of the image at Kanheri which might have indicated the Great Void, Great Wisdom and Compassion.

Scholastic activities were also growing dynamically at Kanheri. The epitaphs found in the vicinity of the cemetery (*Nirvāṇa vīthi*) record the names of teachers along with their scholastic merits. They not only throw light on ancient Buddhist monastic institutions but also give evidence for the first time, of a teachers' tradition in western India. The teachers had different qualifications:

1. Tevijja—The monk who has threefold knowledge.
2. Aṇāgami—The monk who has attained the third stage of 'Arhathood'.
3. Sadabhijñani—The monk who has the power of six knowledges.

This tradition continued even in the early Rāshṭrakūṭa period (the 9th century AD). The inscription of the king Pullaśakti records a donation to the *vihara* and a part of it was meant for the purchase of books.

The scholastic activities at Kanheri were nurtured in *Vassāvāsa*, the monsoon period. There is more significant evidence corroborating the existence of the educational tradition at Kanheri. As mentioned earlier, there are remains of a brick *stupa* in front of the main chaitya at Kanheri. A copper plate of Traikuṭakas (AD 495) was found in the *stupa*, which records the construction of the chaitya of stone and brick by a monk named Buddharuchi, the son of Buddhasri and Puṣyavarman who hailed from Sindhudesh (Sindh in Pakistan). It was dedicated to the venerable Śāradvatiputra, the chief disciple of the Buddha. The orthodox view is that the 'Abhidhamma Piṭaka' was first preached by the Buddha to his mother and then was repeated by him on the banks of the Anotatta lake to Sāriputra. Sāriputra taught the same to his colleagues. Abhidhamma is a totality of the psychological and philosophical teachings of Buddhism. At Kanheri, hardly 200 m to the east of the *stupa*, are the remains of an ancient stone-wall and an inscription (dated to the 2nd century AD), which was constructed to collect the water by a merchant named Pūṇaka from Sopara. This is the earliest dam construction in Maharashtra. It is a beautiful synchronism of literary and archaeological evidence. It is significant that the *stupa* of Sāriputra was erected in front of the main chaitya. There is a *stupa* of Sāriputra at Nalanda, but Kanheri offers earlier evidence.

Thus, Kanheri enjoyed the benefit of the monsoon in the early period and flourished as a great centre of Indian civilization and culture.

References

Bapat, P.V. (1958): *Majhimanikayapali,* (Gen. Ed. Bhikkhu J. Kashyapa), Vol. I, Pali Publication Board, Bihar, Nalanda, pp. 121–136.

Davids, T.W.R. and H. Oldenberg (1984) (ed.): *Sacred Books of the East,* Vol. XIII (1881) and Vol. XX (1885), Oxford Clarendon Press, London.

Gokhale, S. (1991): *Kanheri Inscriptions,* Deccan College Pune.

Lozang Jamspal Santina (1983): *Nāgārjuna's letter to Gautamiputra.* Motilal Banarsidass, Delhi.

Maxmuller, F. (1894): *Sacred Books of the East,* Vol, XLIX *Amitagur Dhyansutra,* trans. by Takakusu, Oxford Clarendon Press, London, p. 159.

Mirashi, V.V. (1955): *Corpus Indicarum,* Vol. IV, Government Epigraphist, New Delhi, pp. 1–2.

Nanjo, B. (1883): *A Catalogue of Chinese Translation of Buddhist Tripitakas,* Oxford Clarendon Press, London, Nos. 124, 136, 327.

Sherman, Lee and Wai-kam-Ho (1959): *Artibus Asiae,* Vol. XXII, Ascona, Switzerland, pp. 121–136.

Sircar, D.C. (1965): *Select Inscriptions Bearing on Indian History and Civilization,* University of Calcutta, Calcutta, pp. 233, 422, 519.

Spink, Walter (1968): Monuments of the Kalachuri Period, *Journal of Indian History,* Vol. XLVI, pp. 263.

Vadekar, R.D. (1940) *Milinadapano,* University of Bombay, Bombay, p. 195.

Vaidya, P.L. (1960): *Mahāyanasutrasangraha,* Vol. I: Mithila Institute of Post Graduate Studies and Research in Sanskrit Learning, Darbhanga, Bihar, pp. 221–253.

Vaidya, P.L. (1960): *op.cit.,* pp. 254–257.

Winternitz, M. (1991): *History of Sanskrit Literature,* Vol. II p. 327, 2nd Edition, printed by Subir Das at Ajanta Offset and Packing Limited, New Delhi.

Indians have a deep religious faith in water. This faith in water will no doubt be reevaluated in the 21st century, a century threatened by global warming. Rajasthan, India. (Photo by Takeshi Takeda)

A possessed woman in India. Coexistence with the gods is a routine in the daily lives of people. Rajasthan, India. (Photo by Takeshi Takeda)

Chapter 21

Monsoon and Religions

YOSHINORI YASUDA

Religion is Born from Climate

■ The Desert and the Forest

As we have seen thus far, the typical climate of the east Monsoon Asia is damp forest. On the other hand western Arid Asia is dry grasses, plains and desert. The climates of the forest in Monsoon Asia and the desert in Arid Asia have also played an extremely vital role in the birth of major religions.

> Whether you believe that all of creation repeats itself in an endless cycle, or whether the heavens and the Earth have a definite beginning and an end, these are the only two theories that humans have. The former view is Buddhist cosmology, and the latter view is Christian cosmology. Buddhism was born in the forests, and Christianity was born in the desert.

This extract has been taken from Suzuki Hideo's famous book, *Shinrin no Shiko, Sabaku no Shiko* (*Ideas of the Forest, Ideas of the Desert,* Suzuki, 1978).

Climate has a major effect on the birth of religions. Buddhism was born in the midst of the forest, and Christianity was born in the desert. The opposing climates of the forest and the desert spoke to a small number of geniuses raised in these climates, giving birth to two major religions—Buddhism and Christianity. Religious geniuses were people whose ears were attuned to the whispers of the climate. A major religion could only be born when this small number of geniuses listened to the climate, and then interacted with it to come up with philosophical ideas to save humankind.

People, however, have somehow become arrogant. They lay emphasis only on the fact that the Buddha was the great founder of Buddhism, and that Jesus was the great founder of Christianity, without any regard for the climate. There is little doubt that Buddhism is the religion founded by the Buddha, and that Christianity is the religion founded by Jesus. However, when I say that Buddhism is the religion of the forest, and that Christianity is the religion of the desert, their faces suddenly light up with suspicion, since the image of the founder alone has been pushed to the fore as a major idol, and worshipped.

Even some of the orthodox Buddhist sects, or the new religions, have forgotten nature and the climate, and magnified only the idol of the founder out of all proportion. To face the global environmental crisis head-on and to ensure that humanity survives, however, it is probable that, in the 21st century when coexistence with nature will be the most important issue, these founder-venerating religions shall die a hasty death.

Religion is a structured thought system of the interaction between humans, nature and the climate. That makes it imperative for us to understand that there is an urgent need today for a peaceful coexistence with nature amidst a population explosion, especially for those of us living at this time, who will face an apocalyptic crisis in 20 to 30 years' time of an acute shortage of food. Must we really wait for apocalypse to be upon us to realize this? The truth is that neither the Buddha nor Jesus, but the forest and the desert are the true founders of religions.

■ Monotheism and the Desert

I wonder if you, the reader, have ever stood in a desert (Fig. 1). Desert nights are enveloped in stillness. All you can hear is the intermittent blowing of the wind across the wasteland. In the absence of the murmur of the wind, however, the desert is enveloped in such stillness that you almost feel that the stars that fill the sky, like a shiny blanket above your head, are making noises as they glitter. The stillness is the stillness of death. If you lose your way in the desert, certain death awaits you. The desert is a truly lifeless place. The only living things moving here are you and the overwhelming force of the stars, spreading out towards you from the horizon. The only audible sound is the murmur of the wind. There are no signs of life around you, and in the climate of the desert, where a star-filled sky is all you can see, identifying the one and only God in the heavens is an inevitable consequence, given human psychology. In addition, the birth of the creation myth—that God created the world—was also an inevitable consequence in this lifeless expanse, and it is but natural to conclude from the foundation of this

cosmology that this world will, at length, return to the desert and end.

In the beginning God created the heaven and the earth. And God said, Let the waters under the heaven be gathered together unto one place, and let the dry land appear: and it was so. And God called the dry land Earth; and the gathering together of the waters he called Seas: and God saw that it was good. And the earth brought forth grasses, and herb yielding seed after his kind, and the tree yielding fruit, whose seed was in itself, after his kind: and God saw that it was good. And God said, Let the waters bring forth abundantly the moving creature that hath life, and fowl that may fly above the earth in the open firmament of heaven. And God said, Let the earth bring forth the living creature after his kind, cattle, and creeping thing, and beast of the earth after his kind: and it was so. And God made the beast of the earth after his kind, and cattle after their kind, and everything that creepeth upon the earth after his kind: and God saw that it was good. (*The Bible*, Old Testament, Genesis, Chapter I.)

In Genesis, Chapter I, in the Old Testament, it is written that God created man, and that he gave man to rule over all the life on Earth. Man was given the right and obligation to rule over all living things on Earth. This gave birth to a cosmology established on a ranking system of God → Humans → Nature. The notion that God

Fig. 1. The desert is an expanse enveloped in stillness with very few living things. The only life-forms encountered in the desert are human and camels. Syria desert. (Photo by Yoshinori Yasuda.)

gave humans to govern the world of life created on this planet was probably first born in a climate where only we humans existed between the heavens that spread out towards us from the horizon as if to cover us, and the lifeless expanse of land. In the tropical rainforests, overflowing with life and drenched in the violence of that life, humans could never have concluded that they could control the world of life.

■ Polytheism and the Forest

In the forest, your line of vision is cluttered with twigs, and the star-filled sky is hard to see. The stars in a sky filled with the water vapour that the forest lets off, do not appear to be glittering; they seem blurred. Forests are truly noisy (Fig. 2). Not only can you hear the hooting of owls and the baying of wolves, you can also hear the gurgling of brooks, and the wriggling, buzzing, scurrying, and squirming of various animals. What makes the forest even noisier is the sound of rustling leaves. In the desert, the only audible sound, that of the wind, transforms itself in the forest into the rustling of leaves, which when magnified ten or even a hundred times over, at times, sounds very eerie.

Forests are teeming with life and it is these very forests with their abundant harvests that have provided food for humans. Unlike in a desert, were you to lose your way in a forest, it would not mean instant death.

Forests are blessed with abundant rich harvests. At the same time they change profoundly with the seasons in a manner that in these forests the drama of the cycle of life and death plays itself out endlessly. Therefore, it is impossible that the notion will ever arise in forests, of a God who created the heavens and the earth, and that it will all end one day. In forests, the notion that can emerge is one of rebirth and reincarnation, which repeats itself in an endless cycle. Suzuki (1978) calls this a cyclical

cosmology, and contrasts it with the linear cosmology of the desert, which proceeds in a straight line to the final end.

However, the forest climate can be savage and rough. At times, people have faced the brunt of its overwhelming savagery. Hence their feeling of gratitude to nature was always complemented with a healthy respect for it. Faced with the forest overflowing with life, and directly in contact with its untamed savagery, how could people possibly think of governing life forms on Earth? All things surrounding humans, from the trees and the animals in the forests to the mountains and rivers, were of equal value to human life, and at times, were even venerated as gods with fearsome powers that far overrode the powers of mortal men.

As Suzuki (1978) has indicated, the conclusion arrived at in these forests overflowing with life was one of 'Live and Let Live', that is, the life that fills the world around you and your own life are equal, and can become part of the one whole. This is the very starting point of animism. It is also the starting point of the system of belief of the theory of Absolute Enlightenment in the Tendai sect of Buddhism, which propounds the belief that 'All things and everything, mountains, rivers, grasses, trees, and the soil of this land can all attain the wisdom of Buddhahood.' The idea from the forest of being just a tiny small part of life among all life forms on the earth including humans, and the idea from the desert that humans are given to rule over all life forms on the earth, are fundamentally very different.

Climatic Change Gave Rise to New World Religions

■ Desertification and the Spread of Monotheism

To a greater or lesser extent, the peoples of the world lived in a world of animism. The climate

Fig. 2. Forests overflow with the sounds of life: the sound of rustling leaves, the sound of murmuring brooks, etc. Beech forest in Aomori Prefecture, Japan. (Photo by Yoshinori Yasuda)

change, however, proved a great opportunity for the establishment and spread of monotheism that would almost replace animism. The most important fluctuation of all was the climate change in the region of the eastern Mediterranean around 1200 BC.

The region that gave rise to the great religious thinkers and philosophers who elaborated a notion of human-centrism that Ito (1985) has named a spiritual revolution broadly coincides with the region affected by urban culture until around 500 BC. Egypt and the region around the Yangtze valley, however, were excluded from the cradle of this spiritual revolution. But why did that happen?

For many years Egypt was governed by the gods of animism, and polytheism. The Yangtze region was similarly governed by animistic Taoism that formed the central philosophy for generations. The spiritual revolution cast out this animism, and gave birth to a human-centric cosmology. Most regions outside of the cradle of the spiritual revolution can be viewed as the world of animism and pantheism.

In the course of the spiritual revolution were born the major world religions, along with ideas that placed maximum value on human intellect and notions of happiness alone. These arose between 1200 and 400 BC. The period between 1200 and 400 BC was one of fierce change with frequent tribal displacement and starvation, as we were visited by the worst climate deterioration in the last 10,000 years, as is clear from the results of the pollen analysis performed by the author and others.

The severe temperature drop in the climate, which began around 1200 BC and was harsh from 900 to 500 BC, cast a very long shadow on the birth of monotheism (Yasuda, 1993), which identifies a single God in man's image in the heavens, a concept that is very different from the world of monotheism with its presentiments of the existence of God in all living things.

From 1200 BC onwards, regions southward from 35°N, such as the lowlands of Mesopotamia, Israel, the Indus plains, and the downstream region of the Nile were all equally assailed by the dry climate and desertification. Meanwhile, Greece and the Anatolian plains northward of 35°N experienced cooling and increased humid climate.

It is my contention that this temperature drop was the last nail in the coffin of Mycenaean culture and the Hittite Empire, which were directly faced with the crises of population pressure, the drying up of forest resources, and the reduction of arable soil (Yasuda, 1993). In addition, it seems that behind the collapse of these major civilizations, there was a major mass migration, known as a 'Sea of People', and the southward migration of nomads such as the Dorians. This wave of human migration shows a tendency to move from north to south, or from inland to the sea. This fact indicates that the human migration of the time was from the north of the continent and from inland, where the climatic conditions were harsher, to the south and to the coast, which were blessed with a better climate. In other words, there is a strong possibility that the worsening of the climate is closely linked with this human migration. In addition, the major perpetrators of this migration were nomads.

It was in 1200 BC that nomadic Israelis began to live in the Canaan valley. I believe that the increased power of the God of Climate, and the identification in heaven of a single God, are both sides of the same coin. Meanwhile, in Greece, the God of Storms, Zeus, emerged as the most powerful of all gods, but polytheism remained the same as ever. I think that this is probably because a wet climate developed in the region northward of 35°N, thereby ensuring the abundant harvests of the land for that time. In northern Greece, at that time there were still dense forests of deciduous oak and European beech, in which the polytheistic gods could live hidden (Fig. 3).

Suzuki (1976) points out that it was the increasing aridity that raised the Supreme God to

Fig. 3. Greece was hitherto covered with dense forests in which the polytheistic gods continued to live. The Parthenon in Athens, Greece. (Photo by Yoshinori Yasuda)

agricultural peasants, due to a striking drying out of the climate from around 1200 BC.

For many years, humankind lived in a world of animism and polytheism. However, a monotheistic God, identified as the one and only true God in heaven, was born, with a completely different quality from this polytheistic world. Moreover, the major revolutions in human spiritual history also occurred in the dry regions of the desert and the plains of western Europe. The hordes of nomads that came charging out of these arid regions of the Steppes and the desert have swept across human history ever since.

■ The God of Storms

It was the God of Storms who became the monotheistic God. A typical example of this is the God of Climate, Baal (Fig. 4). But how did the God of Storms come to occupy such a preponderant place?

The Mediterranean coast receives rainfall in winter. People wait for these winter rains to arrive and announce that the long, dry season has ended and that at last they can plant barley seeds. In November, the arrival of the rainy season is presaged by violent storms. The wind blows fiercely, and rain pours down from the sky torrentially. For the people of the Mediterranean coast, it is these very winter rains, presaged by violent storms, which promise rice harvests for the year to follow. It is the God of Storms who promises these bounteous harvests.

Until this time, the God of Storms was no more than one god in the polytheistic pantheon. However, the laying waste of the land due to the climate change around 1,200 BC caused a sudden spurt in the belief in the God of Storms. Thus the God of Storms, or the God of Climate, Baal, who could control the rhythm of the rains and the rhythm of the skies, came to wield more power than the goddesses of the earth and harvests when the climate became more arid around 1,200 BC.

Another major additional factor was the

the position of the One True God, and that in Greece, where the forests continued to thrive due to the winter rains, polytheism also continued to thrive, and I believe that he is right. In addition, he also points out that the monotheistic God is deeply bound up with the nomads, whereas polytheism has strong links with the agricultural communities. For the nomads who depended on the unreliable rains alone, the God of Storms came to occupy the position of the leader of the gods. Faced with a similar situation, the Israelis had a direct covenant with this God. Thus, behind the establishment of monotheism is the existence of major factors that caused mass migration and the nomads to invade and conquer the villages of

Fig. 4. The God of Climate, Baal, clasping a great snake in his left hand. Discovered from the Teruk site, Syria. (Photo by Yoshinori Yasuda)

movement of the nomads, driven out of the desert by the increasing aridity and their invasion of agricultural villages. The invasion of these nomads also contributed greatly to the establishment of a monotheistic religion. Belief in the God of Climate was originally a practice of the nomads. Driven out of the desert by increasing aridity, the nomads invaded the villages of agricultural peasants, and their culture had a major influence on the culture of the peasants, which was also an important factor in the birth of monotheism. In this way, the seat of the gods moved from the earth to the sky, and was transferred from the female to the male.

The Hebrew God of Storms, Yahweh, became the monotheistic Judaeo-Christian God, and the God Baal, whose believers were in Syria and Lebanon, became Allah, and came to wield a major influence on Islam.

At almost exactly the same time, the God of Storms, Zeus, also came to wield great power in Greece, on the northern shores of the Mediterranean. However, Zeus did not attain the monotheistic character of Israel, and the polytheistic world was maintained in perpetuity. This is because increasing aridity of Greece northward from 35°N was scarcely remarkable, and furthermore, at the time, Greece was still covered with dense forests. As long as the land could support these forests, polytheism was able to continue existing in Greece.

■ Religious Leaders Who Taught the Importance of Self-Sacrifice

Modern *Homo sapiens* possessed more forceful personalities than the rest of humanity, as if to symbolize the times of fearsome change during their birth some 200,000 to 140,000 years ago. They were driven by a fierce desire to live in comfort. To achieve personal comfort, they possessed powerful selfish genes that permitted murder even among their own community. In all probability, this fiercely selfish gene was acquired while trying to survive the era of drastic climatic change at the end of the Late Glacial period. In addition, the agricultural revolution provided an impetus to the selfish gene of modern *Homo sapiens*. We can also say that civilization was born as the ultimate expression of this desire to live in comfort.

Thus, it was the founders of major religions, such as Jesus and the Buddha, who first espoused the theory of self-sacrifice for the comfort of the others, in opposition to the selfish-gene ways of modern *Homo sapiens*. Jesus bore the sins of all humankind and was crucified. The Buddha explained the importance of comfort by giving himself to the starving tiger in the valley. The founders of the major religions explained the importance of sacrificing oneself so that others might live in comfort, in opposition to the raw desires of modern *Homo sapiens*.

The followers of Jesus and the Buddha, however, thought that the message of the teachings that explained the importance of living for the benefit of others applied only to humans. Consequently, the major religions succeeded in saving humans. However, 'others' include trees, grasses, insects, and all the living things on Earth. The majority of Jesus' followers in particular, only paid attention to humans. For this reason, as Christianity spread, forests were destroyed, and the animals in the forests were killed (Fig. 5). The preachers of the gospel, who should have practised their devotions in remote areas to faithfully accomplish Jesus' teachings of self-sacrifice, instead, destroyed the forests and became the advance guard who killed forest animals.

Let me introduce a single example that could establish the connection between the spread of Christianity in Europe north of the Alps in the beginning of the 12th century, and the destruction of the forests. The 12th century has been called the era of great land reclamation, and the opening up of the great forests encroached into the area north of the Alps. It was the Christian missionaries who were the advance guard who destroyed these forests.

To spread Christianity in the world of the forests north of the Alps, it was first necessary to destroy forests, because forests were dens of heathenism. The white robes of the Druids, who tried desperately to save the sacred oak, were stained red with blood, and the church was founded on the stumps of the oak tree. In this way, with the destruction of the forests the animistic gods of the Germans and the Celts were driven out, and the light of Christianity permeated every land in Europe. This also paved the path for the destruction of the European forests (Yasuda, 1991).

To spread Christianity throughout Europe, it was necessary to destroy the forests, but the climate condition that strongly supported the spread of Christianity was the long, harsh winters of Europe. As Watsuji (1935) has pointed out, 'The agony of the gloom of winter makes a perfect resonance with the fear of the desert,' and it is this desolate European winter that made it so easy for Europeans to accept the religious ideas of the lifeless desert.

That the spread of monotheism is indivisibly connected with climate can also be identified by studying the global spread of Christianity following geographical discoveries. From the 15th century onwards, fierce deforestation accompanied immigration to the New World of the Americas. The North American continent, on which the native American had hitherto lived, was a kingdom of forests. However, in a mere 300 years since its colonization by Christians, more than 80% of the forests of the United States was destroyed (Yasuda, 2001).

Monotheism, born in the lifeless wastelands of the desert, has come to bear being ceaselessly pursued by climatic desolation through desertification, because Christians interpret Jesus' teachings of living through self-sacrifice for others as self-sacrifice directed towards only humans.

Protecting Forests is Protecting Buddhism

■ Forests Protected the World of Polytheism

Although Christianity has not spread throughout Japan, why has it been so successful in Korea? This remains one of the major mysteries for theologians today. When reviewing religion through its relationship with climate, one factor that must be considered is the deep connection with the nature of the forests. There are many reasons why monotheism has not spread in Japan, but I think that the most important reason is the

Fig. 5. A saint stamping a dragon and lifting a cross in the church of Rome. (Photo by Yoshinori Yasuda)

deep connection with the fact that as much as 67% of the land is covered with forests. Shinto, which has passed on to the modern day the traditions of animism that have existed since the Jomon Period, still survives. The fact that a religion that venerates snakes, wolves, and foxes as gods, or worships them as messengers of the gods, still survives, must surely be because of the forests. It must surely be because of the existence of a climate of forests in which the giant trees and animals can survive. Shinto erected shrines and protected the groves in which the gods lived to ensure the safeguarding of its dogma. Buddhism was also changed into a Japanese-style 'Buddhism of the Forests,' with a very strong animistic and polytheistic character, upholding the veneration of mountains and the esoteric teaching that, 'All things and everything, mountains, rivers, grasses, trees, and the soil of this land can all attain the wisdom of Buddhahood.'

As a result, Japanese forests were protected and secured as sacred groves and temple lands. The forests protected the gods of polytheism, and these gods in turn protected the forests. However, it appears that to date, neither theologians nor religious adherents have realized the importance of this mutual protection. This is deeply connected with the fundamental question of what is religion? Religion is not simply the teachings of the Buddha in Buddhism or Jesus in Christianity. It is the teachings of the Buddha who breathed in the forests, reflected while communicating with the forests, experienced them, and gained enlightenment. It is the teachings of Jesus who breathed in the desert, and gained experience and enlightenment by communicating with the desert. This is made clear by the anecdotes that when Jesus climbed the bare Hermon mountain, his robe turned white, whereas when the Buddha lived in the cave but could not gain enlightenment, he left the cave and first gained enlightenment at the base of a large tree. Religion is a system of thought born from the interaction between humans, nature and the climate. We cannot talk of religion in isolation from this interaction.

By looking at a giant tree, you can for the first time experience the sense of reverence for the life of the tree (Fig. 6), and when you stand in the desolation of the desert, you can for the first time feel that the world is going to end. Religion is again a legacy of the climate.

It is the theologians and religious adherents, who gain their daily bread through religion, who must reflect upon this fact most deeply now. If the forests of Japan were to be lost, not only would Japanese Shinto disappear, but so would the teachings of Buddhism and the veneration of mountains and the ideas of the natural philosopher. For people involved with Japanese religions, even though protection of the forests is their daily bread, these religious aficionados do not raise their voices stridently for a better protection of the forests. In particular, the world of Japanese Buddhism today thinks only of humans. Despite the fact that Buddhism is the religion of the forest and if the forest disappears, so will Buddhism, there has yet not been even a hint of a movement to protect the forests, and I can only say that it is truly regrettable that there is no such movement in the world of Buddhism. Buddhists in Japan today are required to practise the Buddha's asceticism to rescue nature. In this current climate, Japanese Shinto has at last realized the importance of the sacred forest groves *(Chinjuno mori)* and begun a movement to protect these 1000-year-old forests. The Shinto priests who maintained their silence during the 50 years that followed World War II, have for the first time realized the importance of their religion of the forest when faced directly with the global environmental crisis.

True human happiness cannot be achieved by humans alone. Despite both—beautiful nature and abundant forests—it is only when the animals and plants in these forests live healthily that true human happiness can be attained. Forests continue to remind us of the importance

Fig. 6.   The giant Japanese cedar, wrapped with a
sacred rope in Yamagata Prefecture, Japan.
(Photo by Yoshinori Yasuda)

of the philosophy of renewal and rebirth. Continuing to live serenely amidst this eternal cycle of renewal and rebirth, is the only religion that can guarantee the continued existence of humankind on this small and limited planet. Today, more than in any other era, it is fervently hoped that the importance of polytheism as protected by the verdant forests will appeal to the world, and contribute to human peace and prosperity.

From an 'Era of Conflict of Civilizations' to an 'Era of Fusion of Civilizations'

■ The Era of Fusion of Civilizations

Since religion is a legacy of the climate, it tends to be limited by that climate. The climate limitations of religions which appeared in the desert are particularly crippling given the need to rescue the present and the future, when directly faced with the twin crises of a population explosion and global environmental problems.

One such shackle is the pecking-order system of monotheism. Christian ideas that elected humans to govern over the organic world inspired the courage to rule over nature and to create human kingdoms. In an era of humans being overwhelmed by nature and global nature being still abundant, such ideas guided humankind to happiness. At the same time, however, these ideas gave humans the right to destroy and exploit nature. In addition, when humans carved out kingdoms on Earth, nature was painfully destroyed and impoverished by human hands.

In this era of a global environmental crisis, it is necessary to reinterpret the will of God. Giving humans to govern over all organic creation does not give humans the right to exploit these life forms, but rather to guard the order and harmony of nature, and humans are morally obliged to maintain that order. Furthermore:

And the Lord said unto Noah, Come thou and all thy house into the ark... Of every clean beast thou shalt take to thee by sevens, the male and his female: and of beasts that are not clean by two, the male and his female. (*The Bible*, Genesis, Chapter VII.)

Here, the number seven and the number two are used to clearly delineate between clean and unclean beasts.

This pecking order is applicable not only to the relationship between humans and nature, but also to relationships between humans. Viewing other religions as heresy, Christianity became a warring religion that approved of the Crusades to spread the faith that it considered to be correct. A monotheistic God, recognized as the one and only true God in heaven, could not help being an intolerant religion based on a premise of class rule.

The idea of an 'Era of Conflict of Civilizations' formulated by Huntington (2000) could only have been born from a warlike, monotheistic view of history based on just such a premise of a ruling class. The professor's cultural theory that the 21st century will be one of clash and conflict between the Christian cultural sphere of the West and the Confucian and the Islamic cultural sphere of the East is an extremely dangerous one to advance in an era when, barely twenty years from now, the population on the planet will reach eight billion. This theory is an invitation for another major war to break out, which if handled badly will unfailingly result in the extinction of humankind. To say that this is the advent of the end of the world according to Christianity is no laughing matter. Religions whose essence is conflict and war cannot possibly govern a world as small as the one this will become in the 21st century.

For humanity to survive in such a small and limited world, we must plan for a peaceful coexistence between humans and nature, control desire, but most importantly, plan the peaceful coexistence of people of different races and religions. Now is not the time to create an 'Era of

Conflict of Civilizations,' but rather, an 'Era of Fusion of Civilizations' (Yasuda, 1994).

At this point, it will not be the religion of the desert or the monotheistic cosmology, but the East Monsoon Asian polytheistic cosmology, the religion of the forests and its attendant background, that will be able to play an important role. The religion of the forests is founded on a philosophy of harmony and peaceful coexistence between all the living things in the world.

The world within the forest endlessly repeats the cycle of life, death and rebirth. While the peoples of the desert hold a linear-development view of history, the concept and cosmology of the peoples of the forest is cyclical and repetitive. The religion of the forests is founded on the principle of equality, not on a pecking order. Human life too is no more than a single strand in the world of life on Earth that endlessly repeats the cycle of birth, death, and rebirth.

The symbol of the Monsoon Asia is the dragon. This mythical beast, with the horns of the deer, the face of a camel and a crocodile, the claws of a hawk, the palms of a tiger, and the body of a crocodile or a snake is the legacy of an amalgam of totems from several different tribes (Fig. 7). The dragon is a concrete manifestation of the possibility of the fusion of civilizations. The dragon has been accepted by Buddhism, Confucianism, and furthermore, Shinto, and has changed its form to accommodate their differences (Fig. 8). It is this very fusibility and adaptability that is most sought in the present civilization.

However, in the West dragons were always killed.

■ The Green Cross

The seas of Alexandria are ultramarine in colour similar to the seas of Japan. People go fishing along the coast. There are many fish, and the ports overflow with fishing boats.

What enriches the seas of Alexandria are the vast quantities of nutrients carried along by the Nile river. By contrast, the clear, deep blue waters of the Aegean hold no fish. It is even rare to see a fishing boat out on the waters. What makes the Aegean Sea so barren and poor is that almost no nutrients are provided from the rivers of Greece and Asia Minor. Rivers that flow through lands that have been deforested cannot supply the seas with nutrients. The clear blue Aegean is a dead sea with few living things.

However, Europeans are in love with this clear blue sea, which stirs up a feeling of exoticism in Japanese. It is an ideal sea to swim in, but that is all. There is no sign here of coexistence between humans and nature.

Fig. 7. The dragon is a symbol of power unification in China. A dragon in a palace, Beijing. (Photo by Yoshinori Yasuda)

Fig. 8. The dragon in Japan is a symbol of the Water God. The dragon in Mitsumine shrine, Saitama Prefecture, Japan. (Photo by Yoshinori Yasuda)

Recently, I have come to doubt the wisdom of the idea that swimming in a dead sea where nothing lives is ultimately healthy.

In the very same Mediterranean, the seas of Alexandria are rich with life due to the blessings bestowed by the Nile river. Forests and seas are linked by rivers and maintained by a cyclical system. If the forests die, the rivers become thin and weak, as do the seas. Humans cannot continue to live by ignoring this cyclical system.

Alexandria was the main stronghold of the Ptolemaic Dynasty (Fig. 9). It was from beneath the streets of this city that catacombs were discovered dating from around the 1st century AD.

Qom Shakafa has 99 steps, and was built 30 m underground. The graves were in part immersed in the aquifer.

To the right and left of the entrance to the graves were carved a decorative Medusa, the symbol of Rome, and a cobra, the symbol of Egypt (Fig. 10). The people of Rome considered Medusa to be a guardian of the underworld, and the people of Egypt thought the same of the cobra.

These catacombs display perfectly the fusion of the Roman and Egyptian Civilizations. Even though Egypt was ruled by Rome, Egyptian Civilization was neither destroyed nor buried, but fused instead with the Roman civilization.

Fig. 9. Egyptians also worshipped the cobra. Alexandria, Egypt. (Photo by Yoshinori Yasuda)

Fig. 10. Medusa and the cobra unearthed at the Qom Shakafa site, Alexandria, Egypt. Medusa (upper) and the cobra (lower) suggest a possible fusion of the Roman and Egyptian civilizations.

Both, a cinereous vulture, the symbol of Upper Egypt, and a cobra, the symbol of Lower Egypt, are affixed to the golden mask of Tutenkhamun (Fig. 11). These symbols depict perfectly the fusion of cultures.

Even Christianity, with its policies of conflict and direct confrontation, underwent a fusion with the existing Germanic and Celtic religions, as it spread throughout Europe north of the Alps. The Feast of St. Martin, for example, is a fusion of the Germanic god Woden's feast, and Christianity. Such were the ways in which Germanic harvest festivals and customs were assimilated into Christianity (Veda, 1995). The cross changed into the green cross by coming into contact with the forested landscapes of Europe.

What monotheists should be forced to reflect upon today are these aspects of fusion and adaptability in Christianity. When monotheism renounces its confrontational and warring aspects for those of fusion and adaptability, it will guarantee the first step to world peace. Religions that would contribute positively to the 21st century would be those rich in the ability to fuse and adapt, founded on a principle of equality rooted in an exchange with nature and with the climate.

Fig. 11. The golden mask of Tutenkhamun also suggests the possible fusion of civilizations. (Photo by Yoshinori Yasuda)

In December 1995, I conducted an international symposium in Nara Prefecture, entitled, *Forest and Civilizations* (Yasuda, 2001), during which I accompanied participating researchers from the Christian sphere of civilization, the West, on an excursion to the Wakamiya Festival at the Kasuga Grand Shrine. At midnight on the morning of 17 December, the priests who came out while it was pitch dark to welcome the God of the Mountain, and were engulfed by the whirling sound of the winds, proceeded from the God of the Mountain to the sacred site with twigs embedded in their hair. The Western researchers were all thoroughly impressed and moved by the sight, citing it as an excellent spectacle.

When I saw this, I felt that an interaction with nature and with the climate, was the wellspring of humanity, and that on this point, both Christians and Buddhists thought alike. God is born out of the interaction between humans and nature, and humans and the climate, and if this is the premise at the start, even believers of other religions can understand it perfectly well. Everyone would then arrive at the conclusion that all things being equal, humans are all equally children of nature, and the same is therefore true for the gods embraced by these human children of nature.

Religion as a Tool to Control Desire

■ A Limpid Soul and a Sincere Life

Thus far, I have discussed the climate of the East Monsoon Asia and West Arid Asia, deserts and forests, and monotheism and polytheism. However, humans cultivate religious faith and feel the presence of God in their interaction with nature and the climate, and on this point, both monotheism and polytheism are similar. This is possible only because humanity is all equally children of the earth, and children of nature. This feeling originates in the wellspring of life.

It is believed that in about 20 or 30 years from now, the world will reach a point where an apocalyptic catastrophe may precipitate such a global environmental crisis, that there would be no more room to discuss monotheism versus polytheism. What is even more terrifying is that people have lost the sense of God in their lives, as summed up by the expression, 'Religion is opium.'

After World War II, Japanese education has completely excised religious teaching from historical teaching based on the notion that 'Religion is opium.' As a result, we have lost sight of the means to control our desires both at school and at home.

During Japan's period of high economic growth, Mammonism was rampant. Money was the only issue of concern, and those with money could accomplish anything. What is wrong with making money? Well, who knows whether it is the pursuit of moneymaking that is totally destroying the global environment?

However, it seems evident that a world in which religion, be it monotheism or polytheism, exists as a spiritual tool for controlling human desires, is far more preferable.

In the past, monotheistic Christianity had a dark history of destroying forests, colonizing territories. Contact with their God, however, by going to church every Sunday and listening to the priest's sermon tells the tale of the hidden possibility of Christianity being able to fulfil a major role as a tool for controlling human nature and human desire. Better still, Islam, whose believers pray to Mecca several times a day, as a religion can probably play an even greater role as a tool for controlling desire. It is very easy to appeal to people who go to a mosque and pray to Mecca by borrowing the name of God, to exercise control over their desires so as to save the endangered global environment.

And what, on the other hand, of the Japanese? They go to the Shinto shrine on New Year, marry in a Christian church, and their funerals are held in Buddhist temples. There are, in truth, many young Japanese who write 'No religion' in the box on their passport applications. I have thought several times that we would be better off writing 'Polytheistic' instead of 'No religion.'

What parents and guardians hope for most from modern primary and junior high schools is moral education. That is because today, the great mistake of post-war Japanese education after World War II that completely neglected religious teaching from school education, has clearly manifested itself in problems such as bullying in school and truancy. This is a running amok of young people who had received a school education devoid of any religious education, especially in historical education. They needed a religion. Japanese school education, and historical education in particular, should seriously include religious issues.

For humanity to stand shoulder to shoulder and survive on this small planet, the existence of religion as a spiritual tool to control human desires, and a cosmology founded on that control, will henceforth grow in importance.

■ **A Culture that Sells its Dreams**

What does it mean in the historical humanitarian sense, to say that Japan is the only developed G7 nation that does not belong to the Christian cultural sphere?

When the countries of the West hammer out a clear policy of using confrontation and war to solve their problems, Japan would indicate a different policy of peaceful coexistence and harmony with others. In addition, Japan would use its economic strength and high technology as a means and plan to achieve this policy of peaceful coexistence and harmony with others. At the beginning of the 21st century, this is historically the humanitarian mission of Japan, an economic superpower and a technologically advanced nation.

We experienced during the 50 years after World War II just how much suffering results from attempting to solve problems through forceful confrontation and war.

The former Soviet Union's invasions of Eastern Europe and Afghanistan, America's meddling in Vietnam and Iraq, and the race to develop nuclear weapons between India, Pakistan and North Korea, far from solving problems, embroiled people in the slime of war, destroyed nature, and ruined people's lives. The result of trying to use force to solve problems only results in the loss of valuable human life and abundant nature. And yet the world pursues, as before, the logic of force, based on a cosmology of survival of the fittest.

In 2020, the earth's population will exceed 8 billion, and it is said that it will break the 10 billion barrier in 2050. However, there are limits to the earth's resources. The arable land that yields our food has a top most limit of 2.1 billion hectares, no matter how you look at it.

In the midst of this population explosion, the 21st century will witness cut-throat competition for resources and foodstuffs. Naturally, force will be used as a solution to the problem, and we will see resource wars and food wars. Especially water resource will reach a crisis point. However, at that time, the Western cosmology of seeing war and conflict as a good thing will once again hold sway. At the same time, we should heed the warning that this will also be the first step towards the extinction of humanity.

To avoid this path to extinction, all humankind must discard weapons as Mohandas K. Gandhi (1869-1948) preached and practised. If we do not do that, we will not be able to avoid the path to our own destruction. In addition, we must discard the pecking-order cosmology and the view of history of survival of the fittest, and adopt a cosmology and view of nature and of peaceful coexistence, founded on egalitarianism.

The Western view of history and cosmology alone, predicated upon force and confrontation, has been unable to maintain peace and prosperity on this small planet, which is already burdened with a population in excess of 6

336

billion. We ought to realize that, far from being able to preserve nature, this cosmology of survival of the fittest predicated upon the idea of force and confrontation is clearly limited when it comes to maintaining even simple peace between people.

But even as I say this should a missile come flying across from North Korea, then the short-sighted criticism of 'What would you do then?' is likely to be thrust at me. Certainly, the solution is not that easy. There is still a long, long way to go before all humankind renounces its weapons. However, it is very wrong to theorize that therefore we too must walk the path to becoming a military superpower, as that would entail treading the same dangerous ways that would lead humanity to the pit of extinction. We should remember again the message of Mahatma Gandhi.

Japan, which became an economic superpower in the latter half of the 20th century, ought to first raise high the ideal of the worldwide abolition of weapons. In addition, when superpowers like America and Europe try to undo solutions to problems by using force or conflict, we should search for a new solution by using the alternate logic of peace and harmonious coexistence. Is this not Japan's mission at the beginning of this new millenium, this era of constant ethnic and religious conflict, where modern Western civilization is bogged down with the global environmental problem?

This may all end in tears. Even if, hypothetically, Japanese civilization were to uphold the ideal of peaceful coexistence between different ethnic groups and harmony with nature, and then were to go down because she did not have either the weapons or the ability to fight, then I believe that should be Japan's contribution to humanity.

That is the Buddha's ideal of self-sacrifice. That is Mahatma Gandhi's wisdom of non-resistance and non-violence. Only a civilization that can truly understand the Buddha's ideal and

Mahatma Gandhi's wisdom can create a new civilization for the 21st century, and be a global leader. If Japanese and Indian civilizations were to live the Buddha's idea of self-sacrifice and Mahatma Gandhi's wisdom of non-resistance to guarantee world peace and prosperity, peaceful coexistence between ethnic groups, and the global environment, Japan and India would undoubtedly become the world's pre-eminent leaders in the 21st century. As the master architects (leaders) of a civilization of peaceful coexistence and harmony, their name would shine forever in the annals of human history.

But can we truly hope for such a resolution? The resolution needs to be thrust deep into the hearts and minds of Japanese and Indian politicians, and each and every Japanese and Indian citizen. When each and every Japanese and Indian citizen exhibits the resolve to achieve this ideal, Japan and India will be assured of their position as the leading civilizations of peace and harmony.

A Chinese scholar who was a visiting professor to the International Research Center for Japanese Studies listened to what I had to say, and told me, 'You're just a dreamer.' In all probability, there are many readers who share his feeling. However, I did have this to say to the Chinese scholar: 'Recently, modern Western civilization has sold us its dream of material wealth. In addition, it rules the world. The 21st century civilization must sell the dream of harmony with nature and peaceful coexistence with people of other cultures, other religions, and other races. When you return to China, you should try to sell this dream too.'

References

Huntington, S. (2000) *Japan's choice in the 21st century*, Syueisha, Tokyo, 205 pp.

Ito, S. (1985): *A Comparative Civilization (Hikakubunmei)*, University of Tokyo Press, 258 pp.

Suzuki, H. (1976): *Transcendentalist and climate (Choetsusha to Fudo Taimeido)*, Tokyo, 168 pp.

Suzuki, H. (1978): *Ideas of the Forest and Ideas of the Desert (Shinrin no Shiko Sabaku no Shiko)*, NHK Books, Tokyo, 222 pp.

Ueda, S. (1995): *Festival and God in Europe* (Yoroppa no Kami to Matsuri). Waseda University Press, 387 pp.

Watsuji, T. (1935): *Climate (Fudo)*, Iwanamishoten, Tokyo, 253 pp.

Yasuda, Y. (1991) *An Age of the Mother Goddess (Daichiboshin no Jidai)* Kadokawashoten Tokyo, 240 pp.

Yasuda, Y. (1993): *Climate Modifies the Civilization (Kiko ga Bunmei wo Kaeru)*, Iwanamishoten, Tokyo, 116 pp.

Yasuda, Y. (1994): *Snake and Cross (Hebi to Jyujika)*, Jinbunshoin, Kyoto.

Yasuda, Y. (ed.) (2001): *Forest and Civilisations*, Lustre Press and Roli Books, Delhi, 200 pp.

Chapter 22

From Harmony to Discord to Harmony?
Mankind's Approach to Natural Environment

KIRIT S. PARIKH

Nature in Prehistoric Societies

Primitive man must have been mystified by the world around him. In an instinctive way, however, he must have understood the importance of nature. The fruits he ate and the animals he hunted were nature's gifts. He may have also feared the bigger wild beasts and nature's fury. So he may have started worshipping nature.

Man's understanding of the importance of nature, the need to live in harmony with it and the duty to protect it evolved over the centuries. Some of this is reflected in the approach to life and nature of many tribal and indigenous people today. Even though theirs may not be the lifestyle that we might want to follow, their perceptions are illuminating. As Martin Claude (1993) describes:

> High in the Himalayas, on the earth's tallest mountain, the snow of Sagarmatha (Everest mountain) is melting. The snows from the Mother of the Ocean, the literal translation of the Sanskrit word sagarmatha, move slowly down the mountainside to feed the streams until they rush forward to reach Gangamatha, or Mother Ganges, the sacred river of India. The holy waters flow toward the great ocean. Far away, on another continent, the Mamas, or Kogi high priests, are singing and dancing on their snowfields, praying that when they die they will return to the source of all life—the highest snow peak in the Sierra Nevada of Northern Colombia, known to them as the Land of the Mother.

Like most indigenous people, the Kogi live by what they call 'the Law of the Mother', a complex code of behaviour developed by the Tayrona people in pre-Colombian times that regulates human behaviour in harmony with plant and animal cycles, astral movements, climatic phenomena, and the sacred geography of the mountains.

Indigenous peoples, who number around 300 million today, are the traditional guardians of the Law of Mother Earth, a code of conservation inspired by a universally held belief that the source of all life is the earth, the mother of all creation.

As Julian Burger, author of the *Gaia Atlas of First Peoples* and secretary for 1993 of the United Nations' International Year for the World's Indigenous People, says: 'For the First Peoples the land is the source of life—a gift from the creator that nourishes, supports, and teaches. Although indigenous people vary widely in their customs, culture, and impact on the land, all consider the Earth a Parent and revere it accordingly. Mother Earth is the centre of the universe, the core of their culture, the origin of their identity as a people. She connects them with their past (as the home of the ancestors), with the present (as a provider of their material needs), and with the future (as the legacy they hold in trust for their children and grandchildren). In this way, indigenousness carries with it a sense of belonging to a place.'

As the human population grew, the hunter-gatherers developed agriculture. It is then that communities developed. Ideas may have evolved and become more structured, perhaps more abstract and more philosophical. Some of these later got transformed into religions. At the same time, it was with the growth of population and its concentration in towns that man's actions began to have a visible impact on nature, at least locally. And perhaps a need to have a world view was felt.

Nature in Religious Thoughts

Some of the dominant ideas of the Hindu world view which have bearing on natural environment, are as follows.

The first thought is that God is in all the elements. The Vedic sages who wrote the Upanishads, saw God in everything. Thus the Svetsvatara Upanishad (Mishra, 1997), says:

> The God who is in fire, who is in the water, who pervades the whole universe, who is in medicines, who is in vegetation, we salute that God.
>
> (Chapter 2, Verse 17)

Indians have the same attitude towards all the objects of the universe. We have recognized the divine being present in the *panchabhutas* (the five elements), namely, air, water, fire, earth and space. We have perceived divine existence in trees, medicines, rivers, lakes, mountains and living beings (Banwari, 1992).

Then again as Joshi (2000) observes that in the *Bhagwat Gita*, Lord Krishna says:

Of all unmovable things	I am Himalaya
of all trees	I am Asvattha
of all beasts	I am Lion
of all birds	I am Garuda (Eagle)
of all fish	I am alligator
of all rivers	I am Ganges.

This invests sanctity to everything, inanimate and animate. One has to respect it and worship it and certainly not misuse it.

This is also reflected in what the *Yajurveda* says:

> I salute all trees
> I salute all forests
> and I salute the master of all forests.

The other thought that dominates Hinduism is reincarnation. One's soul is immortal and at death the soul leaves the body and may be reborn as another being which could be anything, an ant, a fish, a scorpion, a lion or any other form of life. It then follows that one should preserve the habitats of all living beings, for who knows what you may be born as in the next incarnation. Thus, in some incarnations, God himself was born in the form of a tortoise, a boar and a horse.

These ideas have developed into a respect for all life forms and preservation of their habitats and nature.

This is also reflected in the custom of many households in India. I grew up in a Hindu, religious but modern, household. At every meal, before anyone ate, some food was set aside to feed stray animals, cows and dogs. A pot was kept outside the home in which the food was placed.

Influenced by these ideas concerning nature and the rights of all living beings, Hindu sages have recognized the limitations of human nature. Since some of these have an important bearing on Sustainable Development (SD), I elaborate on them.

It is recognized that the desire for consumption and greed for possession are limitless. The more you consume, the more you want to consume, but satisfaction does not increase proportionately. As is observed in *Vayupurana*:

'Desire is not satisfied by consumption. Just as a hungry fire fed with butter becomes even stronger, so does one's desire for consumption grow with consumption.'

Similarly greed is infinite. Man's greed knows no bounds. As an old Sanskrit proverb says:

A pauper pines for hundred
The centurion wants thousands
One with a thousand pines for a million
And a millionaire wants nothing less than
To be the lord of the universe.

This potential for endless demand is realized through advertisement and other forms of persuasion. A consequence of ever-increasing demand is that manufacturers build in obsolescence in their products. This increases the material intensity of the economy and also adds to solid waste generation.

The insatiability of want is not just confined to material goods. It is also for power and dominance. Corporate houses want to be ever bigger not just because economies of scale dominate but also because every chief executive wants to be number one.

Thus another important characteristic of present development patterns is dominance by a few large firms with monopoly power.

Recognizing this, Hindu philosophers have emphasized contentment, giving, and limits and discipline in consumption. Thus as Joshi (2000) observes, *Bhagwat Mahapurana* says:

> One is entitled to only that amount of wealth which is enough for one's upkeep; the rest has to be shared and if one does not do so, the same can be taken away.

Similarly *Ishopnishad* says: Enjoy it by giving it up – don't grab what you see (Joshi, 2000 and Parikh, 2001). This idea that one should not take more than one needs exists also in Jain philosophy, and is defined as *Aparigraha*. Mahatma Gandhi's actions illustrate this. Once, at a river, Gandhi took just a small tumbler of water to wash his face. Someone asked him why he was so miserly when the river was full of water that was free. Gandhi said, 'I should not take anything more than what I need, someone else may need it.'

That contentment leads to happiness and not

unlimited consumption is preached by all major religions. See Box 1.

Box 1: Revolt against Consumer Materialism in Religion

Restraint in consumption has been recognized as a virtue throughout the ages by many religions, as is reflected in their texts and teachings.

In Hinduism:
'When you have the golden gift of contentment, you have everything.'

In Islam:
'It is difficult for a man laden with riches to climb the steep path that leads to bliss.'
'Riches are not from an abundance of worldly goods, but from the contented mind.'

In Taoism:
'He who knows he has enough is rich.'
'To take all one wants is never as good as to stop when one should.'

In Christianity:
'Watch out! Be on your guard against all kinds of greed: a man's life does not consist in the abundance of his possessions.'

In Confucianism:
'Excess and deficiency are equally at fault.'

In Buddhism:
'By the thirst for riches, the foolish man destroys himself as if he were his own enemy.'
'Whoever in this world overcomes his selfish cravings, his sorrow falls away from him, like drops of water from a lotus flower.'

Source: UNDP (1998).

Yet environment is under stress. Nature is being plundered at all levels, local, national and global. Why? What has changed?

The Nature of Industrial Societies

The industrial revolution made a quantum change by introducing energy-intensive technology that multiplied man's productivity manifold. This had large economies of scale. Technical progress continues and we have ever-increasing production. An industrial society has made abundant mass consumption possible.

This culture of mass consumption spreading to more and more people demands more and more material resources. It consumes an increasing quantity of energy and generates an even larger number of by-products, which alter nature's balance.

One should recognize that this is not just American culture, it is the industrial culture. All aspire to it, the Chinese, the Indians, the Africans. They will not rest till they have caught up with the US lifestyle. There is the main threat to our earth.

Consumption Patterns: The Driving Force of Environmental Degradation

The present stress on global environment is primarily due to the consumption pattern of the rich in the world. The 24% of the world's population, that lives in industrialized nations consumes more than 70% of most natural resources and also emits 70% of CO_2. See Table 1.

The insatiability of consumption demand can be seen from the consumption of food. See Box 2.

At present it is not population growth but the consumption patterns of the rich that causes the global environmental problem.

It may be possible to produce enough food to meet the demand if one had a stable population of 8 billion persons and the present level of food consumption in industrial countries remained the same. The same cannot be said about consumption of energy intensive services and products. If each of 8 billion persons were to emit 3.26 t of CO_2 per year, as is the present average emission rate of industrialized countries (or 5.4 t as in the US) then the annual gross atmospheric emissions would be 26 billion tonnes (or 43 billion tonnes using US rate) compared to the present gross emissions of CO_2 of 5.3 billion tonnes per year. The rate of CO_2 build up in the atmosphere would be more than 8 to 14 times the present rate. This is obviously not sustainable.

If we could bring down the consumption of all to the level of India today, the threat to climate change would disappear. But I do not recommend it. A life of poverty for all is not what we should aim at. Of course, as the poor emulate the lifestyles of the rich their growing populations would put a larger stress in future. How do we reconcile sustainable development with the aspirations of developing countries?

Actions for Sustainable Development?

New technology, variously known as clean technology, factor 10 technology, zero emission technology could perhaps help us all attain the present US lifestyle without threatening the global ecosystem.

Not only should we not count on it, as it would be too risky, but also it is only a temporary solution. The insatiable desire for consumption would soon outpace the new technology and we would have to confront the same problem again.

Thus, we have to modify consumption patterns. We must set new goals, new role models for others to emulate. How do we attain such a change in attitude?

We may take the following actions:

Recognize rights of others: all living beings, human and non-human.

This will put a restraint on destroying habitats and ecosystems. Consumption of certain products

Table 1. Consumption patterns for selected commodities: distribution among developed and developing countries

Category	Products	World Total (MMT)**	Share		Per Cap. (kg. or m sq.)		ADR* Developed/ Developing	EDR* USA/ India
			Developed	Developing	Developed	Developing		
Food	Cereals	1801	48%	52%	717	247	3	6
	Milk	533	72%	28%	320	39	8	4
	Meat	114	64%	36%	61	11	6	52
Forest	Round wood	2410	46%	55%	388	339	1	6
	Sawn wood	338	78%	22%	213	19	11	18
	Paper, etc.	224	81%	19%	148	11	14	115
Industrial	Fertilizers	141	60%	40%	70	15	5	6
	Cement	1036	52%	48%	451	130	3	7
	Cotton & wool fabrics	30	47%	53%	15.6	5.8	3	6.4
Metals	Copper	10	86%	14%	7	0.4	19	245
	Iron & Steel	699	80%	20%	469	36	13	22
	Aluminum	22	86%	14%	16	1	19	85
Chemicals	Inorganic	226	87%	13%	163	8	20	54
	Organic	391	85%	15%	274	16	17	28
Transport vehicles	Cars	370	92%	8%	0.283	0.012	24	320
	Commercial vehicles	105	85%	15%	0.075	0.0006	125	102

ADR: Average Disparity Ratio—average per capita consumption of developed/developing countries
EDR: Extreme Disparity Ratio—ratio of average per capita consumption in the US/India.

Notes:

* Cereals data are for 1987
* Milk data include cow milk, buffalo milk and sheep milk (1987)
* Meat data include beef, veal, pork mutton and lamb (1987)
* Round wood include fuel wood—charcoal and industrial round wood (1988)
* Paperboards include newsprint, printing and writing papers and other paper + paperboard (1988) *Statistical Year Book 1987; Handbook of Industrial Statistics 1989; International Trade Statistics Year Book 1987* and *UN FAO book of production 1989.*
* Fertilizer consumption data include nitrogen, phosphate and potash fertilizers *Statistical Year Book 1987; Handbook of industrial Statistics 1989; International Trade Statistics Year Book 1987; UN FAO Book of Production 1989.*
* Cotton and Wool Fabric: *Handbook of Industrial Statistics, 1988 UNIDO* (Cotton fabrics + Woolen fabrics; excluded synthetic will alter this figures substantially; USSR & China excluded due to non-availability of data)
* Cotton and wool fabric—total consumption figures in billion-meter square and per capita consumption figures in meter square. *Statistical Year Book 1987.*
* Per capita data are calculated.

Source: Parikh, J. *et al.,* 1991; Parikh; K., 2001.

Box 2: Insatiable Desires: Even the Demand for Food Resources Goes on Increasing

Since a person can only eat so much, the intake of food calories cannot go on increasing indefinitely with income. However, the quality of food consumed does go up and keeps on improving. Better quality food requires more resources. For example, to produce 1 kg. of poultry meat requires 3 to 4 kg of grains as feed. Thus, consumption of 1 kg of poultry meat is equivalent to consumption of 3 to 4 kg of grains. Though these grains may be of different kinds than those consumed by humans directly, the resources required to grow these grains may be comparable.

In the figure, the direct protein intake and the consumable protein content of that intake are plotted against per capita income for groups of countries that fall in different per capita income classes.

It is seen that while there is a satiation in terms of consumed protein, the demand for consumable protein keeps on increasing with income and shows no sign of satiation.

Thus, even for the resources needed to meet basic needs, the rich and the poor compete. The environmental stress caused increases with income.

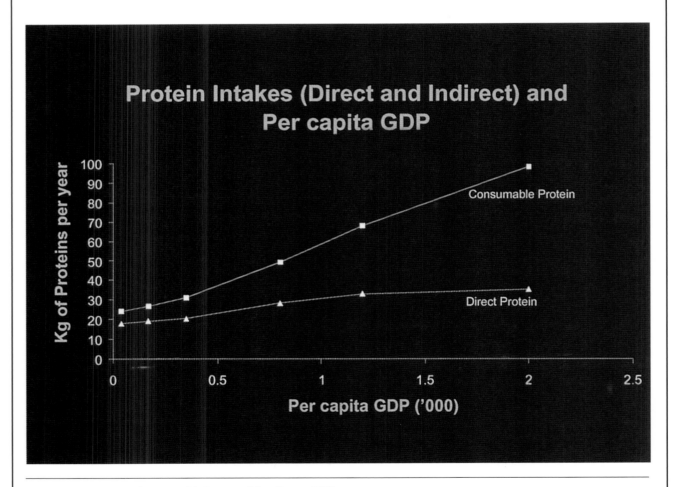

Fig. Protein Intakes—Direct and Indirect and Per Capita GDP.
Source: Parikh et al., (1991) based on data from Linnemann et al., (1979).

will be considered unethical and so its production will cease. We have seen how trade in skins of endangered species and fur coats has virtually ceased through consumer boycott.

This would also mean fairly sharing the environmental space with other human beings. If my share of carbon emission is one tonne of carbon per year, I must limit my consumption to that. This will put a break on certain types of consumption. I would not be able to go on a long drive in a car, but I would be able to take long walks and perhaps enjoy nature more or read or sing or dance.

Empower all people, locally, nationally and globally, with rights to resources and rights to information.

Environmental resources and natural systems have to be looked after everywhere. It cannot be done without the willing participation of all. They should also have knowledge about the state of the environment and what actions and whose actions cause what.

Can such a system function? Would it not lead to a paralysis of decisions and actions? There is a danger that every frivolous, one-point agenda group can stall a participatory decision process. How to avoid this?

The best answer is that one must spread an informed awareness among people. One must integrate concerns of different disciplines and different stakeholders and generate a multi-disciplinary, multi-stakeholder perspective.

Create capacity for such integrative thinking. We need people who can do this; otherwise people will talk across each other and not to each other.

This will require new kinds of research and action institutions. Only when people well-versed in one discipline, work on real life issues with people with other disciplinary perspectives, can effective multi-disciplinarity be generated. Exposure and involvement with different stakeholders can bring richer, integrative and pragmatic insights. Students studying in such an environment can grow into such holistic professionals.

Changing attitudes is a difficult task. However, I think it is not an impossible one. It is not too late to rethink, reorient and restructure our attitudes and lifestyles for a sustainable, equitable, peaceful world living once again in harmony with nature.

References

Banwari, (1992): *Pancavati: Indian Approach to Environment*, translated from Hindi by Asha Vohra, Shri Vinayaka Publications, Delhi.

Joshi, M.M. (2000): Sustainable Consumption: A New Paradigm. Paper presented in the *2nd National Technology Day Lecture* by the Minister for Human Resource Development, Science and Technology and Ocean Development, Government of India, New Delhi.

Martin, C. (1993): Introduction. In Kemf, E. (ed): *The Law of the Mother*, Sierra Club Books, San Francisco, pp. XV-XIX.

Mishra, R.P. (1997), 'The Indian World View and Environmental Crisis in Saraswati,' Baidyanath (ed.), *Integration of Endogenous Cultural Dimension into Development*, D.K. Print World (P) Ltd., New Delhi.

Parikh, J., K. Parikh, S. Gokarn, J.P. Painuly, B. Saha and V. Shukla (1991): *Consumption Patterns: The Driving Force of Environmental Stress*. Report prepared for the UNCED, Indira Gandhi Institute of Development Research, Mumbai.

Parikh, K. (2001): Enjoy It by Giving It Up: Towards Sustainable Development Patterns. In Matsushita K, (ed.): *Environment in the 21st Century and New Development Patterns*, Kluwer Academic Publications, Dordrecht, pp. 271-290.

Prabhupada, A.C.B.S. (1989), *Bhagwat-Gita, As It Is*, Bhaktivedanta Book Trust, Bombay, Ch. 10, pp. 536-542.

UNDP (1998): *Human Development Report 1998*, Oxford University Press, New York, p. 40.

People praying in a Lamaist temple in Tibet.
(Photo by Takeshi Takeda)

The rich water resource supplied by the Asian monsoon has contributed to the development of the civilizations of the Asian monsoon region. A farmer busy planting his field in Guizhou province, China. (Photo by Takeshi Takeda)

Part V

Climate and Civilizations

Rajasthan is located on the northern periphery of the Indian monsoon region, and thus, does not receive the full blessings of the monsoon. Even children must work hard collecting fuel for each day. (Photo by Takeshi Takeda)

Chapter 23

The Rise and Fall of Peruvian and Central American Civilizations: Interconnections with Holocene Climatic Change— A Necessarily Complex Model

ROBERT A. MARCHANT, HENRY HOOGHIEMSTRA AND GERALD A. ISLEBE

Introduction

Within the global change community there has been a growing awareness of the processes of Holocene environmental change, how this impacts our planet, and its human population. Numerous site-specific palaeoecological investigations have produced records of vegetation change that span many thousands of years. Climate changes are invoked to explain vegetation change in older material. More recently, although often spanning several thousand years, the forcing factor of human activity is often used by way of an explanation. The magnitude and spread of this impact is such that the majority of the earth's vegetation cover has been altered or manipulated in some way. The nature of this impact ranges from complete regional deforestation (Marchant and Taylor, 1998) to promoting certain components of the vegetation, such as oil-bearing palms (Vincens *et al.*, 1999). A problem of interpretation arises at times when there is a combination of possible forcing factors, anthropogenic and natural factors operating in conjunction. However, unravelling the complexities of human-environmental interactions are central to understanding past changes in our environment, the mechanisms behind this change and what bearing these may have on current and future responses. Within this chapter we review the archaeological, documentary, linguistic and sedimentological evidence for cultural and environmental change in two case studies (Fig. 1). To compare changes from different sites, chronological control is required. This is mainly derived from

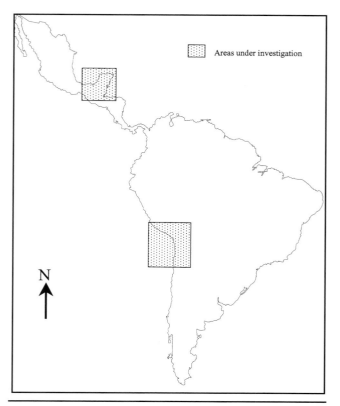

Fig. 1. Map of Latin America indicating the location of the two areas presented: coastal and Andean Peru, and the area settled by the Maya.

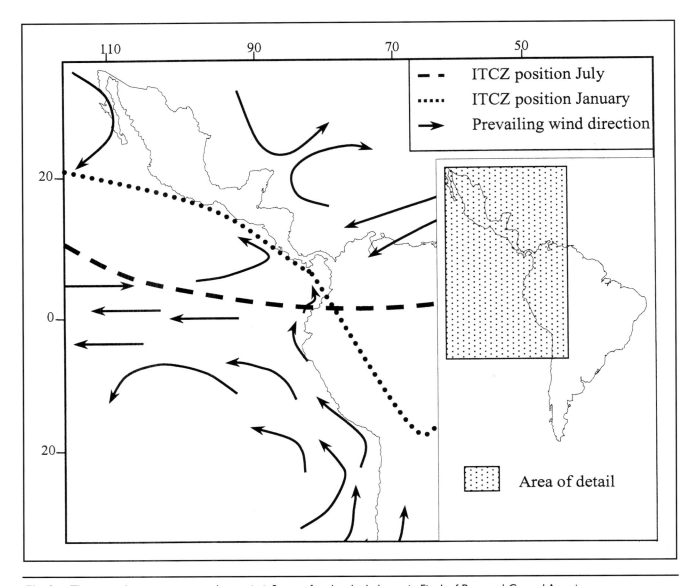

Fig. 2. The main climatic systems and oceanic influence for the shaded area in Fig. 1 of Peru and Central America.

radiocarbon-dated sediments or artefacts. All radiocarbon dates were standardized by conversion to calibrated ages using the Calib 4.3 computer software (Stuiver *et al.*, 1998; http://radiocarbon.pa.qub.ac.uk/calib/calib.htm) from which the most central date is used. Using these data, we investigate possible inter-dependencies between Holocene environmental and cultural change in northwestern Latin America. A brief background to the Latin American environment is presented, indicating the dominant climate systems (Fig. 2), the land-ocean interactions inherent in the climate system and other non-climatic controls that can drive climate change. Using information incorporated within the sedimentary record, the Latin American environment is shown to be quite dynamic throughout the Holocene. We then present the approach used to investigate past cultural-environmental connections; particularly emphasizing the shift away from the over-simplified and widely debunked deterministic view that appears to have become entrenched within palaeoecological study. Archaeological and palaeoecological data for two parts of northwestern Latin America are combined to investigate the nature of the response of separate cultures, and demonstrate possible adaptive strategies adopted in response to changing environmental conditions and how they may evolve.

Environmental Background for Northwestern Latin America

■ The modern climate

Hastenrath (1991) has documented the climate of northwestern South America. The dominant climate system is described by the annual migration of the Inter-Tropical Convergence Zone (ITCZ) or the meteorological equator (Fig. 2). This climate system is characterized by the migration of the equatorial rainfall belt, corresponding to the annual migration of the belt with maximum solar isolation. Due to the influence of the westerlies from the Pacific, and the sharply rising topography of the Andes, the ITCZ has a sinusoidal profile over northern south America (Fig. 2). Here the seasonal intensification of the northeast trade winds dominates the climate system. Due to the proximal location of the Pacific-based moisture source and the steeply rising ground of the Andes mountain chain, precipitation is highest west of the Andes. The third important climate influence brings moisture derived from the Atlantic Ocean, precipitation being high on the eastern slopes of the Andes; the concave nature acting as a natural receptacle for moisture derived from the Atlantic Ocean (Fjeldså, 1993). A strong environmental influence relates to the dramatic topographic differences in the region, the Andes mountain chain being described by approximately 7000 m of altitudinal change. Applying a lapse rate of 6.6 C per 1000 m (van der Hammen and González, 1965), this

altitudinal rise equates to a temperature change of nearly 30°C, and as such results in a significant change in environment recorded over a relatively small area. For example, there are transitions from cool high-altitude grasslands to 'temperate' forests at middle altitudes and diverse tropical rainforests within a few kilometres.

■ Latin American 'monsoon'

Monsoons are commonly associated with the tropical areas where there are seasonal shifts of moisture from sea to land, such as the well-documented monsoon for East Africa and West India arising from the Indian Ocean (Caratini et al., 1994). Although the domain of the classical monsoon is limited from 20°W to 180°E longitude, a similar cycle, although no-less dramatic but more predictable, occurs in Latin America. The added complexity to the South American climate system is concerned with the El Niño Southern Oscillation (ENSO). The climate of the tropical Pacific basin, extending from the western Americas across to Australia, New Zealand, and northeast Asia, oscillates at irregular time intervals (3–7 years) between an El Niño phase with warm tropical waters upwelling off Pacific coastal South America and a La Niña phase with cold tropical waters dominating. As climates, particularly rainfall patterns, are driven by temperature differences between land and ocean, the influence of changing oceanic sea surface temperatures (SST) on the coastal South American environment can be dominant, and have a strong influence elsewhere. The ENSO phenomenon has attracted the attention of the global change community, particularly due to the well-documented economic and cultural impacts, both locally in South America and globally, that are felt within a wide latitudinal band about the equator. Although not a classic monsoon system, the events being episodic rather than annual, ENSO events are the largest coupled ocean-atmosphere phenomena resulting in climatic variability on inter-annual time scales (Godínez-Domínguez *et al.*, 2000).

Complicating these macroclimate systems are factors that cause climate change but do not adhere to strict cycles such as the changes caused by volcanic eruptions. One key eruption in the recent past has been the Huaynaputina eruption in Peru that is recorded at 1600 cal. yrs. AD (Briffa *et al.*, 1998) that was of such force that volcanic glass derived from this eruption has been found in Antarctic Ice core layers (Palais *et al.*, 1990). The main impact of such an eruption would have been aerosol loading into the atmosphere, resulting in changed temperature and moisture balance. The spatial extent of this climatic perturbation would be a function of the location of the eruption and the ambient atmospheric circulation. For example, where the ITCZ has a sinusoidal profile, as over the Northern Andes, this would concentrate any climatic change impact within a relatively restricted area (Zielinski, 2000), with the impacts felt in an elliptical pattern orientated along a north-south axis more than an east-west axis.

Holocene Environmental Dynamics in Northwestern Latin America

Given our understanding of present-day climate dynamics, we know climates of South America are in a continual state of flux. Over the past forty or so years there has been a growing understanding of how Post-glacial climatic events have played a major role in the evolution of Neotropical vegetation. A particularly good understanding of how our planet reflects, and has responded to environmental changes can be obtained from plants; each individual plant species being enclosed within an environmental envelope. By over-lapping individual environmental envelopes, large vegetation communities (biomes) develop. As the environment changes, so will the associated vegetation composition and distribution reflect this change. Additionally, human interaction with plants has been increasingly intertwined with cultural development. For example, in excess of 100 plants were under cultivation before the arrival of the Europeans in the 15th century (Piperno and Pearsall, 1998). Along with being tightly constrained by their environmental envelopes, plants are well suited to unravelling past environmental changes as they leave behind signals such as plant macrofossils and pollen. By accessing archives, normally lake or bog sediments that contain such plant fossils and placing these within a time frame provided by radiocarbon dating it is possible to reconstruct plant communities of the past and determine how the vegetation at a single site has changed over time (Behling *et al.*, 1999; Islebe *et al.*, 1996). However, the problem of equifinality must be considered where a range of interpretations is often available to explain changes in a site-specific study. We know that vegetation patterns result from a factor-complex of ecological, environmental and human-induced variables. This is further complicated by understanding how accurately the pollen reflects the vegetation, the relationship between pollen and the parent taxa not being one-to-one (Van't Veer and Hooghiemstra, 2000; Marchant and Taylor, 2000). Thus, palaeoenvironmental reconstructions from single sites form complex and detailed records of environmental change, but in isolation they are not suitable to decipher regional to supra-regional scale environmental shifts. Indeed, there can be a danger in over-interpreting such records, particularly while implying regional environmental change when only the local is recorded. Climate changes, and how these impact on the vegetation, are rarely spatially uniform. We therefore need to analyse and interpret data from numerous sites to determine the spatial pattern to this variability

(Prentice and Webb III, 1998). To enable such investigations of pollen data at a wide range of spatial and temporal scales, fossil pollen data are retained within the Latin American Pollen Database (http://www.ngdc.noaa.gov/paleo.html) where they can be accessed and analysed simultaneously.

The two main changes in climate are either in temperature or moisture. Additional changes in seasonality also occur although available data is not sufficient to investigate these at present. Unlike the large temperature changes associated with full glacial conditions, the temperature changes associated with the Holocene are thought to be relatively small, for example, the middle Holocene hypsithermal period was possibly only about 1°C warmer relative to today (Marchant *et al.*, 2001a). Servant *et al.*, (1993) suggest that dry periods were more frequent during the middle Holocene, moisture flux along the South American coast being quite responsive to change (Markgraf, 1989; Markgraf *et al.*, 2000). Indeed, shifts in moisture, either in the distribution or intensity, may have been more important in driving environmental changes over the Holocene than temperature (Marchant *et al.*, 2001 b; 2002 a). One of the main archives used to determine moisture shifts is fluctuation in lake levels (Markgraf *et al.*, 2000). To explain moisture changes over the Holocene, we need to stress the importance of convective moisture sources, which could have responded rapidly to increased continentality as a result of rising sea levels. To determine more precisely the influence of sea level changes and ocean circulation in driving moisture changes, modelling studies are needed that can address 'What if?' scenarios (Foley *et al.*, 1996) such as those focused at a similar period on Saharan Africa (Kutzbach *et al.*, 1996). These are able to combine with databased investigations and provide the link and verification between observed and modelled reconstructions (Marchant *et al.*, 2002 b).

Cultural-Environmental Inter-Relationships: A Tricky Business

- **The rise and fall of determinism in linking cultural and environmental change**

In this section we outline our approach for investigating cultural inter-relationships. The relationship between environment and culture has been debated since Mason (1896) noted correlations between environmental conditions and the level of human development. Environmental determinism underpinned Childe's (1926) hypothesis that agricultural origins in the Levant Valley of the Near East were stimulated by climatic warming following the last ice age. Leakey (1931) linked hunter-gatherer settlement patterns to 'pluvial' and 'interpluvial' periods in the Eastern Rift valley of Africa. However, these correlations were constructed from a relatively restricted platform of largely qualitative archaeological and palaeoenvironmental data, much without independent chronology. The argument for environmental determinism to explain environmental causation of cultural changes runs roughly as follows. Vegetation changes recorded within a pollen sequence indicate that climate changes led to a shift in abundance or availability of resources. This results in a shift in attention to other resources, migration leading to changes in technology, subsistence strategy settlement pattern, and social and political structural change. Reversing this argument leads us to think that for any adaptive shift there must be an environmental change to precipitate it. This fails to take into account a whole array of other non-environmental variables (Robertshaw, 1988). Furthermore, correlating environment and culture by invoking adaptation does not tell us anything about how, or why, a particular cultural manifestation came into being. Indeed, the details of environmental-cultural correlation can be easily misconstructed

(Robertshaw, 1988); single causes of cultural change, such as those driven solely by the environment, were rejected in anthropology years ago (Gnécco, 1999). To move from the domain of description, which in some cases has proved to be erroneous, to the domain of explanation we must switch from the synchronic to the diachronic (Robertshaw, 1988). This move will necessitate the incorporation of numerous interactions within the populations, with adjacent populations, and with the backdrop of environmental change, how this is recorded and responded to by the adjacent populations. Rather than investigating how populations responded to climate change, or shifts in the physical environment, we can ask how populations responded and adapted to their perceptions of change in the physical world (McIntosh, 2000). Therefore, the wealth of deterministic explanations that abound in palaeoecological literature should be treated sceptically, particularly as more research is stressing the problems of involving climate change as a driver of human activity and vice-versa.

■ Methods of reconstructing past cultural change

While our understanding of varying climate conditions over the Holocene has improved dramatically over the past decade, our knowledge on the human use of this changing environment has not developed to the same degree (Aldenderfer, 1999). For example, within palaeoecology new methodologies and greater numbers of studies have allowed for regional reconstructions which have given additional insights into the character and impact of climate forcing mechanisms (Marchant et al., 2001 a). As with the synthesis of the information derived from the accumulated sediments, a range of techniques combine to reconstruct a picture of how the composition and distribution of populations has changed over the Holocene.

These range from direct analysis of past occupation layers revealed by archaeological investigations and the associated artefacts such as pottery (Herrera et al., 1999). Additional sources of information to reconstruct past human occupation comes from the modern composition and distribution of ethnic groups (Schurr, 2000), linguistic analysis, investigations of individual starch grains identifiable as manioc (Manihot esculentia), yams (Dioscorea spp.) and arrowroot (Maranta arundinacea) on assemblages of plant milling stones dating from 6000 to 4000 cal. yrs. BC which indicate ancient and independent emergence of plant domestication in the lowland Neotropical forest (Piperno et al., 2000). Other information on past human activity comes from a range of fingerprints on the present-day ecological composition of vegetation communities. For example, the present vegetation of northeastern Guatemala is predominantly tropical semi-evergreen forest interspersed with patches of savanna; some researchers believe this patchwork is a relict of Maya land-use practices (Leyden, 1987). We also incorporate historical evidence, where available, although little documentary evidence survives from before the incursions of the Spanish in the middle 1500s. However, care must be taken to incorporate these direct records in light of other available information; propaganda being rife in historical times. Similarly to the record of environmental change, the Holocene is characterized by recognizable cultural developmental stages; these allow observable transitions in the various sources of evidence and reconstructed over the Holocene.

Despite this plethora of techniques providing insights into the distribution and practices of past human populations, most knowledge of pre-Hispanic civilizations, and their impact on and interactions with the environment continues to come from archaeological excavations. Although these are quite widespread and cover a range of time periods, the spatial and temporal distribution, as also the palaeoenvironmental information, is focused on specific periods and places. As new securely dated material becomes available our existing understanding is also being reassessed. For example, the understanding of the prehistoric, and particularly the Pre-Inca Andean cultures, has shifted notably in recent decades, one of the key developments being the understanding that a strong economic base, and associated societal structure including religion, social stratification and controlled agricultural production was in existence from at least 50 cal. yrs. BC (Morris, 1999).

Case Studies of Cultural-Environmental Dynamics

Two examples from Latin America, located in Peru and Central America (Fig. 1), are investigated to suggest how early civilizations may have responded to Holocene environmental changes. These case studies are chosen as they compare and contrast nicely, representing lowland and highland ecosystems, describing different periods of the Holocene (early to late Holocene respectively), have different cultures with different resource bases and are both influenced by El Niño events. These combine to offer exceptional insights in human-environmental interactions developing under a range of different environments.

■ **Early to middle Holocene environment and culture in coastal and Andean Peru**

The first case study focuses on Eastern Peru (Fig. 3) where there are a number of archaeological and palaeoenvironmental records, the latter being notable for a high-resolution ice-core record of regional environmental change derived from the high Andes (Thompson *et al.*, 1984; 1985; 1994). Although some radiocarbon dating of past occupation layers from the western flanks of the

Peruvian Andes indicates inhabitation in excess of 22,000 cal. yrs. BC, this is very controversial. A more reliable suite of dates indicate human occupation back to 12,000 cal. yrs. BC in the lowlands, and 11,000 cal. yrs. BC at high elevations with entry routes into the region being coastal in origin (Aldenderfer, 1999; Baied and Wheeler, 1993). The timing and direction of this initial colonization is the focus of ongoing debate, this being partly fuelled by numerous archaeological sites being submerged by rising sea levels; sea levels being some 80 m lower than the present-day during the early Holocene (Adams *et al.*, 1999). Indeed, it is not a coincidence that many coastal sites date from 5000 cal. yrs. BC, the time when seas reached modern levels (Caronato *et al.*, 1999). Rising sea levels would have had serious implications for the changing use of marine resources and the ensuing archaeological records. For example, the archaeological site of Quebrada Tacahuay, Peru, dating from 11,000 cal. yrs. BC to 4500 cal. yrs. BC, contains some of the oldest evidence of maritime-based economic activity in the New World. Recovered materials include a hearth, lithic cutting tools and flakes, and abundant processed marine fauna, primarily seabirds and fish (Keefer *et al.*, 1998). The development of early complex societies on the east coast of Peru is further evidenced by the numerous temple structures that predate agriculture (Moseley, 1972). Many of these are dated between 4800 and 1000 cal. yrs. BC, this being earlier than anywhere else in the archaeological record (Kornbacker, 1999). These temples, and presumably the political-religious system they supported, appear to have collapsed relatively rapidly (Burger, 1981), with the last one to demise (1000 cal. yrs. BC) showing signs of strengthening, presumably against mud-flows. Such events are usually associated with increased rainfall around this period (Burger and Gordon, 1998).

It is widely assumed that modern hunter-gatherer societies lived, until very recently, in isolation from food-producing societies and

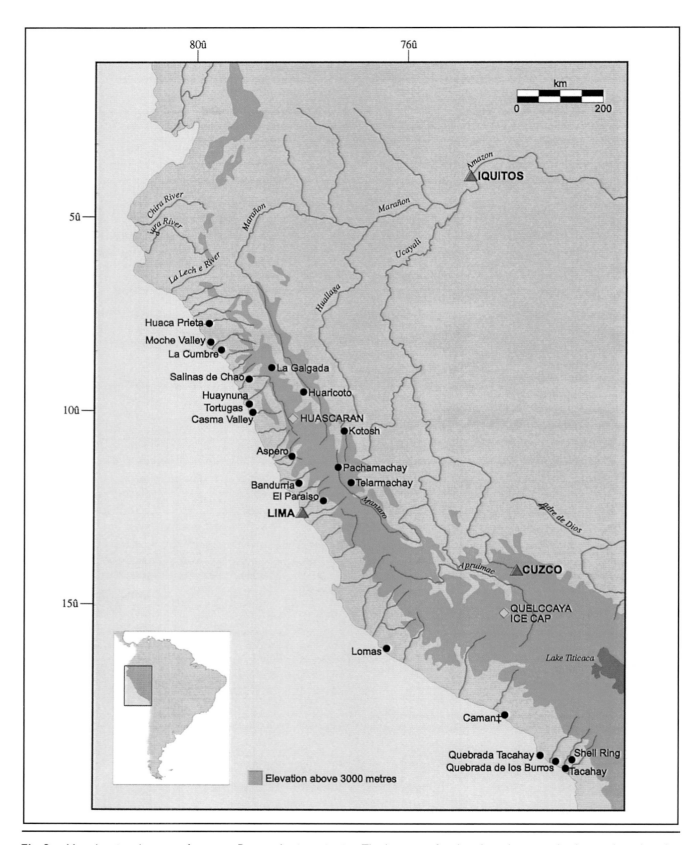

Fig. 3. Map showing the area of western Peru under investigation. The location of archaeological sites, and palaeoecological studies mentioned in the text are shown. The extent of land above 3000 m also shows the close proximity of this to the coast.

Labels within the map:

- km 0 — 200 (scale bar)
- Amazon
- IQUITOS
- Chira River
- ura River
- Marañon
- Marañon
- La Leche River
- Ucayali
- Huallaga
- Huaca Prieta
- Moche Valley
- La Cumbre
- La Galgada
- Salinas de Chao
- Huaricoto
- Huaynuna
- Tortugas
- Casma Valley
- HUASCARAN
- Kotosh
- Aspero
- Pachamachay
- Bandurria
- Telarmachay
- El Paraiso
- LIMA
- Mantaro
- Apruimac
- CUZCO
- Madre de Dios
- QUELCCAYA ICE CAP
- Lomas
- Lake Titicaca
- Camant‡
- Quebrada Tacahay
- Shell Ring
- Quebrada de los Burros
- Tacahay
- Elevation above 3000 metres

states, and that they practised neither cultivation, pastoralism, nor trade—this view being supported by writings of explorers, missionaries, and until very recently anthropologists (Headland and Reid, 1989). Andean Civilizations developed in a range of different environments; the northernmost Peruvian sites fall into two distinct cultural complexes as determined by characteristic tool finds—a unifacial tradition and bifacial tradition thought to represent different technological solutions to similar environmental contexts in a primarily subsistence economy (Aldenderfer, 1999). In the Jauja-Huancayo basin of central Peru there is a long history of semi-nomadic camelid pastoralism dating back 7000 years or more, this economy being well-suited to the Puna ecosystem (Browman, 1974). However, this system is under a number of climatic limitations, the annual rainy season being of crucial importance for pasture regeneration, crucial to sustaining hunting and herding activities. The rains vary from year to year in their inception, duration, location and total fall. In the Peruvian Andes older patterns of nomadic wandering associated with hunting and gathering were supplanted about 1800 cal. yrs. BC by sedentary agricultural villages (Browman, 1974). The earliest agricultural communities are associated with the Chiriapa culture (1800–350 cal. yrs. BC) (Binford et al., 1997), this being characterized by dryland farming that continued after the emergence of the large state at Tiwanaku about 560 cal. yrs. BC. Tiwanaku was the capital of an extensive empire covering some 600,000 km² (Morris, 1999). In the Tiwanaku core area raised-field cultivation in wetland areas started after 600 cal. yrs. BC and became widespread from 1000 cal. yrs. AD (Binford et al., 1997). The Lake Titicaca region became a major centre for Pre-Incan agricultural activity—hardy tubers and the unique cold-adapted chenopod grains (*quinoa* and *caniwa*) being readily cultivated (Thompson et al., 1988). The extent of the agriculture is thought to extend to 800,000 hectares between 3850 and 3800 m; making this the most extensive single area of Pre-Hispanic agriculture in South America (Thompson et al., 1988).

Therefore, Peru was characterized by a wide range of economies, from hunter-gatherer and pastoralists of montane areas to increased sophisticated agricultural communities. Small indigenous societies are likely to have been involved in minor food production and/or trade with large food-producing societies (Headland and Reid, 1989). This trade is partly evidenced by finds of obsidian from a near-shore environment. These tools came from highland sources some 130 km inland indicating that people either travelled to the highlands themselves, or traded with people who did (Sandweiss et al., 1996). A possible stimulus to the growth of trade between maritime hunting and gathering communities to those operating inland and/or vice versa would be a sudden disruption of the population/resource balance. Such a perturbation is suggested for coastal Peru where the Humbolt current yields an extremely resource-rich coastline, except during El Niño events (Yesner, 1980). The people taking advantage of such a rich resource would have to develop some kind of device to get over a decline in food availability. The development of the temples (and the associated religion), and trade with groups from the Peruvian Highlands may provide such a device. To explore these views further we should first look at the variability of the Holocene environment for this area.

The early Holocene was characterized by a drying environment as documented by a drop of 60 m in the water level of Titicaca lake (Cross et al., 2000). These changes correspond to the precession-related northward shift of the ITCZ, and associated moisture, rather than ENSO, as currently defined, which is thought not to have operated in the early Holocene (Markgraf, 1989). A dry warm climate characterized the Central Peruvian Andes between 6000 and 2600 cal. yrs. BC (Hansen et al., 1994). Farther north on the Bolivian Andes, a dry phase is recorded from approximately 4500 cal. yrs. BC (Abbot et al., 1997). δ¹⁸O measurements from an ice core from highland Peru record middle Holocene climatic warming from 7200 to 4250 cal. yrs. BC, with maximum aridity from 5500 to 4200 cal. yrs. BC (Thompson et al., 1995). On the eastern Andes of northern Peru, the dessication signal appears much weaker (Hansen and Rodbell, 1995), possibly due to continued influence from Amazonia-generated moisture. Following middle Holocene aridity, the shallow southern basin of Titicaca lake was completely desiccated prior to 1800 cal. yrs. BC (Abott et al., 1997); a lake level increase being recorded between 2000 and 1600 cal. yrs. BC (Cross et al., 2000). This period of environmental change appears to have a global character with the strongest signal being recorded in the tropics. For example, this is recorded as a dry period in Africa (Marchant et al., 1997; Vincens et al., 1999) and a wet shift in South America (Behling and Hooghiemstra, 2000; Berrio et al., 2002). More recently, the Quelccaya ice core record shows four dry periods (540–610 cal. yrs. AD, 650–730 cal. yrs. AD, 1040–1490 cal. yrs. AD, and 1720–1860 cal. yrs. AD). Although the intervening period would have been relatively wet, two were particularly so (760–1040 cal. yrs. BP and 1500–1720 cal. yrs. AD) (Thompson, 1992). The Quelccaya ice cores, while showing naturally occurring events, also record some human activity. For example, fine particles in the cores dated at 900 and 600 cal. yrs. AD indicate increased agricultural activity corresponding to a relatively dry period (Thompson et al., 1988).

One of the main climatic mechanisms to impact the Peruvian resource base, particularly through moisture variability, is changeable ENSO activity. Such impact would particularly influence the Pacific coast, being recognizable to Peruvian societies. For example, normal variation in precipitation is ± 20% but is also described by very wet (270%) and very dry (19%) years (Browman, 1974). ENSO events were absent or significantly different between 6800 and 3800 cal. yrs. BC (Sandwiess et al., 2001), the timing of this appearing to coincide with the establishment of modern climatic conditions and sea level. Intense ENSO activity did not start until after 1600 cal. yrs. BC (Marchant et al., 1999). For example, a particularly strong period of ENSO activity has been dated between 1700 and 100 cal. yrs. BC, this being associated with alternating droughts and/or heavy rains (Sandwiess et al., 2001). The consequent effect of this ENSO activity on the biological food chains in the Peruvian cold current, and disturbances to in-shore fishing communities would have been serious (Manzanilla, 1997). Archaeological data from some twenty sites along the Pacific coast (Fig. 3) indicate a large molluscan assemblage within middens that show a faunal response to changing oceanic conditions between 4700 and 1500 cal. yrs. BC (Sandwiess et al., 2001). For example, in coastal Peru, El Niño events significantly reduce mollusc populations (Moore, 1991). The middle Holocene onset of ENSO, and the associated impact of increased climatic variability on coastal resources, may have facilitated transition or connection to highland locations. Such connections can be seen as the temple mounds around Titicaca lake, these being indicative of intensive cultivation in raised fields, a system inaugurated by the Pre-Inca civilizations (Kolata, 1986). Human and camelid offerings at Akapana, Tiwanaku's main pyramid site, are dated to about 900 cal. yrs. AD. This may be related to a period of environmental change recorded between 600 and 1000 cal. yrs. AD (Paulsen, 1976). Strong rains and floods had a considerable effect on the Chimu agricultural system, so that intensive cultivation strategies were adopted, particularly raised fields (Moore, 1991). The substantial 'decline' in regional agriculture after 50 cal. yrs. BC, that has been attributed to a loss of soil fertility and/or climate change is not detected in the palaeoecological record (Chepstow-Lusty et al., 1996). Elsewhere, agricultural terracing may have been adopted in response to variable moisture supply, such as recorded from 560 to 590 cal. yrs. AD (Keys, 2000). Additionally, coastal Peru was affected by a series of disastrous El Niño years around 890 cal. yrs. AD (Chepstow-Lusty *et al.*, 1996) that would have produced relatively dry environmental conditions in Central America.

Thus, the area of Peru under investigation has, and continues to be, an area characterized by a dynamic environment—the only constant being change. One of the main drivers of such change comes from ENSO variability. Such a variable environment places a premium on mobility and diversity of resources, therefore the development of secondary subsistence modes such as connections between Andean and coastal dwellers are likely to have a long history. However, connections cannot be just confined to times of food shortage and it is likely there developed an exchange of goods including food and locally available goods such as obsidian. For example, camelid meat from pastoral communities living at high altitude was transportable following preservation by drying into strips called Ch'arki (Stahl, 1999). The relationship between these two groups is unlikely to have been one solely concerned with food. It is likely there was a myriad of connections at the material, social and spiritual levels. The nature of these connections can only be inferred as these aspects of former cultures are not readily preserved within the archaeological record, and modern analogues are not available.

■ Late-Holocene environmental and cultural background in the Maya lowlands

The second example comes from the lowland area of Yucatán and Guatemala; the area inhabited by the Maya (Fig. 4). Similarly to the Peruvian case study, this area has a wealth of archaeological information enabling past cultures and their economic base to be reconstructed. This area has been intensively studied, partly due to the richness of the artefacts left behind by the Maya and partly because there was large-scale collapse of the population around 900 cal. yrs. AD, this taking place over a relatively short time (50–100 years) (Leyden, 1987). The Maya stem from about 50 cal. yrs. BC with the Classical period lasting upto 1150 cal. yrs. AD; this culminating in a large population that had developed an intricate system of agriculture, ritual and social cohesion. Exponential growth through the Classical period led to large settlements, increased mono-culture, and consequent expansion into the less fertile Yucatán Peninsula (Fig. 4) (Hsu, 2000). For example, isotopic analyses of faunal dogs and deer remains at Colha indicate increased C4 plant (maize) diet and herbivory over time, reflecting a subsistence shift at the end of the Preclassic period (White *et al.*, 2001). The main associated impact on the environment of this expansion has been forest clearance that may have resulted in associated soil erosion. However, regional deforestation is known to have begun much earlier, initial deforestation being dated at 3350 cal. yrs. BC in Belize (Alcala-Herrera *et al.*, 1994), and 3900 cal. yrs. BC from Guatemala (Islebe *et al.*, 1996). Disturbance indicators in Belize, such as *Ambrosia*, Asteraceae and *Rumex* reach their highest abundance after this period, crop plants (*Zea mais*; Amaranthaceae-Chenopodiaceae (e.g. *Chenopodium quinoa*) also being present (Alcala-Herrera *et al.*, 1994). However, large-scale deforestation is really only attributed to the Classic period when population density reached some 200–300 people per square kilometre (Morely *et al.*, 1983). This latter period is coincidental with the timing of marl deposition at Cobweb Swamp, Belize, this being topped by peat indicating that swamp vegetation had recolonized the site following a period of erosion and ensuing sediment inwash to the swamp (Alcala-Herrera *et al.*, 1994). Elsewhere in Mexico, exploitation of increasingly marginal lands appears to have resulted in extensive soil erosion, reaching a maximum about 1040 cal. yrs. AD (McAuliffe *et al.*, 2001). While severe soil loss appears to have occurred within the Peten lakes region to the north, soil loss in the Petexbatún areas appears relatively minimal, where effective land-use management, such as terracing, was able to prevent widespread erosion (Beach and Dunning, 1995; Dunning and Beach, 1994).

Many reconstructions of the collapse of the Classical Maya political, social and economic systems around 1050 cal. yrs. AD emphasize anthropogenically induced failure (Santley *et al.*, 1986). Other people suggest the collapse was not so catastrophic and the magnitude of population decline was not uniform throughout the Mayan region (Rice and Rice, 1984). The debate as to the nature and duration of the catalyst(s) that precipitated the rapid population decline continues to rage on, partly driven by varying responses from different regions. However, there is a consensus that several years of unstable climate may have been a contributory factor in undermining an otherwise stable agricultural system (Gunn *et al.*, 1995; Jacob, 1995), particularly in more mesic locations. Although this is likely to have been of similar duration and magnitude across the region, the 'on-the-ground' environmental response would have been quite diverse. Within some areas, the Maya appear to have been able to adapt to changes. Thus, the popular model of human-induced self-destruction of the complex society is not readily applicable throughout the Mayan area to explain the population collapse. For example, in the Petexbatún area political and economic conflict appear to be causal factors in societal failure (Inonmata, 1997). Recent research indicates the process of agricultural failure was regional, and did not occur throughout the Mayan area with the resultant landscape a mosaic of field and forest (Dunning *et al.*, 1998). Today, northeastern Guatemala is predominantly tropical seasonal forest interspersed with patches of savanna (Fig. 4); this patchwork possibly reflective of Maya land-use practices (Leyden, 1987). The impact of environmental change, particularly the proposed shift to drier environmental conditions, would have been most strongly recorded where there were intensive land-use practices and/or population pressure was high. In the southern areas of the Maya territory (Fig. 4), the combination of low population densities and sophisticated agricultural techniques allowed continuous land use without the coincident widespread destruction reported from areas to the north (Emery *et al.*, 2000). This is supported by palaeobiological examination of human remains that indicates nutritional deterioration or health problems over time (Wright and White, 1996). Indeed, isotopic evidence from deer within the Petexbatún area indicates there was no deterioration in the availability of crops, depopulation not being extreme enough to affect the diet of these common foraging animals (Emery *et al.*, 2000).

To understand the possible contribution from climate dynamics in this area we must investigate the regional evidence for Holocene climate change. However, separating out human-induced and climate-induced changes can be problematic as the environmental impacts from both forcing factors are interchangeable. Furthermore, a uniform response to a single climate-forcing factor throughout the area inhabited by the Maya that is geologically, ecologically, and climatically diverse (Fig. 4) would similarly be diverse. The middle Holocene is characterized by a relatively xeric environment for much of northwestern Latin America (Marchant *et al.*, 2002b). In contrast to northwestern South America, the circum-Carribean area records a shift to dry conditions much later; where it occurs at approximately 780 cal. yrs. BC (Bradbury *et al.*, 1981; Bush *et al.*, 1992; Curtis *et al.*, 1998). This is followed by a cooling climate from the middle Holocene that may have allowed the Mayan Civilization to prosper. The cooling of this area would have forced the malaria-carrying mosquito to migrate further to the south, allowing farming

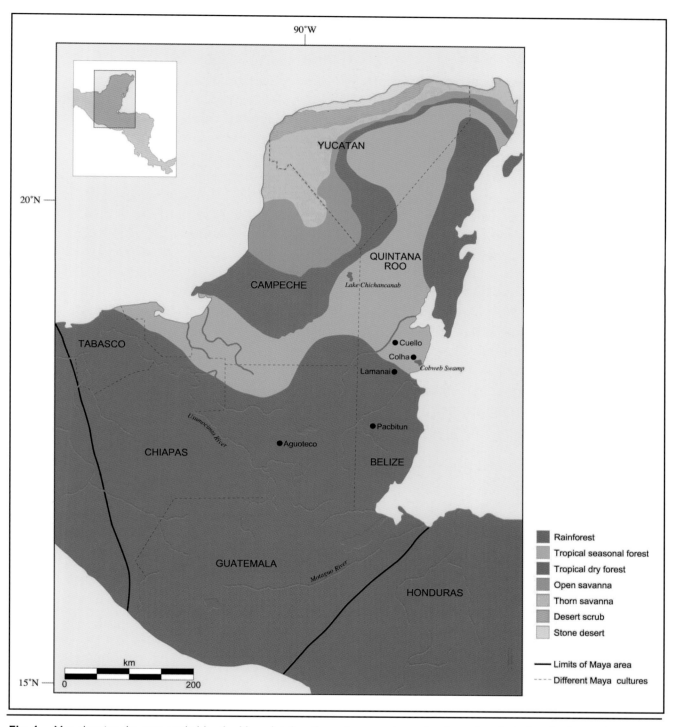

Fig. 4. Map showing the area settled by the Maya detailing the main vegetation units, the archaeological, and palaeoecological sites mentioned in the text.

expansion and construction of cities (Hsu, 2000). A well-dated multi-proxy study from Chichancanab (Hodell *et al.*, 1995) indicates a significant shift to a relatively dry environment about 980 cal. yrs. AD. Taken in isolation this record cannot support the case for regional drying, in combination with a number of records from Colombia (Marchant *et al.*, 2001 a),

363

Guatemala (Islebe *et al.*, 1996) and Mexico (Metcalfe *et al.*, 2000); these indicating a phase of relatively rapid climate shift about 980 cal. yrs. AD. Reduced precipitation is recorded in a number of Central Andean areas and more extensively western South America, for example a dry period is recorded by a low stand at Titicaca lake (Binford *et al.*, 1997). Although the resolution of many palaeoarchives is not sufficient to fully characterize this period, it seems that a series of strong El Niño years around 1050 cal. yrs. AD (Chepstow-Lusty *et al.*, 1996) resulted in relatively dry environmental conditions in the area inhabited by the Maya. This event may also have coincided with a period of increased solar activity as a part of a 208-year duration cyclical variation in solar activity (Hodell *et al.*, 2001). Within the Yucatán Peninsula where the ambient environment is already xeric, this appears coincident, and possibly precipitative, of the cultural collapse of the Maya about 900 cal. yrs. AD (Hodell *et al.*, 1995). However, such a model cannot be applied throughout the area. Environmental stress centred within the Yucatán Peninsula may have induced migration and possible political tensions in areas to the south. Again, the impact that such political tensions may lead to are difficult to reconstruct from the fragmented records on past human and environmental dynamics. Such problems are the focus of the next section.

Environmental Adaptation—the Response of Human Populations to Environmental Change

This section will indicate the possible interconnections between environmental and cultural change. Holocene environmental conditions in northern South America, in common with most places of the world, have alternated between different climate modes—the only constant appears to be change. Adaptive strategies used to cope with different environments have been associated with diminishing territorial mobility and the appearance of new technologies for processing previously unused, or harder to get, resources (Gnécco, 1999). It appears that climatic events resulting in ecosystem changes would have played an important role in the evolution of these strategies (Gnécco, 1999). Particularly important would have been relatively sudden changes, occurring over a single generation which would have created difficulties or opportunities. Indeed, the full implications of these sudden pulses of environmental change have barely begun to be considered (Adams *et al.*, 1999).

Both the coastal and highland areas in Peru are climatically very sensitive. Climate would have played, as today, an important role for cultures living in these areas. Dry periods during the Holocene in Andean Peru are associated with crop failure and, in extreme cases, famine, whereas wet periods are associated with floods and accompanying loss of infrastructure and crops. This situation leads to the highland and coastal cultures flourishing out of phase with each other (Paulsen, 1976). To begin any investigation into the role of it is to the environmental threshold for food production. In the Peruvian example, there is such a varied array of economic bases with a concomitant array of environments that changes would have been responded to in a manner very different to the Maya. Environmental thresholds are dynamic and will be crossed through time as climates change, populations grow, cultures and their technologies develop, and resources are depleted or substituted (Seltzer and Hastorf, 1990). Passing these thresholds is likely to have been more rapid when there was a relatively large population, in relation to the resource base. As populations grow and become more complex, their ability to buffer change may become more complex as more connections are made, both within the society and with adjacent populations (Fig. 5). This growth in

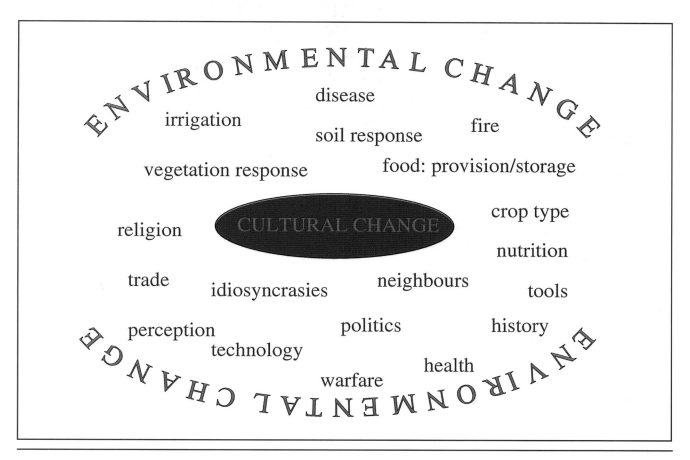

Fig. 5. Model of growing complexity in cultural environmental interactions. When cultures become more developed the connections and complexities increase both within the population and with connections outside.

cultural complexity, and the associated ability to buffer deleterious effects of climate changes, will incorporate some, if not all of the components indicated in Fig. 6. The range of possible inter-linking factors between environmental change and the culture become more numerous through time. An example of this increased complexity is shown by developing state-administered systems, 'urban' populations being able to exploit many widely separate habitats creating elaborate networks for the trade and redistribution of geographically restricted products such as shells, obsidian and food (Peterson and Peterson, 1992); shells of giant snails efficiently substituted stone scrapers to work wood (Prous and Fogaça, 1999). Basic subsistence produce like corn and beans also became part of the complex system of redistribution to support growing populations in

urban centres (Peterson and Peterson, 1992). Ethnic groups did not occupy partitioning territories within which a given type of economy was developed. There would have been an overlap, combinations and connections that would respond to a shift in climatically controlled resource availability in a number of ways. The close correlation and synergistic links between climate change, vegetation response, and cultural development are present although these are intertwined with technology, history, cultural practice, religion, perception, individual and group idiosyncrasies (Fig. 6); these affecting the way a society and its members respond to change (Sandwiess *et al.*, 2001).

Notwithstanding doubts about uni-directional 'cause and effect' relationships between environmental and cultural change, which

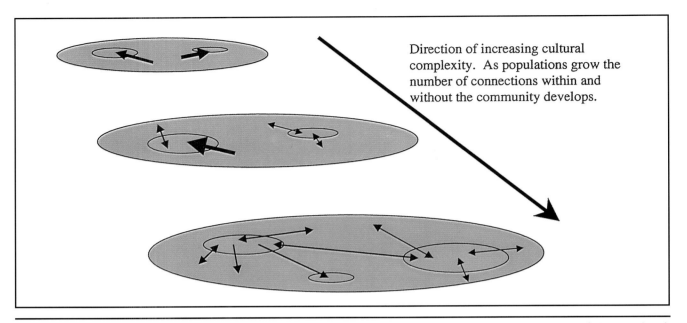

Fig. 6. Model of cultural environmental interaction and the various influences that play a part in this development. Culture is placed centrally with environmental change all encompassing. Intermediate between these two are a range of factors that will determine how environmental change may translate into a cultural impact.

environmental determinism demands, there are strong synchronicities between archaeo-historical and palaeoenvironmental data (Proctor, 1998). Correlations between climatic and cultural change can be found throughout the world (van Geel *et al.*, 1996). However, for every example of a positive correlation a negative one can be found. These 'non-conformities' do not necessarily stem from the non-impact of climatic change but could arise because the community being impacted may be able to respond to, and possibly buffer, the impacts by adaptation. A thorough understanding of past cultural-environmental dynamics is key to fully understanding present-day human interactions and landscape impacts and will allow us to contextualize the present-day landscape. Although the amount of data has greatly increased over the last decade many gaps remain, the evidence being biased towards certain key sites and periods. From the available data, it is apparent that long-term agricultural activities and exploitation of natural resources had resulted in the substantial modification of natural habitats (Fig. 7). Such early practices, combined with

insensitive land management, resulted in the heavily degraded landscape observable today at numerous locations (Fig. 8). This common depiction of human impact being deleterious to the environment is not always the case. A different depiction (Fig. 9) shows an apparently productive agricultural system that is better suited to the environment and results in a more sustainable land-use (Fig. 10). Indeed, in some areas protection may have been afforded to certain areas, this protection continuing through to the present day, under the guise of protected area legislation. As cultures became more complex certain areas would have been concerned with aspects of the environment other than for agricultural production. For example, extensive trade in Quetzal feathers by the Aztec may have resulted in many cloud forests and associated flora and fauna being disturbed considerably over several hundreds of years (Peterson and Peterson, 1992). Other activity may have encouraged the local growth of naturalized species such as cashew (*Anacardium occidentale*) that appears to have spread

Fig. 7. Depiction of early land-clearance practices (after Mendoza de Caltha, 1643). Such scenes of deforestation and land-use make for powerful images of anthropocentric control on the environment.

367

Fig. 8. An area of felled tropical seasonal rainforest from approximately 100 km northwest of Rio de Janeiro, Brazil. The ensuing land management focuses around grazing, increasingly by goats; such practice resulting in extensive soil erosion, land degradation and increasing marginal land use. (Photo R. Marchant)

Fig. 9. A relatively harmonious view of agriculture with plentiful crops and an ensuing content population (from Poma de Ayala, 1583).

following forest clearance (Roosevelt *et al.*, 1996). In numerous locations archaeological, enthnobotanical and palaeoecological evidence indicates that the lowland forests, once thought to be virgin, were at least in some parts, settled, cut, burned, and cultivated repeatedly during Prehistoric times. Indeed, substantial segments of present patterns of biodiversity may relate to such past human activities (Tuomisto and Ruokolainen, 1997) and how these interact with a dynamic environment. Understanding past land-use systems, associated social structure and how these were able to precipitate and respond to changes in the environment are crucial for strengthening traditional land-use practices. However, the connections between climate and culture remain largely reflective of the given environmental setting and require case-by-case study with high-temporal resolution of investigations and precision of the dating. For example, understanding the evolution of the ENSO system, and its influence on Prehistoric societies in the Pacific basin, requires annually resolved palaeoenvironmental records as preserved within coral ecosystems (Cole *et al.*, 2000).

Throughout our synthesis we have presented a range of information from two different schools of natural science and indicated the possible inter-relationships between independent findings. Combining information from these disciplines, with independent methodologies and interpretations of data, will enhance our understanding of past environmental variability, and how cultures respond to this. As a caveat to such interdisciplinary endeavours, caution should be levied to the marriage of

Fig. 10. A highly developed system of agricultural terraces from the Rukiga Highlands, Southwest Uganda. Agricultural practice of inter-cropping in narrow strips has developed over a long period. The crops are harvested at different times and have different rooting depths preventing soil loss in an area of steep topography and high rainfall. (Photo R. Marchant)

371

archaeological and palaeoenvironmental data as there are numerous problems of sampling biases and paucity of data found in both disciplines (Robertshaw, 1988) these multiplying up when investigated in combination. In the areas we investigated the intercentury-scale palaeoclimatic and archaeological data range from the locally quite excellent to the problematic or non-existent. In addition to increasing the resolution, there is need to apply as yet under-utilized components of sediments that show some potential as proxies of human impact such as fungal spores, grass cuticles, isotopic analysis (phytoliths), seeds, charcoal and geochemical proxies. This synthesis has indicated areas, and periods, where our interpretations are particularly contentious due either to a lack of information or conflicting data sets. These areas could be targets for new initiatives, when combined with a spatial awareness of similar cultural shift in other parts of the tropics (Taylor *et al.*, 2000), and it may be possible to make some tantalizing links combining theories of climate change, environmental and cultural response.

Conclusion

The different systems represented here, like all human-environmental systems, had recurrent crises and subsequent adjustments. The impact of environmental change on these systems would vary spatially depending on the ability of the culture to respond to such change. Diverse agricultural strategies, including connections with outside communities for trade appear more responsive to macro-scale environmental change than reactively restricted agricultural systems.

The Holocene has witnessed strong environmental and cultural transitions. Within this picture of a changing environment was the influence of sea levels as they approached modern levels and the associated establishment of the ENSO. This latter component of the climate system appears to have been an increasingly important driver in determining the environment of northwestern Neotropical areas. Given the widespread occurrence of strong pulses of climate change, such as those centred about 1800 cal. yrs. BC and 1050 cal. yrs. AD, the characterization of these should be a target for future investigation. Indeed, further work is necessary to add new primary data on cultures and environmental change *in situ*. These data can be used to model and test the stresses and opportunities presented by different climatic regimes.

There is little doubt that the physical, biological and climatic environment has influenced the nature and development of human culture, although this influence is mediated via an increasing number of complex interactions within society. Periods of significant environmental change appear to have led to 'stress' on established societies often resulting in cultural and technological changes, these changes occurring in a complex and not uni-directional manner. To understand the human dimension within global change science is of immense societal, political and economic relevance. Indeed, we must take into account the long-term relationships between environment and human activity, both past and present. To fully understand interactions between environments and cultures we must combine a spatial perspective with site-specific studies that combine archaeology and palaeoecology *in situ*.

References

Abbot M.B., G. Seltzer, K.R. Kelts and J. Southon (1997): Holocene palaeohydrology of the tropical Andes from lake records. *Quaternary Research*, 47: 70–80.

Adams, J., M. Maslin and E. Thomas (1999): Sudden climate transitions during the Quaternary. *Progress in Physical Geography*, 23: 1–36.

Alcal-Herrera, J.A., J.S. Jacob, M.L.M. Castillo and R.W. Neck (1994): Holocene palaeoasalinity in a Maya wetland, Belize, inferred from the microfaunal assemblage. *Quaternary Research*, 41: 121–130.

Aldenderfer, M. (1999): The Pleistocene/Holocene transition in Peru and its effects upon human use of the landscape. *Quaternary International*, 54: 11–19.

Baied, C.A. and Wheeler, J.C (1993): Evolution of high Andean puna ecosystems: environment, climate, and cultural change over the last 12,000 years in the central Andes. *Mountain Research and Development*, 13: 145–156.

Beach, T., and N. Dunning (1995): Ancient Maya terracing and modern conservation in the Peten rainforest of Guatemala. *Journal of Soil and Water Conservation*, 50: 138–145.

Behling, H., J.C. Berrio and H. Hooghiemstra (1999): Late Quaternary pollen records from the middle Caquetá river basin in central Colombian Amazon. *Palaeogeography, Palaeoclimatology, Palaeoecology*, 145: 193–213.

Behling H. and H. Hooghiemstra (2000): Holocene Amazon rainforest-savanna dynamics and climatic implications: high-resolution pollen record from Laguna Loma Linda in eastern Colombia. *Journal of Quaternary Science*, 15: 687–695.

Berrio J.C., H. Hooghiemstra, H. Behling, P. Botero and K. van der Borg (2002): Late Quaternary savanna history of the Colombian Llanos Orientales from Lagunas Chenevo and Mozambique: a transect synthesis. *The Holocene*, 12: 35–48.

Binford, M.W., A.L. Kolata, M. Brenner, J.W. Janusek, M.T. Seddon, M. Abbott and J.H. Curtis (1997): Climate variation and the rise and fall of an Andean civilisation. *Quaternary Research*, 47: 235–248.

Bradbury, J.P., B. Leyden, M.L. Salgado-Labouriau, W.M. Lewis Jn., C. Schubert, M.W. Binford, C. Schubert, D.G. Frey, D.R. Whitehead and F.H. Weibezahn (1981): Late Quaternary environmental history of Lake Valencia, Venezuela. *Science*. 214: 1299–1305.

Briffa, K., P.D. Jones, F.H. Schweingruber and T.J. Osbourne (1998): Influence of volcanic eruptions on northern hemisphere summer temperature over the past 600 years. *Nature*, 393: 450–455.

Browman, D.L. (1974): Pastoral nomadism in the Andes. *Current Anthropology*, 15: 188–196.

Burger, R.L. (1981): The radiocarbon evidence for the temporal priority of Chavín de Huántar. *American Antiquity*, 46: 592–602.

Burger, R.L. and R. B. Gordon (1998): Early Central Andean metalworking from Mina Perdida, Peru. *Science*, 282: 1108–1111.

Bush, M., D.O. Piperno, P. Colinvaux, P.E. Oliveria, L.A. Krissek, M.C. Miller and W.E. Rowe (1992): A 14,300-yr palaeoecological profile of a lowland tropical lake in Panama. *Ecological Monographs*, 62: 251–275.

Caratini, C., I. Bentab, M. Fontugne, M.T. Morzadec-Kerfourn, J.P. Pascal and C. Tissot (1994): A less humid climate since ca. 3500 yr BP from marine cores off Karwar, Western India. *Palaeogeography, Palaeoclimatology, Palaeoecology*, 109: 371–384.

Caronato, A., M. Salemme and J. Rabassa (1999): Palaeo-environmental conditions during the early peopling of southernmost South America (Late Glacial-Early Holocene, 14–8 ka BP). *Quaternary International*, 53: 77–92.

Chepstow-Lusty, A.J., K.D. Bennett, V.R. Switsur, A. Kendall (1996): 4000 years of human impact and vegetation change in the central Peruvian Andes—events paralleling the Maya record. *Antiquity*, 70: 824–833.

Childe, V.G. (1926): *The most ancient Near East.* Routledge and Kegan Paul, London.

Cole, J.E., R.B. Dunbar, T.R. McClanahan and N.R. Muthiga (2000): Tropical Pacific forcing of decadal SST variability in the Western Indian Ocean over the past two centuries. *Science* 287: 617–619.

Cross, S.L., P.A. Baker, G.O. Seltzer, S.C. Fritz and R.B. Dunbar (2000): A new estimate of the Holocene low-stand level of Lake Titicaca, central Andes, and implications for tropical palaeohydrology. *The Holocene*, 10: 21–32.

Curtis, J.H., M. Brenner, D.A. Hodell, R.A. Balser, G.A. Islebe and H. Hooghiemstra (1998): A multi-proxy study of Holocene environmental change in the Maya lowlands of Peten, Guatemala. *Journal of Palaeolimnology*, 19: 139–159.

Dunning, N.P., and T. Beach (1994): Soil erosion, slope management, and ancient terracing in the Maya lowlands. Latin American *Antiquity*, 5: 51–69.

Dunning, N.P., D. Rue, T. Beach, A. Covich and A. Traverse (1998): Human environmental interactions in a tropical watershed: the palaeoecology of Laguna Tamarindito, El Peten, Guatemala. *Journal of Field Archaeology*, 25: 139–151.

Emery, K.F., L.E. Wright and H.N. Schwarcz (2000): Isotopic analysis of ancient deer bone: biotic stability in collapse period Maya land-use. *Journal of Archaeological Science*, 27: 537–550.

Fjeldså, J.A. (1993): The avifauna of the *Polylepis* woodlands of the Andean highlands: the efficiency of basing conservation priorities on patterns of endemism. *Bird Conservation International*, 3: 3–55.

Foley, J.A., I.C. Prentice, N. Ramankutty, S. Levis, D. Pollard, S. Sitch and A. Haxeltine (1996): An integrated biosphere model of land surface processes, terrestrial carbon balance and vegetation dynamics. *Global Biogeochemical Cycles*, 10: 603–628.

Gnécco, C. (1999): An archaeological perspective on the Pleistocene/Holocene boundary in northern South America. *Quaternary International*, 53: 3–9.

Godínez-Dominquez, J.R. Rojo-Vazquez, V. Galvan-Pina, B. Aguilar-Palomino (2000): Changes in the structure of a coastal fish assemblage exploited by a small scale gillnet fishery during an

El Niño-La Niña event. *Estuarine Coastal and Shelf Science*, 51: 773–787.

Gunn, J.D., W.J. Folan, H.R. Robichaux (1995): A landscape analysis of the Candelaria watershed in Mexico: insights into palaeoclimates affecting upland horticulture in the southern Yucatán peninsula semi-karst. *Geoarchaeology*, 10: 3–42.

Hansen B.C.S. and D.T. Rodbell (1995): A glacial/Holocene pollen record from the eastern Andes of Northern Peru. *Quaternary Research*, 44: 216–227.

Hansen, B.C.S., G.O. Seltzer and H.E. Wright Jr. (1994): Late Quaternary vegetational change in the central Peruvian Andes. *Palaeogeography, Palaeoclimatology, Palaeoecology*, 109: 263–285.

Hastenrath, S. (1991): *Climate dynamics of the tropics*. Dordrecht, Kluwer, 488 pp.

Headland, T.N. and L. Reid (1989): Hunter-gatherers and their neighbours from prehistory to the present. *Current Anthropology*, 30: 43–66.

Herrera, R.S., H. Neff, D. Glascock (1999): Ceramic patterns, social interaction and the Olmec: neutron activation analysis of early formative pottery in the Oaxaca Highlands of Mexico. *Journal of Archaeological Science*, 26: 967–987.

Hodell, D.A., J.H. Curtis and M. Brenner (1995): Possible role of climate in the collapse of Classic Maya civilization. *Nature*, 375: 391–394.

Hodell, D.A., M. Brenner, J.H. Curtis and T. Guilderson (2001): Solar forcing of drought frequency in the Maya lowlands. *Science*, 292: 1367–1370.

Hsu, K.J. (2000): *Climate and Peoples: A Theory of History*, Orell Fussli, Zürich.

Inomata, T. (1997): The last day of a fortified Classic Maya centre: archaeological investigations at Aguateca, Guatemala. *Ancient Mesoamerica*, 8: 337–351.

Islebe, G.A., H. Hooghiemstra, M. Brenner, J.H. Curtis, D.A. Hodell (1996): A Holocene vegetation history from lowland Guatemala. *The Holocene*, 6: 265–271.

Jacob, J.S. (1995): Ancient Maya wetland agricultural fields in Cobweb Swam, Belize: construction, chronology and function. *Journal of Field Archaeology*, 22: 175–190.

Keefer, D.K., S.D. de France, M.E. Mosely and J.B. Richardson III, D.R. Santerlee and A. Day-Lewis (1998): Early maritime economy and El Niño at Quebrada Tacahuay, Peru. *Science*, 281: 1833–1835.

Keys, D. (2000): *Catastrophe: A quest for the origins of the modern world*. Ballantine, 343pp.

Kolata, A.K. (1986): The agricultural foundations of the Tiwanaku State: a view from the heartland. *American Antiquity*, 51: 748–762.

Kornbacher, K.D. (1999): Cultural evolution in prehistoric coastal Peru: an example of evolution in a temporally variable environment. *Journal of Anthropological Archaeology*, 18: 282–318.

Kutzbach, J., G. Bonan and S. Harrison (1996): Vegetation and soil feedbacks on the response of the African monsoon to orbital forcing in the middle Holocene. *Nature*, 384: 623–626.

Leakey, E.C. (1931): *The Stone Age culture of Kenya Colony*. Cambridge University Press.

Leyden, B.W. (1987): Man and climate in Maya lowlands. *Quaternary Research*, 28: 407–414.

McAuliffe, J.R., P.C. Sundt, A. Valiente-Banuet, A. Casas, J.L. Viveros (2001): Pre-Columbian soil erosion, persistent ecological changes, and collapse of a subsistence agricultural economy in the semi-arid Tehuacan Valley, Mexico's 'Cradle of Maize'. *Journal of Arid Environments*, 47: 47–75.

McIntosh, R.J. (2000): The peoples of the Middle Niger. In R.J. McIntosh, J.A. Tainter and S. K. McIntosh (eds.): *The way the wind blows: climate, history and human action*. Colombia University Press.

Manzanilla, L. (1997): The impact of climate change on past civilizations, a revisionist agenda for further investigations. *Quaternary International*, 43: 153–159.

Markgraf, V. (1989): Palaeoclimates in Central and South America since 18,000 BP based on pollen and lake level records. *Quaternary Science Reviews*, 8: 1–24.

Markgraf, M., T.R. Baumgartner, J.P. Bradbury, H.F. Diaz, R.B. Dunbar, B.H. Luckman, G.O. Seltzer, T.W. Swetnam and R. Villalba (2000): Palaeoclimate reconstruction along the Pole-Equator-Pole transect of the Americas (PEP 1). *Quaternary Science Reviews*, 19: 125–140.

Marchant, M., D. Hebblen and G. Wefer (1999): High resolution planktonic foraminiferal record of the last 13300 years from an upwelling area off Chile. *Marine Geology*, 161: 115–128.

Marchant, R.A., D.M. Taylor and A.C. Hamilton (1997): Late Pleistocene and Holocene history at Mubwindi Swamp, southwest Uganda. *Quaternary Research*, 47: 316–328.

Marchant, R.A. and D.M. Taylor (1998): A Late Holocene record of montane forest dynamics from south-western Uganda. *The Holocene*, 8: 375–381.

Marchant, R.A. and D.M. Taylor, (2000): Numerical analysis of modern pollen spectra and *in situ* montane forest—implications for the interpretation of fossil pollen sequences from tropical Africa. *The New Phytologist*, 146: 505–515.

Marchant, R.A., H. Behling, J.C. Berrio, A.M. Cleef, J. Duivenvoorden, H. Hooghiemstra, P. Kuhry, A.B.M. Melief, B. van Geel, T. van der Hammen, G. van Reenen and M. Wille (2001a): Mid- to late Holocene pollen based biome reconstructions for Colombia: a regional reconstruction. *Quaternary Science Reviews*, 20: 1289–1308.

Marchant, R.A., H. Behling, J.C. Berrio, A. M. Cleef, J. Duivenvoorden, H. Hooghiemstra, P. Kuhry, A. B. M. Melief, B.

van Geel, T. van der Hammen, G. van Reenen and M. Wille (2001b): A reconstruction of Colombian biomes from modern pollen data along an altitudinal gradient. *Review of Palaeobotany and Palynology*, 117: 79–92.

Marchant, R.A., H. Behling, J.C. Berrio, A. Cleef, J. Duivenvoorden, B. van Geel, T. van der Hammen, H. Hooghiemstra, P. Kuhry, B.M. Melief, G. van Reenen, and M. Wille, (2002a): Colombian vegetation derived from pollen data at 0, 3000, 6000, 9000, 12,000, 15,000 and 18,000 radiocarbon years before present. *Journal of Quaternary Science*, 17: 113–129.

Marchant, R.A., Boom, A. and Hooghiemstra, H. (2002b): Pollen-based biome reconstructions for the past 450,000 yr from the Funza-2 core, Colombia: comparisons with model-based vegetation reconstructions. *Palaeogeography, Palaeoclimatology, Palaeoecology*, 177: 29–45.

Mason, O.T. (1896): Influence of environment upon human industries or arts. In *Annual Report of the Smithsonian Institute for 1895*, Washington: 639–665.

Metcalfe, S.E., S.L. O'Hara, M. Caballero and S.J. Davies (2000): Records of Late Pleistocene-Holocene climate change in Mexico—a review. *Quaternary Science Reviews* 19: 699–721.

Moore, J.D., (1991): Cultural responses to environmental catastrophes: post-El Niño subsistence on the prehistoric north coast of Peru. *Latin American Antiquity*, 2: 27–47.

Morely, S.G., G.W. Brainerd and R.J. Sharer (1983): *The Ancient Maya*. Stanford University Press, Stanford, CA.

Morris, A. (1999): The agricultural base of the pre-Incan Andean civilizations. *The Geographic Journal*, 165: 286–295.

Moseley, M.E. (1972): Subsistence and demography: an example from prehistoric Peru. *Southwestern Journal of Anthropology*, 28: 25–49.

Palais, J.M., S. Kirchner and R.J. Delmans (1990): Identification of some global volcanic horizons by major element analysis of fine ash in Antarctic ice. *Annals of Glaciology*, 14: 216–220.

Paulsen, A.C. (1976): Environment and empire: climatic factors in pre-historic Andean culture change. *World Archaeology*, 8: 121–132.

Peterson, A.A. and A.T. Peterson (1992): Aztec exploitation of cloud forest: tributes of liquidambar resin and quetzal feathers. *Global Ecology and Biogeography Letters*, 2: 165–173.

Piperno, D.R. and D.M. Pearsall (1998): *The origins of agriculture in the lowland neotropics*. Academic Press, San Diego, 400 pp.

Piperno, D.R., A.J. Ranere, I. Holst, P. Hansell (2000): Starch grains reveal early root crop horticulture in the Panamanian tropical forest. *Nature*, 407: 894–897.

Prentice, I.C. and T. Webb III (1998): BIOME 6000: reconstructing global mid-Holocene vegetation patterns from palaeo-ecological records. *Journal of Biogeography*, 25: 997–1005.

Proctor, F. (1998): The meaning of global environmental change: retheorizing culture in human dimensions research. *Global Environmental Change*, 8: 227–248.

Prous, A. and E. Fogaça (19999: Archaeology and Pleistocene-Holocene boundary in Brazil. *Quaternary International*, 53: 21–41.

Rice, D.S. and P.M. Rice (1984): Collapse intact. Postclassical archaeology of the Peten Maya. *Archaeology*, 37: 46–51.

Robertshaw, P.T. (1988): Environment and culture in the Late Quaternary of Eastern Africa: A critique of some correlations. In J. Bower and D. Lubell (eds.): *Prehistoric cultures and environments in the Late Quaternary of Africa*. Cambridge University Press.

Roosevelt, A.C., M. Lima da Costa, G. Lopes Mochado, M. Michnab, N. Mericer, H. Valladas, J. Feathers, W. Barnett, M. Imazio da Silveira, A. Henderson, J. Sliva, B. Chernoff, D.S. Reese, J.A. Holman, N. Toth, and K. Schick (1996): Palaeoindian cave dwellers in the Amazon: the peopling of the Americas. *Science*, 272: 373–383.

Sandweiss, D.H., K.A. Maasch, R.L. Burger, J.B. Richardson III, H.B. Rollins and A. Clement (2001): Variations in Holocene El Niño frequencies: climate records and cultural consequences in ancient Peru. *Geology*, 29: 603–606.

Sandweiss, D.H., J.B. Richardson III., E.J. Reitz, H.B. Rollins, H.B. and K.A. Maasch (1996): Geoarchaeological evidence from Peru for a 5000 years B.P. onset of El Niño. *Science* 273: 1531–1533.

Schurr, T.G. (2000): Mitochondrial DNA and the peopling of the New World. *American Scientist:* 88, 246–253.

Santley, R.S., T.W. Killion and M.T. Lycett (1986): On the Maya collapse. *Journal of Anthropological Research,* 42: 123–159.

Seltzer, G.O. and C.A. Hastorf, (1990): Climatic changes and its effect on prehispanic agriculture in the Central Peruvian Andes. *Journal of Field Archaeology*, 17: 397–414.

Servant, M., J. Maley, B. Turcq, M. Absy, P. Brenac, M. Furnier, M.P. Ledru (1993): Tropical forest changes during the Late Quaternary in African and South American lowlands. *Global and Planetary Change,* 7: 25–40.

Stahl, P.W. (1999): Structural density of domesticated South American camelid skeletal elements and the archaeological investigation of prehistoric Andean Ch'arki. *Journal of Archaeological Science,* 26: 1347–1368.

Stuiver, M., P. J. Reimer, E. Bard, J.W. Beck, G.S. Burr, K.A. Hughen, B. Kromer, F.G. McCormac, J. van der Plicht and M. Spurk (1998): http://radiocarbon.pa.qub.ac.uk/calib/calib.html. *Radiocarbon*, 40: 1041–1083

Taylor, D., P. Robertshaw, and R.A. Marchant (2000): Environmental change and political economic upheaval in precolonial western Uganda. *The Holocene*, 10: 527–536.

Thompson, L.G. (1992): Ice core evidence from Peru and China.

In: Bradley, R.S., P.D. Jones, (eds.): *Climate since AD 1500*. Routledge, London and New York, 517–548.

Thompson, L.G., E. Mosley-Thompson, B. Morales Arnao (1984): Major El Niño Southern Oscillation events recorded in stratigraphy of the tropical Quelccaya ice cap. *Science*, 226: 50–52.

Thompson, L.G., E. Mosely-Thompson, J.F. Bolzan and B.R. Koci (1985): A 1500-year record of tropical precipitation in ice cores from the Quelccaya ice cap, Peru. *Science*, 229: 971–973.

Thompson, L.G., M.E. Davis and E. Mosely-Thompson and K.B. Liu (1988): Pre-Incan agricultural activity recorded in dust layers in two tropical ice cores. *Nature*, 336: 763–765.

Thompson, L.G., M.E. Davis and E. Mosely-Thompson (1994): Glacial records of global climate: A 1500-year tropical ice core record of climate. *Human Ecology*, 22: 83–95.

Thompson, L.G., E. Mosely-Thompson, M.E. Davis, P.E. Lin, K.A. Henderson, B. Cole-Dai, J.F. Bolzan and K. Liu (1995): Late glacial stage and Holocene tropical ice core records from Huascarán, Peru. *Science*, 269: 46–50.

Tuomisto, H. and K. Ruokolainen (1997): The role of ecological knowledge in explaining biogeography, and biodiversity in Amazonia. *Biodiversity and Conservation* 6: 347–357.

Van`t Veer, R. and H. Hooghiemstra (2000): Montane forest evolution during the last 650,000 years in Colombia: a multivariate approach based on pollen record Funza-1. *Journal of Quaternary Science*, 15: 329–346.

van der Hammen, T. and E. González (1965): A Late glacial and Holocene pollen diagram from Cienaga del Vistador Dep. Boyacá, Colombia. *Leidse Geologische Mededelingen*, 32: 193–201.

van Geel, B., J.M. Buurman, H.T.M. Waterbolk (1996): Archaeological and palaeoecological indications of an abrupt climatic change in The Netherlands, and evidence for climatological teleconnections around 2650 BP. *Journal of Quaternary Science*, 11: 451–460.

Vincens, A., D. Schwartz, H. Elenga, I. Ferrera, A. Alexandre, J. Bertaux, A. Mariotti, L. Martin, D.J. Meunier, N. Nguetsop, M. Servant, S. Servant-Vildary and D. Wirrman (1999): Forest response to climate changes in Atlantic Equatorial Africa during the last 4000 years BP and inheritance on the modern landscapes. *Journal of Biogeography* 26: 879–885.

White, C.D., M.E. Pohl, H.P. Schwarcz and F.J. Longstaffe (2000): Isotopic evidence for Maya patterns of deer and dog use at Preclassic Colha. *Journal of Archaeological Science*, 28: 89–107.

Wright, L.E. and C.D. White (1996): Human biology in the Classic Maya collapse: evidence from paleopathology and paleodiet. *Journal of World Prehistory*, 10: 147–198.

Yesner, D.R. (1980): Maritime hunter-gatherers: ecology and prehistory. *Current Anthropology*, 21: 727–750.

Zielinski, G.A. (2000): Use of palaeo-records in determining variability within the volcanism-climate system. *Quaternary Science Reviews*, 19: 417–438.

Chapter 24

Droughts and Fossils: A Hypothetical Perspective

VIJAY SATHE

Overview

Drought can be a major factor in terminating life assemblages from a landscape. It is starvation more than dehydration that is the main cause of death during a drought. It is a known fact that under such conditions all types of animals tend to concentrate around any available water source in the vicinity. Animals are even known to travel long distance in search of water bodies. Eventually it may turn into death on mass scale representing taxonomically a diverse skeletal assemblage around the water bodies.

This study is an attempt to raise a model and build a hypothesis regarding the interrelation between changing climatic conditions due to drought and the fossil skeletal assemblages that are concentrated in the vicinity of waterholes.

Introduction

Until recently, the status of Vertebrate Palaeontology in India has been largely limited to descriptive taxonomy. The methodological advances such as taphonomy, dental histology and genetic mtDNA studies have demonstrated that the faunal study can be made more meaningful by understanding the bio-ecology of the palaeoenvironment. Today, there are over 500 known Quaternary fossil vertebrate localities, spread all over extra peninsular and peninsular India. The fossil material from the localities is

spread within a time span ranging from about 2 million years to 20,000 years BP. Most of these localities are situated on abandoned palaeochannels of ancient rivers. These localities have yielded the fossilized skeletal remains of proboscideans, artiodactyls, perissodactyls, reptiles and various members of rodentia and other orders.

The taxonomic details of fossil assemblages and their litho-stratigraphical provenance dominate publications. Nevertheless, taphonomic studies of Siwaliks and Eocene fauna of north-west India have significantly contributed to our understanding of palaeoenvironment. However, the review of the literature reveals that this aspect (cause of mortality) is yet to be adequately appreciated with regard to fossil assemblages from peninsular river valleys. The mode of death and cause of mortality are fundamental to the future course of death assemblage (thanatacoenoses) into the formation of fossil record. For a clearer picture of the taphonomic history of an assemblage, it is necessary to identify the cause of death of the animals. The cause of death in general, catastrophic and attritional in particular, are such aspects that have never been looked into, leaving a major lacuna in the methods of the taphonomic researches in India. Researches outside India however have paid attention to the impact of situational oddities leading to mass mortality (Haynes, 1988; Conybeare and Haynes, 1984).

The present study discusses the implications of drought for vertebrate fossil assemblages using a hypothetical model based on mortality of present day large herbivores in African wildlife sanctuaries. The observations are drawn using analogy with modern counterparts. Eventually this announcement brings in a new dimension of taphonomic investigations with regard to Indian Vertebrate Palaeontology.

Drought: Definition and Its Consequences on Local Biology

Although Tannehill (1947) has defined drought as the 'outcome of a deficient rainfall for an extended period—a season, a year or several years, threatening the survival of flora and fauna in the region'; no single definition of drought is likely to be universally acceptable in view of the very subjectivity of this phenomenon. It can be defined according to three criteria, viz., meteorological, agricultural and hydrological, all attributed to the environmental stress on modern ecosystems. The resultant destabilization of the food chain accelerates starvation and death of the herbivores, causing an acute shortage of food for predatory animals and birds alike.

It differs significantly from other natural hazards (e.g., floods, tropical cyclones and earthquakes). Since the effects of droughts are cumulative over a period of time and may linger for years even after their termination, the precise duration of drought is difficult to determine. Thus Tannehill (1947) often refers to drought as a creeping phenomenon.

Drought perpetuates starvation more than dehydration due to disappearance of vegetation. It is a known fact that under such conditions all types of animals tend to concentrate around any available water source in the vicinity. Animals are even known to move from their original habitats over considerably long distances in search of water bodies.

Prolonged starvation would eventually turn into death on mass scale representing taxonomically a diverse skeletal assemblage in the vicinity of water bodies. Naturally these assemblages show a larger quantum compared to assemblages resulting during normal times. However, the state of preservation would reflect a bias with reference to the percentage of individual species. Such assemblages can certainly help us in forming important parameters in the process of identifying characteristic alterations of an ecosystem caused by drought. Severity of drought can be assessed by the successive degeneration of plants and animals, which depend on the water support directly or indirectly for survival. Alteration of a landscape by mild to severe drought takes places in three stages leaving a great impact on the flora and fauna. According to Shipman (1975), these stages can be explained as follows:

Stage I: Drought is mild, with rainfall being less than the normal. As a result there is a visible damage to vegetation. Seasonal water bodies and streams dry up. Migratory megavertebrates move out to better pastures and more water. Animals unable to move out are subject to abnormal periods of dryness leading to elevated rates of mortality, more than the normal attritional levels noticeable especially among younger individuals.

Stage II: On account of continuous, deficient rainfall, severity of drought is enhanced multifold. This results in the shrinking of perennial rivers and lakes. Calcrete develops in alkaline soils. Evaporites precipitate and cracks develop in the dried-up beds of rivers and waterholes. Response of animals to drought conditions depends upon their frequency of visits to waterholes. Animals habitually visiting waterholes once a day tend to congregate around it irrespective of their normal habitat preference. Vegetation in the vicinity of waterholes is consumed and only the animals with minimum need of water move out to forage better pastures away from the congregated waterholes. In this process animals that are unable to sustain extreme heat and lack of vegetation die of starvation. It is here that the carnivores emerge on the scene to consume food, made available as a result of scarcity of a vital source of energy for herbivores. Carnivores tend to eat soft tissue parts. For sheer abundance of animal meat, predators deviate from their routine bone chewing and the resulting skeletal assemblage is generally minimally damaged by carnivore activity.

Stage III: The stage of extreme drought is identified with continued absence of normal rainfall over a prolonged period. The drying up of lakes and rivers lead to mass death of aquatic fauna. Large herbivores unable to move to better pastures throng around shrinking pools of water and eventually die struggling to drink. This leads to a temporary extinction of plants and animals within the given locus. Low soil moisture and organic contents could induce desertification. If similar conditions prevail for a longer duration, the landscape is permanently altered into a desert or semi-desert. It should be noted here that logically the progression of the desertification process should reflect a fall in the frequency of fossil records. However, the rainfall following Stages I and II may result in a large run off because of the lack of vegetation. Passage of water is hindered by accrete, increasing water loss caused by the run off. Such a situation also results in a spillover from seasonal and perennial rivers with transportation of large quantities of sediments. Under both circumstances of causes either by run off or spillover, large quantities of embedded fossils also get transported.

Despite inherent difficulties involved, it is essential that all stages of drought be recognized by means of detailed taphonomic analyses to address the problem of palaeoecological interpretations with greater precision. The

assemblages belonging to Stage I may or may not be isolated from the assemblages formed by either catastrophes or natural death. But the severity of droughts at Stages II and III leaves signature marks such as predominance of younger populations, bones with weathering and breakages. However, there is a distinct escalation of such taphonomic marks witnessed in Stage III. So far there are very few identifiable drought assemblages, which have been recognized to belong to the Stage III of drought assemblage (Romer, 1961).

As mentioned earlier, death is predominantly due to starvation than thirst. The level of strontium in bones is a potential indicator of the animal's diet. It is one of the trace elements whose representation in fossilized bones of animals can be a meaningful application in understanding the diet of ancient herbivores. However, there are two variables, viz., the strontium content in the fossil bone and the level of strontium in the parent soil. Keeping the level of environmental strontium stable is a prerequisite for a precise determination of its presence in bones (Parker and Toots, 1980). Catastrophic mortality due to drought is likely to leave signatures of prolonged starvation, owing to depletion of vegetation cover, which will be reflected in the levels of strontium/calcium and stro. Sr in their bones.

With regard to the relationship between drought and mortality and its bearing on past life assemblages, following palaeontological and geological criteria were proposed by Shipman (1975). This has far-reaching implications of drought-related environmental stress in the formation of a vertebrate fossil record. The assessment of representation of a fossil record is thus substantiated by several parameters to confirm the depositional history of death assemblage in a given area.

1. The age profile and distribution of death assemblages (thanatacoenoses) indicate

catastrophic mortality. Stages I, II and III indicate a consecutively increasing number of younger individuals while during Stages II and III, it would be confined to sub-adults and adults.

2. The skeletons are fairly well preserved because scavengers did not have to disturb the bones as an abundance of meat supply was created because of mass death. Another factor leading to better preservation is that the joints tend to be preserved intact as the lack of moisture dries the soft tissues, making it difficult to disarticulate skeletons.

3. Animals left without a habitat preference gather around available waterholes. High dependence on water in majority of species makes them stay on in the close vicinity of already shrinking waterholes.

4. In extreme drought conditions, aquatic and semi-aquatic fauna die with greater concentration in small clusters.

5. Fine grained fossil yielding strata have mud cracks.

6. Association of calcrete and evaporites with fossils that are uncommon with other non-fossil strata.

7. Overlying sediments indicating rapid deposition.

Materials and Methods

Fossil assemblage from the faunal complex of Tadola, Ganjur, Wangadari and Dhanegaon in the Latur and Beed Districts in the Manjra valley, Maharashtra, was critically examined to identify the signatures (if any) of catastrophic mortality (drought) and subsequent carcass processing prior to burial. The Pleistocene formations in this area are one of the richest fossil-bearing horizons only next to the Siwaliks of the northwest and Narmada valley in Central India. The animal community is represented by well to poorly preserved skeletal remains of primitive and true elephants, horses, large and medium-sized bovids, hippopotamus, deer, crocodiles and turtles. Despite the allochthonous nature of fossil assemblage, attempts to identify the original habitat of these past animal populations have met with considerable success (Sathe, 1991).

The elephant mortality in waterholes due to severe drought in National Hwange Park, Zimbabwe in 1982 led to the death of a large population of elephants at a natural water source. The catastrophic mode of death in natural water sources, age structure of the assemblage and taphonomic processes responsible in modifying the death assemblage were explained as useful analogues for palaeoecological interpretations of fossil proboscidean assemblages (Conybeare and Haynes, 1984).

This case study forms the basis of the hypothesis presented in this study indicating that taphonomic features of the fossil assemblages should be helpful in deciphering the occurrence and recurrence (if any) of drought episodes. Following Conybeare and Haynes (1984) and Shipman (1975), an assessment of the age structure of the fossil assemblage and the evidence of bone modification highlight the mode of representation of the assemblage. The observations drawn are preliminary in nature and are being augmented with more of fossil history of skeletal assemblages from the Quaternary formations of Godavari and Ghod in Peninsular India. They need to be tested to see whether the deductions based on analogues satisfy the conditions, labelled as palaeontological and geological criteria, to identify drought as a probable cause of death of the animal populations.

Discussion

■ Relative abundance of taxa

The fossil localities are multispecies and the types of animals represented in all the fossil localities in the Manjra valley include large herbivores, small herbivores, predators, scavengers and small mammals. The NISP or number of identified specimens and their state of preservation points to several agents of modification that were operating in the area since the Late Pleistocene period.

■ Age determination

Age structure of the assemblage presents a mixed result with majority being sub-adults to adults, known to be the second and third targets of catastrophic mortality especially with reference to the three stages of successive severity of drought. However, it has to borne in mind that owing to obvious modifying factors and/or agents of taphonomy, pulverisation and disintegration of skeletal elements of lower bone density in younger individuals is likely to destroy the bones which may be retarded if the bones do not get exposed to prolonged sub-aerial weathering. In this light it will be premature to assume that the life assemblage did not have any sizable representation of the younger individuals.

■ Mode of skeletal representation

While looking for patterns in bone collections that reflect behavioural and ecological data, a mixed picture emerges that partly satisfies the premise of catastrophic mortality, especially due to drought. The differential weathering, higher frequency of spiral breakage, abrasion, common occurrence of trampling and a near total absence of caudal vertebrae, limited fragmentation of vertebrae and ribs, and a total absence of carnivore activity are some of the interesting features that appear to emerge in a patternized way in most of the taxa represented (Fig.1). The maximum representation of breakage followed by abrasion explains prolonged exposure prior to burial with an interim phase of fluvial transportation. A few sites within the faunal

complex represent skeletal elements belonging to the Voohries' Group III which point to a majority of lag elements such as crania, jaws, etc. The general absence of micovertebrates is yet another line of evidence which may be attributed to their poor prospects for survival. However, a recent discovery of a porcupine tooth from Wangdari makes renewed investigations for microvertebrates imperative. Association of calcrete with fossil-yielding horizons is a common feature while the occurrence of fine-grained fossil horizons with mud cracks and the episodes of rapid deposition of overlying sediments cannot be confirmed with certainty.

The preliminary observations presented here have eventually initiated a new line of investigations in Indian Quaternary Palaeontology, especially with regard to megavertebrates. There are many unanswered questions like how best we can differentiate between catastrophic and natural die offs, which

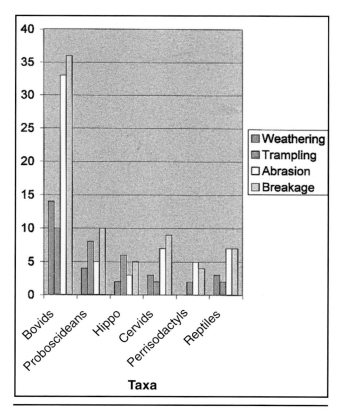

Fig. I. Bone modification in fossil assemblage from Manjra valley.

could have occurred several times over the period of the past several thousand years' history of accumulation? Time resolution of fossil assemblages is a burning issue that needs to be addressed in the light of the secondary context of their provenance. There is a possibility of blending together of assemblages originating from separate episodes of death and burial, giving rise to a bone bed which is indistinguishable and yields little information on that particular slice of time that witnessed episodes of drought or any other catastrophic cause of death. This necessitates the need for an extensive re-examination of field assemblages and laboratory collection to arrive at more significant observations regarding the mortality patterns and their causes that would add to reliable collaborative data for reconstructing palaeo-environment.

Acknowledgement

The author is thankful to Dr Shubhangana Atre, Deccan College, Pune for the thought-provoking discussions that eventually resulted in the preparation of this paper.

References

Conybeare, A., G. Hayens (1984): 'Observations on Elephant Mortality and Bones in Water Holes', *Quaternary Research* 22: 189-200.

Hayens, G. (1988): 'Mass Deaths and Serial Predation: Comparative Taphonomic Studies of Modern Large Mamal Death Sites', *Journal of Archaeological Science* 15: 219-235.

Parker, R.B., H. Toots (1980): 'Trace Elements in Bones as Palaeobiological Indicators', in: Behrensmeyer, Anna K. and A.P. Hill (eds.) *Fossils in the Making: Vertebrate Taphonomy and Palaeoecology*, Chicago University Press, Chicago, pp. 197-207.

Romer, A.S. (1961): 'Palaeozoological Evidence of Climate (Vertebrates)', in: Nairn, A.E.M. (ed.) *Descriptive Climatology*, Interscience Publishers Inc., New York, pp. 183-206.

Sathe, V. (1991): *Quaternary Palaeontology and Prehistoric Archaeology of Manjra valley, Maharashtra*, Unpublished Ph.D. Thesis, University of Pune, Pune.

Shipman, P. (1975): 'Implications of Drought for Vertebrate Fossil Assemblages', *Nature* 257: 667-668.

Tannehill, I.R. (1947): *Drought: Its Causes and Effects*, Princeton University Press, New York.

Chapter 25

Human Response to Holocene Climate Changes—A Case Study of Western India between 5th and 3rd Millennia BC

VASANT SHINDE, SHWETA SINHA DESHPANDE AND YOSHINORI YASUDA

Introduction

The word 'climate' is understood as the condition of a place in relation to various phenomena of the atmosphere, including temperature, moisture, precipitation, and wind, that characteristically prevail in a particular region and affect animal or vegetable life, i.e., the environment which, in turn, affects and influences human responses in the form of cultural changes and adaptations. Different regions have differing environments resulting in different cultural groups or human responses and Steward (1955) has justifiably stated that similar environments may have similar cultural responses from human societies. The origin, development and decline of cultures in the Indian subcontinent right from prehistoric times have been to a larger extent determined by climatic conditions especially by the Indian summer and winter monsoons.

The Indian monsoons are one of the most significant and attractive natural phenomena as they have an unparalleled fundamental impact on eco-diversity and food production. The Indian subcontinent is characterized by two monsoon systems: the summer monsoons coming from the southwest Indian Ocean region between June to September and the winter monsoons emanating in the northwestern region of the subcontinent from December to February (Fig.1). The Indian

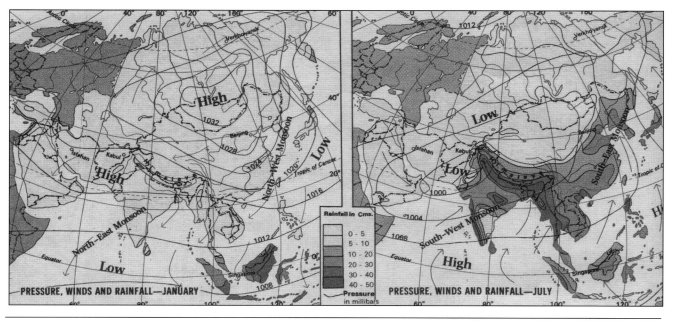

Fig. I. Map showing movement of winds of summer and winter monsoon.

Ocean Monsoon, which responds to global climatic changes, is characterized by circulation of the prevailing moisture-bound winds causing rainfall over the south Asian landmass. The Southwest Monsoon (summer) rainfall (75–80% of the total annual rainfall) is higher than that of the Northwest. There are two streams of the Southwest monsoon; the Bay of Bengal stream arriving earlier and the Arabian Sea stream later. The Southwest monsoon is followed by the retreating Northeast monsoon, which covers the eastern half of the subcontinent and is characterized by frequent cyclonic storms emanating from the Bay of Bengal. The Northwest winter monsoons emanate in the Mediterranean plateau giving winter rainfall in that belt and then move eastwards towards the northwestern part of the Indian subcontinent crossing the Hindukush ranges. The regions of Afghanistan, Pakistan and northwest and north India experience winter showers important for their much needed winter crops. The Southwest and Northwest monsoons thus help generate monsoon circulation that feeds the major water courses (rivers) of the subcontinent (Searle, 1995) and have been doing so for the past 8–14 million years giving rise to different cultural groups in the regions which to a large extent have depended on the climatic and monsoonal changes within the area.

The economy of the Indian subcontinent even today is dependent on rain-fed agriculture to a very large extent especially for the food required to feed the millions of people inhabiting this region. The agriculture and the crops grown for the past six to seven thousand years have depended on the two streams of summer and winter monsoon. The summer rains help the *kharif* crops like paddy, bajra, maize, and types of millets while the winter rains are responsible for a healthy *rabi* crop of wheat, pulses, sugarcane, mustard, sesame, cotton, etc. The region of Western India which forms part of this research falls today within the marginal area affected by the monsoon and is hence dependent on both

winter and summer rainfall. Change or disturbance in either of the two leads to major economic and cultural crisis as is seen from the rise of the Mesolithic and Chalcolithic cultures in this region.

Some of the main morphological changes in the present landscape and drainage system of India occurred between 13,000 yrs. BC and 2000 yrs. BC and forced the human societies of this period to undergo major changes in their living pattern, either advancing or declining. Since the beginning of Holocene (10,000 yrs. BC) the palaeoclimate of Western India has undergone a series of dramatic changes. The beginning of the Holocene was characterized by a shift from a cold and dry Ice Age climate to a warmer and wetter climate with various intermittent semi-arid and arid phases. The Savannah type vegetation shifted to shrubs and occasional trees during the Holocene in this region. These changes continued right through the Chalcolithic period and stabilized only at its end into the conditions prevalent today. Climatic conditions are believed to be a major factor in the rise and decline of the Chalcolithic cultures including the Harappans. This change in climatic conditions could also be one of the triggering factors that led to the theorized shift of the Chalcolithic people from the western parts of the subcontinent to the eastern and southern parts.

Ecologically this region of Western India (Fig. 2) is characterized by environmental uniformity that includes fertile black cotton soil and a semi-arid climate. The annual average rainfall is about 700 mm. The important crops grown in this region include wheat, sugarcane, mustard, millet and maize. The semi-arid region supports typical vegetation such as a variety of grasses, thorny bushes like *ker* (*Capparis decidua*) and trees like *khejdi* (*Prosopis spicigera*) and *babul* (*Acacia arabica*). There is also abundance of pastureland around most of the sites in this region. Thus, the most important factors, which appear to have attracted the Mesolithic pastoralists and incipient

Fig. 2. Map showing physical features of western India including the Mewar region of Rajasthan.

farmers and later Chalcolithic farmers, were the proximity of a perennial source of water, abundant pasturage and fertile soil.

The period between 5th and 2nd millennium BC in Western India witnessed a rise in human population as the Mesolithic hunting-gathering communities began settling down with agriculture as the main source of subsistence in a sedentary economy because of a variety of demographic, economic and environmental pressures. The emergence of an agricultural way of life whether it originated, as Childe (1936) believed, in a nuclear region to the west or as a result of internal evolution and development as David Clark (1972) believed, gradually moved the population from widely segregated spatial environments to concentrated and limited areas leading to the formation of communities with their own cultural components or characteristic features. Interaction has been considered as an important aspect of cultural evolution and it played an important role in the transformation of the hunter-gatherers into semi-nomadic

pastoralists and incipient slash and burn agriculturists with direct exchange systems and a proliferation of ideas and techniques between neighbours.

Western India and its Environment

The region of Western India, which constitutes the states of Gujarat, Rajasthan and western parts of Madhya Pradesh or the Malwa region, includes the Malwa plateau, southeast Rajasthan, northern Rajasthan, western Rajasthan including the desert parts, Saurashtra, North Gujarat, and Kutch (Fig. 2). The archaeological sites that are covered under this study include Bagor, Ahar, Gilund, Balathal, Ganeshwar-Jodhpura in Rajasthan, Loteshwar, Langhnaj, Nagwada, Padri, Prabhas Patan in Gujarat and the region of Malwa plateau in Madhya Pradesh (Fig. 3). The Kutch part of Gujarat is marked by salty marshlands and geophysically consists of small dissected plateaus and scrap lands and a long seaboard indented by large inlets like the Gulf of Cambay and the Rann of Kutch and tidal floodplains and saline marshes, while the rest of the state including Saurashtra and North Gujarat is covered with ample agricultural and pasture land. North Rajasthan is

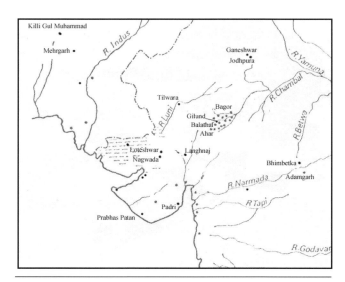

Fig. 3. Map giving details of the distributions of archaeological sites in western India, including the Mewar region of Rajasthan mentioned in the text.

marked by the hilly tracts of the Aravallis with large copper deposits while the southeast Rajasthan region is marked by fertile black-cotton soil and monsoonal rivers like Banas and its tributaries. Western Rajasthan, part of which forms the Thar Desert, has no perennial streams and the only river of importance is the Luni with its various tributaries. In the remaining portions of western Rajasthan, rainwater gathers in shallow depressions some of which have sufficient permanence to give rise to salt lakes such as Sambhar and Didwana. Rajasthan is rich in mineral resources both metallic and non-metallic (*c.f. Encyclopedia Britannica,* 1999) including copper in the Khetri belt, zinc, silver and lead in the southeastern Aravalli belt. The Malwa region is rich in flora and fauna even today with a good monsoonal climate and perennial rivers like the Narmada and Chambal and their tributaries.

All of western India, except western Rajasthan and Kutch in Gujarat falls within the semi-arid environment, with seasonal rivers dependent on rainfall. The climate is characterized by hot and dry summers followed by monsoon and winter season. Temperature varies from 15°C in the winter to 42°C maximum in summer. The average rainfall is 106 cm, which is unpredictable and hence has no correlation with the actual amount every year thus affecting the basic mode of subsistence in the region, i.e., agriculture. The desert part, however, experiences extreme climate, from below zero degrees in winter to 47°C in summer. Though the entire region is fertile because of the presence of black-cotton soils, it falls in the semi-arid zone, which is characterized by low and unpredictable rainfall and shrub forests. But with the appropriate water sources, the fertile soil supports a variety of crops notably wheat, barley, millet, sorghum, sesame, green grams, black gram, mustard, cotton, banana, etc. The foothills of Aravallis in Mewar and the Satpura ranges in southern central India have ample pasturelands and they also support

dense tropical forest of this region. A variety of wild animals such as tiger, panther, wild buffalo, spotted deer, black buck, *nilgai* and wild dog inhabit these forests. As the region lies beyond the influence of the Southwest Monsoon the people living in these areas practise a pastoral mode of subsistence.

Palaeoclimate of Western India

The environment and landscape of the Pleistocene in this area as experienced by the Mesolithic and Chalcolithic people was different from today's. During the Pleistocene man's effect upon his environment was marginal but the Holocene stimulated more technologically resourceful groups conditioned to change and able to exploit new opportunities as they moved into new ecozones and developed. These people not only affected their environment but to varying levels attempted to control it by burning forests for cultivation.

A number of attempts have been made in the reconstruction of the palaeoclimatic sequence, including the palaeo-monsoon right from the Quaternary to the Holocene period with pioneering work being done by the much quoted Singh *et al.,* (1971). As this work was done prior to the discovery that ^{14}C amounts actually varied during different times and the subsequent development of calibration charts, many conclusions based on these uncalibrated dates are actually off by significant lengths of time (Sinha, 2003). The actual evidence for the climatic and environmental changes during the Holocene in western India is derived from the palaeobotanical materials extracted from saline lakes like Didwana, Lunkarnasar and Sambhar in Rajasthan (Singh *et al.,* 1971 & 1990) and is listed below with the Singh *et al.,* (1971 & 1990) dates given in uncalibrated 'BP' and the calibrated dates in 'BC' (from Possehl, 1994). The pollen evidence suggests that during the last glacial period to 13,000 years BP there existed hyper-arid conditions with a treeless savannah grassland environment and a weak summer monsoon with increased winter rains. Between 12800 BP and 9380 BP (12800 BP to cal. yrs. 8085 BC) as shown by rising lake levels and temperatures there was more moisture circulating in the atmosphere leading to the development of a shrub savannah grassland biome. During the next period from 9380 BP to 7460 BP (cal. yrs. 8085 BC to cal. yrs. 6100 BC) the trend continued leading to a tree and savannah biome. However by 6010 BP (cal. yrs. 4719 BC), a declining trend in the precipitation developed and has continued till modern climatic conditions with a semi-arid and shrub savannah biome with occasional acacia trees were reached after 4180 BP or cal. yrs. 2577 BC (Sinha, 2003).

Enzel *et al.,* (1999) worked on the Lunkaransar dry salt lake of Rajasthan and the result of this work revealed that the early Holocene witnessed many minor climatic fluctuations. The lake underwent dramatic fluctuations around 6300 BP (cal. yrs. 5000 BC) when the lake levels rose to a high and with minor fluctuations continued till 5500 BP (cal. yrs. 4200 BC) when it reduced abruptly and then dried completely by 4800 BP (cal. yrs. 3500 BC). This climatic data completely negates the idea that improved climatic conditions led to the rise of the Indus civilization as it was during a dry and semi-arid environment that the culture flourished in India and Pakistan. Further they move on to say that it was not the summer monsoon that was responsible for the increase in lake levels but a higher winter precipitation, which could be the potential source for changed hydrological conditions in the middle Holocene period.

Based on the data from the lake deposits in Rajasthan, Krishnamurty *et al.,* (1981) concluded however that:

Before 8000 BC—Severe Aridity
8000–7500 BC—Relatively Wet
7500–3000 BC—Relatively Dry
3000–1700 BC—Sudden Increase in Wetness

1700–1500 BC—Relatively Dry
1500–1000 BC—Relatively Wet
1000–500 BC—Arid

The authors (see also Shinde, 2001) and the team including scientists from the International Center for Japanese Study, Kyoto, Japan have also worked on the existing climatic problem and recent work done in the Sambhar lake by them has yielded similar results:

Cal. yrs. 6200 BC– cal. yrs. 4100 BC—Wet Phase
Cal. yrs. 4100 BC– cal. yrs. 3800 BC—Dry Phase
Cal. yrs. 3800 BC– cal. yrs. 2200 BC—Wet Phase
Cal. yrs.2200 BC–till Present—Wet Phase begins to decline

Based on the above data, it is clear that there is no consensus among scholars regarding the climatic conditions of the Holocene in western India (Fig. 4). In this situation, we need to take into account some of the theories of origin of agriculture leading to settled lifestyles. Due to an abundance of food supply around 10,000 BP there was an explosive increase in human population with an increase in the number of Mesolithic sites in the subcontinent (Dhavalikar, 1988). The climate enabled them to flourish but the relative change towards dryer conditions around the middle Holocene forced both humans and animals to settle down around congenial environments which led to the rise of domestication of plants and animals or the Neolithic Revolution as suggested by Childe (1936), not only in the west but in several such favourable ecological niches (Clark, 1972). It has been suggested by scholars that the hunting nomadic population even today would prefer their existing lifestyle than shift to an agricultural mode of life as the labour input and the acquired result in the former is proportional to their requirements (Cohen, 1977) and hence the Mesolithic population was forced into food producing for various reasons yet unknown to us.

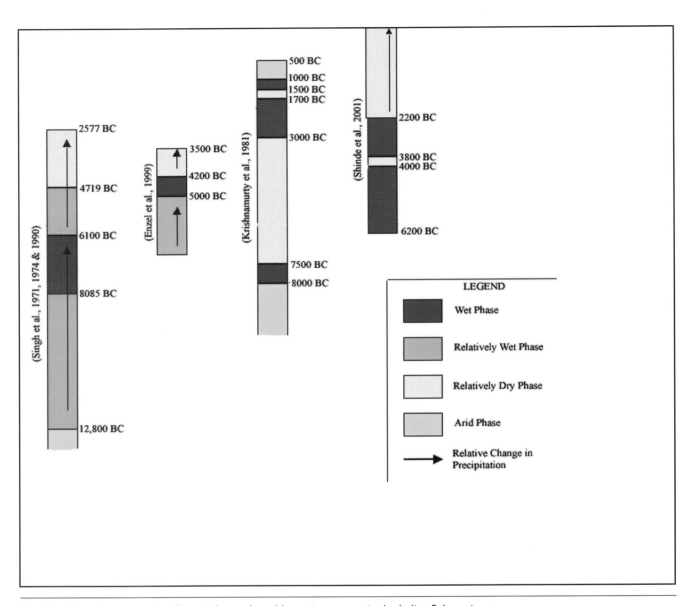

Fig. 4. Chart showing palaeoclimatic data gathered by various teams in the Indian Subcontinent.

Hunting-Gathering to Agriculture— A Cultural Process in Western India

This climatic sequence discussed above can be applicable to the larger semi-arid region that makes up western India even though there is lack of consensus as to what happened in the past. It is argued that these climatic fluctuations were responsible, to a certain extent, for the origin of early farming communities and for cultural changes that took place in the middle Holocene.

It is believed that an abundance of plant and animal food possibly led to an explosion in the population of the Mesolithic hunter-gatherer around 10,000 years ago. This is evident in the sudden increase in Mesolithic settlements all over the subcontinent as can be seen by the large number of settlements in Rajasthan, Malwa and Gujarat except the Kutch. However, the changing environment around the middle of Holocene as pointed by most palaeo-climatologists, population pressure and depleting resources forced these

Mesolithic groups to adopt an agrarian lifestyle and settle in restricted but congenial environments that provided better resources.

Most scholars believe that the domestication of plants and animals spread to the major part of western India from the site of Mehrgarh located near the Bolan Pass in Baluchistan where it first appears around 7000 BC (Shinde, 2002). A few sites in this region like Mehrgarh and Kili Gul Mohammed have produced interesting evidence in respect to the development and transition from hunting-gathering to agriculture. The site of Mehrgarh itself is considered to be a nuclear site for the domestication of plants and animals and the diffusion of agriculture to other parts of India is traced back to it. Recently the region of southeastern Rajasthan has produced evidence of local transition from hunting-gathering to origin of domestication of plants and animals. As mentioned above the climate of western India to a large extent was affected by the winter monsoons (Enzel *et al.*, 1999) and thus the region of northwest Pakistan where the nuclear sites are located was affected much earlier than the western part of India which also receives some amount of summer monsoon from the Southwest and thus could support the increased Mesolithic population till about the 5th millennium BC when the climate must have forced the people to settle down in congenial environments with the domestication of both plants and animals.

The earliest Oasis theory' propounded by Childe for the rise of domestication today seems more and more plausible based on the available evidence from the region of our study. Recent studies carried out by scholars in the Middle East and northeast Africa (Hassan, 2002) suggest the rise of domestication as an answer to a spell of droughts between 7500–6000 years BP and its spread is attributed to cultural interaction. A series of developments all over the world include:

- development of specialized and mixed economies
- ability to develop innovations at a relatively fast rate
- environmental changes or climatic kickers'
- rise of cultural nodes or centres of interaction in the climatically sensitive zones (arid or sub-arid regions) referred to as ekotropic regions'
- within the ekotropic regions are cultural troughs', i.e., localities with regular flow of water and plants or buffer zones where animals and humans converge in the absence of other such nodes of congregation and overcome occasional food scarcities and develop in the meantime innovative strategies to cope with disasters. These regions are also characterized by high plant and animal diversity providing the opportunity for selecting cultigens.

These factors seem to have contributed to the rise of domestication of plants and animals in the region of western India especially southeast Rajasthan which provided all the necessary factors leading to the convergence of plants, animals and humans around a source of water which was also rich in plant and animal diversity. This provided opportunities for selecting cultigens. The evidence in this respect is best documented at the site of Bagor in the Mewar region of Rajasthan.

The Mesolithic site of Bagor (7423E & 2525N) (Fig. 5) located on the left bank of Kothari river, a tributary of the Banas river, lies on a large and a prominent sand dune locally known as *Mahasati,* about 1 km east of the modern village. The site of Bagor also lies in the centre of the Mewar plains in the shadow of the Aravalli hills. The plain has an undulating rocky surface about 500 m above sea level with a gentle slope to the northeast. Much of it is covered by open woodland of *khejdi (Prosopis spicigera), babul (Acacia arabica), dhak (Butia frondosa)* and *khajur.* It falls in the semi-arid environmental zone located on the fringe of a small chain of mountainous land on the eastern and southeastern side and an alluvium plain to the north and west of the site. This area is dotted with stabilized and unstabilized sand dunes and the Mesolithic people had selected these stabilized sand dunes for their settlements.

PASTU
QUART

KOTHARI RIVER

Bagor

AGRICU
LA

Fig. 5. A view of the site of Bagor and the surrounding ecological condition.

The sand dune itself is composed of windblown sand and the habitation material occurs throughout the sandy deposit, thus attesting that the dune was under active formation when the site was inhabited. The dune is located on an elevated rocky outcrop hardly 200 m away from the bank of the river beyond the normal reach of floodwaters and this elevated patch of land has a strategic location commanding a view in all directions. The Kothari river takes a huge meander near the present village, about 1.5 km to the south of the ancient site and continues further towards the southwest and encircles it. It should be noted that in semi-arid regions where rivers are seasonal, most settlements are located on such meanders and considering the location of the site of Bagor it appears that this pattern started during the Mesolithic period and continues till the present. Locating a site on the meander has a number of advantages, especially as the formation of deep pools in such meanders provides water for consumption in the dry season, attracts animals and provides aquatic flora and fauna that can form part of the diet. River meanders with their pools of fresh water therefore form a source of assured water supply for both human and animal life and provide ample opportunities for subsistence. In addition, the meander partially encircles the settlement and provides natural protection on three sides to its inhabitants, a fact that the Mesolithic people may have considered before establishing the settlement at this spot.

There are a couple of other determining factors, which might also have been considered by the Mesolithic people at Bagor. The site is located on the junction of arable and pastureland providing an ideal location for a community practising incipient agriculture and pastoral livelihood (Fig. 5). It seems that the area provided the Mesolithic population with the wild cultigens of the cultivable grass which they collected and utilized for their food requirements and slowly observed the seasonal changes and innovated the process of domesticating these wild cultigens and settled in the region as permanent communities with incipient cultivation and herding besides some amount of food collection. The other side of the river bank has a number of rocky outcrops with a thin cover of coarse red soil ideal for the growth of pasture and even today is one of the important sources of pasture in this region. It is believed that the newly evolving pastoralists and agriculturist Mesolithic community at Bagor had started domesticating sheep, goat and cattle and thus would have required ample pastureland to be located near the habitation area. In addition, in the Mewar region most of the rivers including the Banas rarely spread waters beyond their channels and thus only a few patches of alluvium are found along rivers even though they generally flow through a flat landscape. Fortunately the alluvium land on the left bank of the river at the site of Bagor is one of the most fertile in the region a fact that was exploited by the Mesolithic people who were practising incipient agriculture in this region.

Another important attraction for the Mesolithic people might have been the availability of suitable raw material required for manufacturing tools and equipments required in their daily life at the site (Fig. 6). Quartz, the primary raw material used for their tools is available on the opposite bank of the river from the ancient site in the form of rocky outcrops that contain chunks and nodules of it. In addition, there are also quartz veins running in the northeast-southwest direction all along the opposite bank; however the material found in the veins may not have been commonly used by the people for manufacturing tools as they are found in the form of thin bands whereas those found in the cavities of the schist outcrops are in the form of nodules. At the site nearly 97% of the tools are made of quartz while a small percentage (3%) was made of chert and is confined to the upper part of the Mesolithic level. The raw material required for chert tools is not locally present and

Fig. 6. Photograph showing availability of raw material required for manufacturing stone tools around major Mesolithic sites in the Mewar region of Rajasthan.

may have been acquired by the people from a source away from the site that needs to be identified.

The site is spread over an area of 200 m east-west and 150 m north-south and rises to a height of 6 m above the surrounding plain while the actual cultural deposits are located roughly over an area of 80 m by 80 m. A vertical excavation was undertaken around the intact portion of the mound roughly at the highest point. The stratigraphy that was identified during the 2001–02 excavation more or less matches the stratigraphy of the earlier excavation (Misra, 1973); however, the cultural remains were confined to the upper 70 cm deposit. The earlier excavation had revealed the presence of Mesolithic, Chalcolithic and Early Historic sequence (Misra, 1973) while the recent excavations brought to light the remains of only the Mesolithic period, which has been subdivided into two phases—Aceramic and Ceramic (Fig. 7).

■ Phase A, Aceramic Mesolithic

The lower 25 cm of the habitational Layer 3 constitutes the Aceramic phase of Mesolithic with quartz tools, debitage and bone fragments devoid of ceramics. This phase represents the earliest

Fig. 7. Excavations at the site of Bagor have revealed the structural remains in both, the Aceramic and Ceramic Mesolithic phases. This picture shows the structural remains in the Ceramic Mesolithic phase.

structure found at the site. It is difficult to discern the shape of the dwelling but considering the stone alignment it appears to be circular in shape (Fig. 8). No proper well-made floor levels are associated with the structure, but it is represented by a hard and compact surface intentionally made to that form. One post-hole was noticed near the southwest corner of the stone alignment indicating the presence of a superstructure supported on wooden posts. The inner portion of the structure was rammed hard and smoothened and there appears to be a stone alignment along the periphery, possibly to prevent rainwater from entering the structure. These people seem to have used small flat stones available in the vicinity to suit their various purposes. The stones may have

been used for supporting the posts as a number of these were found near the post-hole. On the surface of the floor was found a large amount of debitage with some tools and charred fragments of animal bones, found lying flat on the surface.

Another similar structure contemporary with the earlier was found at the same level though its exact plan cannot be determined but has produced some interesting evidence for manufacturing tools and food processing equipment. The evidence associated with tool manufacture consists of a considerable large core of quartz with debitage around it at the western end and a small line of stones close to the core. At a distance of 15 cm to the south was found a rubber stone made of fine grain sandstone and to its east at a distance of 25 cm was found another similar heavily used rubber stone. Both lay on the floor of the structure and could therefore be associated with the activities of the structure. On the surface of the floor were found scattered tools of quartz and relatively well-preserved animal bones. The interior of the structure has a relatively high concentration of cultural material as compared to the outer side. The evidence of two structures in the Aceramic Mesolithic phase indicates the primary nature of the site right from the beginning and it is not unlikely that they stayed here for a considerable length of time, but because of lack of relevant evidence these structures cannot be interpreted as permanent structures.

■ **Phase B, Ceramic Mesolithic**

The Ceramic Mesolithic phase is confined to Layer 2 at the site. The evidence from this phase indicates the continuation of the blade industry and structural activity without any drastic change except the appearance of relatively large number of potsherds. It is, therefore, quite likely that the ceramic production at the site was introduced sometime in the middle phase of the Mesolithic period. This phase dated by AMS is placed earlier

Fig. 8. Various structural levels excavated in a trench at Bagor. The stone clusters in places may indicate presence of circular huts.

than cal. yrs. 4500 BC, and hence its pottery is the earliest in this region. It is coarse, red, brittle and handmade on a slow turntable, has grass and sand tempering, is ill-fired and in some instances is decorated with deep incised criss-cross patterns (Fig. 9). This is the beginning of the incised decorated ceramic in this region, which is different from the known Chalcolithic ceramics of the western Indian region. The Chalcolithic ceramic decorated with incised patterns appears to have been derived from this. Similar pottery has been reported from the Mesolithic phase in the Belan valley and this needs to be studied for comparative analysis.

Two structural phases have been identified in the Ceramic Mesolithic phase and the first phase is represented by a patch of well-rammed hard floor (Fig. 7), but unlike the previous aceramic phase there is no stone alignment on the periphery. However, the nature of the floor is the same as the previous structural phase and the average thickness of the floor is 15–20 cm. This structure appears to be a domestic cum manufacturing unit as is clear from the contents. The evidence of the manufacture of stone tools consists of a couple of fragments of cores, quartz raw-material, debitage and finished tools scattered all over the floor area. The evidence for dwelling consists of a relatively high concentration of animal bones on the northern side and a saddle quern with two grinding stones close to it on the southern side. The distribution of pottery fragments is even but with a slightly higher concentration on the northern side. The floor of the structure appears to extend on both the southern and northern sides. It is excavated

Fig. 9. Typical pottery from the Ceramic Mesolithic phase at Bagor. The pottery is red in colour, coarse, ill-fired and handmade.

over an area of 2.5 m by 2 m. The last structural phase of this Mesolithic level is represented by only a well-rammed floor level with an even distribution of pottery and bones on its surface excavated at a depth of 33 cm below the surface. Based on the ethnographic evidence both the structures of the Aceramic and Ceramic Mesolithic could have had walls of grass and conical thatched roofs.

The Aceramic phase has been dated to cal. yrs. 5680 BC and the beginning of the ceramic phase is dated to cal. yrs. 4490 BC by AMS dates. The evidence of flimsy structures, coarse pottery, some food processing equipment and tools suggests that the Mesolithic people had a semi-sedentary life, where they occupied the site for a considerably lengthy period but probably moved to another place for a certain period in their annual cycle. Results of the analysis of botanical remains collected from the site are awaited.

Both the phases of the Mesolithic at Bagor have produced large amounts of finished and unfinished tools, indicating local production. As the name itself suggests these lithic tools are tiny ranging between 40-20 mm while some of them are even smaller between 5-10 mm and were used as composite tools by hafting them on a bone or wooden handle. Most of these tools are made of quartz because of its easy availability all over the subcontinent though chert and chalcedony are a better option but difficult to obtain (Fig. 10). This tool industry at Bagor and Gilund are truly microlithic in nature and geometric in shape with mass production of micro blades and bladelets

Fig. 10. Mesolithic cores and wate flakes of quart and chert found at the site of Bagor, indicating local production.

and their conversion into various tool forms including rhomboids, lunates, crescents, points, arrowheads, etc. (Fig. 11). This tool industry is most suitable for hunting. At the site nearly 97% of the tools are made of quartz while a small percentage (3%) was made of chert and is confined to the upper part of the Mesolithic level. The raw material for the chert tools is not locally present and may have been acquired by the people from a source away from the site that needs to be identified.

The presence of food processing equipments suggests the presence of a number of species of wild and domesticated fauna and some evidence has also been produced during the course of excavation and further investigations are on to identify the species of both. A large number of animal bones have been found *in situ* on the working and living surfaces identified during the excavations and these include cattle, sheep-goat, a variety of deer and other wild species.

The site of Langhnaj and other such Mesolithic sites in north and central Gujarat also indicate similar evidences of a microlithic aceramic and later a ceramic culture. At Langhnaj, phase I is Aceramic Mesolithic dated to around 2500 BC and the second phase has been described by the excavator as a Chalcolithic phase dated to 2000 BC and after (Sankalia, 1965). This date is based on the similarity of copper and Black-and-Red Ware with the artefacts of the Ahar culture dated to 2000 BC (Sankalia *et al.,* 1969). The single date from the site is from a mixed stratum and cannot be used

Fig. 11. Typical microlithic geometric and non-geometric tools of quart found at the site of Bagor.

to date the site objectively, and the dates will probably go back relatively closer to that of Bagor in case of further research (Sinha, 2003).

However, on the basis of relative dating and comparison of the material from the site of Bagor, Phase I at Langhnaj (Sankalia, 1965) can be equated with the Aceramic phase at Bagor but could be slightly later in date as it is possible that the Mesolithic assemblage took some time to move down from the Bagor region. Based on this probable assertion, Phase II at Langhnaj can be asserted as the Ceramic phase similar to Bagor dated to around 3000–3500 BC as the Ahar culture has its beginning around the same time (a few dates from the site of Balathal go back to 3700 BC) and the site of Langhnaj has yielded copper and Black and Red Ware both pointing towards contact with the southeastern regions of Rajasthan. Thus, though the first phase is completely Aceramic Mesolithic, the second phase Ceramic Mesolithic has some amount of Chalcolithic contact especially with the Ahar region on the basis of pottery and copper which can even lead to indirect contact with the north Rajasthan Ganeshwar culture (Sinha, 2003).

Thus, on the basis of the above-mentioned sites and others like Adamgarh, Bhimbetka in Madhya Pradesh, the Vindhyan and eastern Mesolithic and Neolithic cultures it can categorically be argued that the region was inhabited by a Mesolithic population around the 6th millennium BC. Domestication evolved slowly and pottery was introduced around 4500 BC or even earlier. Also the region of southeast Rajasthan especially the site of Bagor has yielded the earliest date for the origin of pottery from where it seems to move into the regions of the Ahar culture and the Ganeshwar Jodhpura area.

The region of Gujarat especially North Gujarat has also yielded evidence of Mesolithic cultures with pottery but the area was probably influenced by the Black-on-Red Ware using people from Sind and Afghanistan (Sinha, 2003) and the Black-and-Red Ware of the Ahar culture that came in later.

Loteshwar in North Gujarat (Mehasana district) gives evidence of the earlier beginnings of the Anarta tradition (Sonawane and Ajithprasad, 1994), even before the Harappans. The site has an undated Aceramic Mesolithic stratum followed immediately by a Chalcolithic deposit with pit dwellings, pottery (gritty red, fine red and burnished grey/black), steatite beads, shell bangles and beads. The Chalcolithic level here has an early date of 2921 and 3698 BC and indicates that they evolved directly from the Mesolithic communities with internal development and possibly outside influence from the regions of Sind and Rajasthan.

The evidence of Mesolithic habitation from sites like Loteshwar and Langhnaj indicates the presence of human traditions in the region and their interaction with the incoming pastoral groups from the Amri-Kot Diji region and to a certain extent the incorporation of the people within their own cultural group.

Origins and Development of Village Life

The introduction and origin of the Chalcolithic cultures in western India was once attributed to Harappan influence, which in itself was the result of a continued development begun at Mehrgarh around 7000 BC. However, in the last decade the first hint of indigenous development of village life in western India has been found in Saurashtra, North Gujarat and at the site of Balathal going back to the middle of 4th millennium BC. Hence based on the evidence from Bagor it has been hypothesized that this area could be another primary zone where domestication and village life may have evolved.

The recent evidence from North Gujarat, Saurashtra in Gujarat and Balathal in Mewar region of Rajasthan suggests that the Chalcolithic village community came into being much before the Harappan period. It is obvious therefore that the Harappans did not play any significant role in the origin of Chalcolithic village cultures in

Gujarat and Central India (Shinde, 2000). The beginnings of village life in these regions go back to the last quarter of the 4th millennium BC as the radiocarbon dates from the sites of Loteshwar in North Gujarat, Padri in Saurashtra and Balathal in Mewar region would indicate. The Chalcolithic cultures here, in fact are contemporary with the Pre/Early Harappan cultures of northern and western Rajasthan, Sind and northwestern Indian subcontinent.

The region of Gujarat indicates a local evolution of village society, it seems to have been influenced from the Sind region especially on the basis of its similar Black-on-Red ceramic type. Small excavations carried out at Loteshwar, 17 km northeast of Shankheshwar in Mehasana district of Gujarat, has produced interesting evidence of the beginning of settled life that is earlier than the Harappan period. The Chalcolithic deposit overlies the Mesolithic material. The thin habitation deposit has revealed numerous large pits, some of which could even have been pit-dwellings, were found filled in with pottery of the Black-on-Red type (Gritty Red Ware), animal bones, clay lumps, etc. The other material equipment associated with the Chalcolithic culture includes micro steatite beads, microlithic tools and some crudely made shell bangles and beads. The site of Padri in Saurashtra also shows the moorings of a village community with Black-on-Red pottery indicating influence from the Sind region which is not impossible as the pastoral groups from very early days might have known the rich pastures of North Gujarat and Saurashtra.

The excavations at Balathal (Udaipur District, Rajasthan) roughly 42 km northeast of Udaipur have produced evidence in respect to the origin of the early farming community of Central India (Misra *et al.,* 1997). The evidence from this site is particularly interesting as the pottery is unique-both the incised ware and the Black-and-Red Ware with an earlier date for the former from the site of Bagor. A considerable thick deposit at the base of

the settlement (Phase A) at Balathal (around 1 metre) has produced evidence of the origin of the Chalcolithic culture and a gradual development of the village community in ascending order (Fig. 12). Excavation to the natural level in a limited area has demonstrated a gradual growth of the settlement from the modest beginning at the site. The people who established settlement on the bedrock constructed mud and wattle-and-daub structures with well-made floors plastered with cow dung. Some of the characteristic Chalcolithic wares of this region such as thick and thin Incised Red as well as Black-and-Red were introduced right from the beginning. However, they are

Fig. 12. Stratigraphic sections at Balathal showing complete history of the site. The lower one metre shows the beginning of the village life and the upper three metres the developed phase. The declined phase is visible towards the end of the section.

coarse, thick in section, inadequately fired and the majority of vessels are handmade. Shapes such as wide-mouth deep carinated bowls, small narrow-mouth jars and storage jars with beaded rim, the fossil types of the Chalcolithic phase in this region, are present right from the beginning. A gradual development is seen in these different wares in terms of technology and quality of vessels (Shinde, 2001) from the early to Mature Chalcolithic.

The earliest phase at Nagwada in North Gujarat has been identified as part of the Amri-Nal-Kot Diji complex on the basis of Gritty Red Ware and the typical Amri pottery from the burials (Sonawane and Ajithprasad, 1994). This indicates that sometime after village life was introduced, the region came under the sphere of influence from northwest part of the Indian subcontinent. Similar evidence also has been reported from the site of Padri near the Gulf of Cambay (Shinde, 1998) and Prabhas Patan near Somnath in Gujarat (Dhavalikar and Possehl, 1991). At the basal levels in Padri dated to 3300 BC is found the evidence of rectangular mud structures with well-made floors plastered with cow dung having low mud walls, coarse handmade Black-on Red painted pottery and evidence of semi-precious stone bead and copper tool manufacture technology. A gradual development is seen in the material culture and towards the end the overall development around 2500 BC results in the culture attaining full maturity. The culture at the basal levels at Prabhas Patan dated to 2900 BC and termed as Pre-Prabhas, is characterized by the presence of plain Black and Red and painted Black-on-Red pottery. We have very limited information available for this culture, as the excavation carried out at the site was very small in nature.

The excavations at Balathal and Gilund in Rajasthan and Padri in Gujarat have demonstrated a gradual development and prosperity in the material culture of the Chalcolithic people from 3000 BC and by 2500 BC there is a drastic change in the lifestyle of the people. The cause for this change is both internal development as a result of agricultural maximization and external influence from the developed Harappans who were in constant contact with the Chalcolithic people for their requirements of varied raw materials. This contact resulted in several shared cultural characteristics especially visible in the Chalcolithic cultural pattern and though they do not reflect a single cultural developmental continuum, they do indicate interaction. At about 2500 BC there is a rapid transformation at Balathal and Gilund, and they become well-planned settlements (Figs. 13 and 14). At Balathal structural complexes of stone and mud-brick are organized along a central road and subsidiary lane; Gilund gives the impression of a citadel and lower town plan with outer fortifications, public structures and large residential complexes (Fig. 15) indicating a socially stratified and politically organized society.

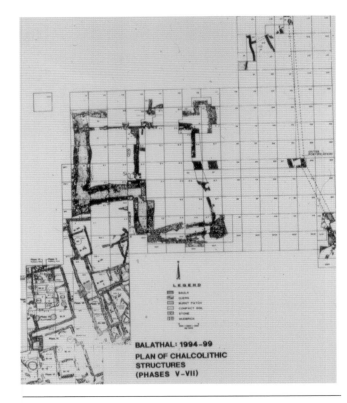

BALATHAL: 1994-99
PLAN OF CHALCOLITHIC
STRUCTURES
(PHASES V-VII)

Fig. 13. A plan showing remains of a developed settlement in the Mature Chalcolithic (devloped phase) at the site of Balathal.

Fig. 14. Structural remains unearthed in the Mature Chalcolithic (developed phase) at the site of Balathal.

Fig. 15. Remains of the parallel wall mud-brick structure uncovered in the Mature Chacolithic phase at the site of Gilund.

Though there is no evidence of a planned settlement at Padri, it develops into an important settlement specializing in salt manufacture around 2500 BC. The site played a significant role in the economic organization of the time. Thus, there is evidence in western India for a gradual development of technology, the circular huts were replaced by rectangular dwellings of mud or occasionally mud bricks and the coarse handmade pottery was replaced by fine and wheel made pottery (Shinde, 2000). By 2500 BC the Chalcolithic settlement is fully transformed, with a prosperous and well-developed trade mechanism, craft specialization and external contact and the beginnings of an urban veneer due to internal and external trade contacts (Shinde, 2002). Around this time we also see the development of the Harappan culture into a flourishing civilization in western and northwestern India and Pakistan. As the climatic data is sketchy and contradictory, we are not sure about its role in the development of village-based cultures into semi/proto or fully urban and developed societies around 2500 BC. There were certainly a number of other factors apart from the climate, which contributed to the cultural process.

Decline of the Chalcolithic

Around 2000/1900 BC the mature Chalcolithic and the Harappan civilization experienced a crisis leading to its gradual decline. Climatic fluctuation is considered to be one of the factors for the decline. A number of possibilities can be postulated such as drying up of the climate, extensive and repeated flooding combined with the shifting of river channels due to sedimentation and tectonic movements. The deterioration in climatic conditions resulting in the desiccation of lakes and changes in vegetation in Rajasthan is dated to around 2000 BC. This event has analogues in major parts of Eurasia and thus can be considered in terms of global climate changes. Reconstruction of the palaeoclimate in China, Mongolia, Western and Central Europe, reveal a similar situation of relative dryness as seen in the western part of India.

The advent of a relative dry phase in western India around 2000 BC had an adverse impact on human cultures. Excavations carried out at Balathal and Gilund in the southeastern part of Rajasthan has shed considerable light on the Late Chalcolithic phase, which was affected by the change in climate. The overall decline in the prosperity and living standard evident in the material culture can possibly be attributed to dry climatic conditions. This has been reflected in their structures, ceramic assemblages and material equipment. The large, spacious and complex structures of the previous phase were replaced by small, carelessly made and haphazardly located mud structures (Figs. 16 and 17). There is hardly any evidence of use of mud-bricks in this phase. The evidence from Balathal suggests that there was continuity in the construction method and building material. However, they are small in size, either single or double-roomed. The evidence from Gilund indicates construction of small mud structures, either rectangular or squarish in shape. The fine delicately modelled pottery from the mature phase is carelessly made, coarse, ill-fired probably as a necessity without aesthetics. The use of copper seems to have reduced with increasing dependence on stone tools like the early phase as the metal was expensive and not available everywhere but had to be imported

Fig. 16. Structural remains from the Late Chalcolithic from the site of Gilund showing decline.

Fig. 17. Structural remains from the Late Chalcolithic at Balathal showing decline.

from north Rajasthan. Towards the end, the people seem to have adopted a semi-nomadic subsistence pattern in a large part of western India, as the presence of circular wattle-and-daub structures would suggest. There is also evidence of movement of the Ahar culture into the Malwa plateau, which provided the people with water from the perennial rivers like the Chambal and Narmada and fertile land and pasture for their herds. At various sites there is evidence of assimilation of the Ahar people with the already existing population while at others they established their own identity. However here too the culture seems to have deteriorated from its proto-Urban status of the mature phase and has a more rural village based economy (Shinde and Sinha, in Press; Sinha, 2003).

Most of western India including Gujarat seems to have undergone some amount of economic decline but Gujarat, especially Saurashtra seems to have recovered from the economic degeneration caused by the decline of the Harappans and the desiccation of the climate by establishing direct trade relations with the west and by cultivating the newly arrived African varieties of millets which were more suited to the arid climate of the region. The sites here too indicate some amount of decline where the urban features are replaced by a village-based but industrial setup though features like fortifications and public buildings seem to exist as seen at Rojdi (Possehl and Raval, 1989). However in North Gujarat we see similar features as in southeastern Rajasthan where village communities shift to a semi-nomadic pastoral subsistence, moving with their herds in search of pasture and livelihood and probably moving into the southern reaches of Malwa and Maharashtra.

Discussion/Conclusion

The above discussion of the available climatic data shows clearly that there is no consensus among scholars regarding the conditions of the climate in the Holocene period. However, the rise and development of various stages of cultural evolution throughout the Holocene shows that the cultural development was not nearly as dependent on the climatic and environmental changes as has been put forth by earlier scholars. Nature has been part of the developmental process throughout the evolutionary history of mankind, but we also need to keep in mind that necessity is the mother of invention, whether it is due to cringes created by the deteriorating environment, increasing population or the need to innovate as a result of developing ideas among mankind. Like Wheeler put forth, ideas have wings and float in the air and are adopted by people who are ready for a cultural development and culture change in order to progress. The studies of various cultural groups show that climate and environment can either expedite or regress the speed of development and cultural evolution but are not the only factors for the development or decline of any culture. Steward (1955) has mentioned that 'culture represents an adjustment to the natural environment', but does not dictate completely its process of evolution or devolution. Carrithers (1990), has very significantly observed that cultural diversity is the result of human capacity to create, maintain and alter social forms over time.

Mankind's progressive evolution is seen right from the Pleistocene period along with favourable and unfavourable climatic cycles, which do not seem to have hindered in any way the process of evolution. Similarly, the Holocene witnessed a number of climatic and environmental changes but the process of cultural evolution continued and at a rather faster pace than ever before. Within a couple of hundred years man had transcended his roaming life with the hunting-gathering lifestyle and was settling down around congenial environments, and from once being totally dependent on nature, was innovating ways to control it. Domestication of the plants and animals was his first step towards this and since then has never looked back to the extent that today the environment is at the mercy of mankind which could have negative effect on human societies.

Systematic research within the Indian subcontinent in the past fifty years has brought to light two such ekotropic zones where man evolved to control the disparities of nature; Mehrgarh and Kili Gul Mohammad in the Kachhi plains in Baluchistan and Bagor in southeast Rajasthan indicate the indigenous and gradual evolution of an urban civilization and proto-urban cultures from a hunter-gatherer band between 7000 and 2500 BC. The excavations at Bagor, Balathal and Gilund have shown that the early farming people are indigenous to the region, were contemporaries of the Harappans and in fact there is evidence to suggest that their cultures evolved simultaneously with them. This co-habitation and the process of development as seen with its beginning at Bagor around 5000 BC continued till 3000 BC, and in the Indian subcontinent witnessed the rise of the Pre and Early Harappan cultures. Small settlements with an agro-pastoral economy and hunting and gathering developed. With increasing prosperity and surplus, various crafts such as ceramic manufacture, copper technology and public architecture had an opportunity to evolve. As new areas were colonized and new settlements were both built and encountered, an exchange network developed and expanded amongst them and this facilitated the procurement of goods not available locally in exchange for products locally produced. Archaeological evidence suggests that sites of this period had a strong agricultural economic base while dietary bone analysis reveals the existence of pastoralism with a predominance of cattle followed by sheep and goat supplemented by wild animals and fish. With the shift in subsistence and the development of sedentary settlements, a marked change in the social organization occurred as cultural requirements entail adaptation to economic and ecological environments and needs. The increasing need for co-operation and organization of activities relating to subsistence led to division of labour, more efficiency and surplus required administrative mechanisms and authority for storage, allocation and redistribution. Authority also plays an important role in the administration of pastures, water sources and all other collectively held and produced goods. Surplus allowed more time and the ability to support skilled and specialized crafts not possessed by all members of the society thus enhancing trade and exchange potential. These socio-economic changes led to the development of the chiefdom (Sinha, 1998 & 1999) and urban societies essential for restructuring and administering these required changes.

From the foregoing discussion, the cultural process in western India from the beginning of the 3rd millennium to the end of the 2nd millennium BC becomes conspicuous. The process of evolution and progress is very clear throughout from 10,000 BC onwards with or without the favourable climate and environment. The evolutionary trend clearly indicates the fact that the progressive cultural trend was the need of the time and that the cultures evolved and progressed because the human mind was ready enough to take a leap forward and innovate to evolve. The process of development is cyclic with evolutions and declines as part and parcel or rather two sides of the same coin. Climate and environment participate in this process by hindering or helping the rate of growth but they do not control the process in its totality. The global picture around the 2nd millennium BC indicates a declining trend along with the decline in the environment and the climate which might have furthered the process of degeneration but it wasn't the sole cause for it as the cultural development of at least 8000 years had completed its cycle and was already on its way to devolution. Further we need to keep in mind that from the beginning of the Holocene till about 2000 BC the trend was progressive whether the climate as

Fig. 18. Drilling survey conducted in Sambhar lake. (Photo by Takeshi Takeda)

deciphered by the climatologists was favourable or unfavourable, so obviously it was not the climate alone that set the evolutionary trend. The work on climate reconstruction undertaken by ALDP in Rajasthan (Fig. 18) needs to be extended in other parts of the subcontinent. We really need larger data on climate to determine its exact role in the origin, development and decline of any culture.

REFERENCES

Carrithers, M. (1990): Why humans have culture? *Man*, 25: 189–206.

Childe, V.G. (1936): *Man Makes Himself*, Watts, London.

Cohen, M.N. (1977): *The Food Crisis in Prehistory: Over Population and the Origins of Agriculture*, New Haven, London.

Clarke, D.L. (1972): *Models in Archaeology*, Methuen, London.

Dhavalikar, M.K. (1988): *First Farmers of the Deccan*, Ravish Publishers, Pune.

Dhavalikar, M.K., and G.L. Possehl (1992): The Pre-Harappan period at Prabhas Patan and the Pre-Harappan Phase in Gujarat, *Man and Environment* 17(1): 71–8.

Encyclopaedia Britannica, 1999.

Enzel, Y., L.L. Ely, S. Misra, R. Ramesh, R. Amit, B. Lazar, S.N. Rajguru, V.R. Baker and A. Sandler (1999): High-Resolution Holocene environmental changes in the Thar Desert, Northwestern India, in *Science* 284.

Krishnamurty, R.V., D.P. Agrawal, V.N. Misra and S.N. Rajaguru (1981): Palaeoclimatic Influences from the Behaviour of Radiocarbon Dates of Carbonates from Sand Dunes of Rajasthan, *Proceedings of the Indian Academy of Sciences (Earth Planet Science)* 90: 155–60.

Hassan, F. (2002): Holocene environmental changes and the transition to agriculture in southwest Asia and northeast Africa, *Origins of Pottery and Agriculture*, Y. Yasuda (ed.), Roli Books and Lustre Press, Delhi, India.

Misra, V.N. (1973): Bagor- A late Mesolithic settlement in North-west India, *World Archaeology*, V-1:92–100.

Misra, V.N., Vasant Shinde, R.K. Mohanty, Lalit Pandey and Jeevan Kharakwal (1997): Excavations at Balathal, Udaipur District, Rajasthan (1995–97), with special reference to Chalcolithic architecture, *Man and Environment* XXII (2): 35–59).

Possehl, G. L. and M. H. Raval (1989): *Harappan Civilization and Rojdi*, Oxford & IBH and the American Institute of Indian Studies, Delhi.

Sankalia H.D. (1965): *Excavations at Langhnaj*: 1944–63 Part I, Deccan College, Pune .

Sankalia, H.D., S.B. Deo and Z.D.Ansari (1969): *Excavations at Ahar (Timbavati)*, Deccan College, Pune.

Searle, M. (1995): The rise and fall of Tibet, *Nature* 374:17–18.

Shinde, V.S. (1998): Pre Harappan Padri culture in Saurashtra: the recent discovery, *South Asian Studies*, 14:174–182.

Shinde, V.S. (2000): The origin and development of the Chalcolithic in Central India, *Indo Pacific Prehistory Association Bulletin 19*, (Maleka Papers, Vol 3).

Shinde V.S. (2001): Chalcolithic phase in Western India (including Central Indian and Deccan Region), *Recent Trends in Indian Archaeology*, K. Paddayya (ed.), ICHR Publication, New Delhi

Shinde, V.S. (2002): Emergence, development and spread of agricultural communities in South Asia, *Origins of Pottery and Agriculture*, Y. Yasuda (ed.), Roli Books and Lustre Press, Delhi, India.

Shinde V.S., S. Sinha Deshpande (submitted to *Antiquity*): *Development of urbanization in Mewar, India in the middle of Third millennium BC*.

Shinde, V.S., Y. Yasuda and G. Possehl (2001): Climatic conditions and the rise and fall of Harappan Civilization of South Asia, *Monsoon* 3: 92-94.

Singh, G. (1971): The Indus Valley culture seen in context of post-glacial climatic and ecological studies in northwest India, *Archaeology and Physical Anthropology in Oceania*, 6 (2): 177–189.

Singh, G. *et al.,* (1974): Late quaternary history of vegetation and climate of the Rajasthan Desert, India, *Philosophical Transactions of the Royal Society of London*, Vol. 267: 467–501.

Sinha Deshpande, S., (1999): *Chalcolithic social organization in Central India: A case study of Balathal*, Puratattva No. 29: 50–59, 1998–99.

Sinha, S., (1998): *Study of Chalcolithic social organization in Central India with special reference to Balathal*, M.A. Dissertation, Deccan College.

Sinha, S. (2003): *A Study of Cultural Interactions in Western India in the 3rd and 2nd Millennium BC*, unpublished Ph.D Thesis, Deccan College, Pune.

Sonawane, V.H. and P. Ajithprasad (1994): Harappa culture and Gujarat, *Man and Environment,* 19 (1–2): 129–39.

Steward, J.H. (1955): *Theory of Culture Change*, Urbana: University of Illinois Press, 35.

Contributors

ANIRUDDHA S. KHADKIKAR
Geology and Palaeontology Group
Agharkar Research Institute
G.G. Agharkar Road
Pune-411 004
Maharashtra
India

BAS VAN GEEL
Research group Palynology and Palaeo/
Actuo-ecology
IBED-ICG, University of Amsterdam
Kruislaan 318
1098 SM Amsterdam
The Netherlands
Tel: +31-20-5257844
Fax: +31-20-5257878
E-mail: vangeel@science.uva.nl

BHASKAR C. DEOTARE
Department of Archaeology
Post-Graduate and Research Institute
Deemed to be a University
Deccan College
Pune-411 006
Maharashtra
India

CATHRIN BRÜCHMANN
GeoForschungsZentrum Potsdam (GFZ)
P.B. 3.3-Sedimente und Beckenbildung
Telegrafenberg
D-14473 Potsdam
Germany

GERALD A. ISLEBE
El Colegio de la Frontera
Sur Unidad Chetumal
Apartado Postal 424, CP77000
Chetumal, Quintana Roo
Mexico

GEORG SCHETTLER
GeoForschungsZentrum Potsdam (GFZ)
P.B. 3.3-Sedimente und Beckenbildung
Telegrafenberg, D-14473 Potsdam
Germany

HENRY HOOGHIEMSTRA
Hugo de Vries-Laboratory
Department of Palynology and Palaeo/Actuo-
ecology
University of Amsterdam
Kruislaan 318, 1098 SM Amsterdam
The Netherlands
Tel: +31-20-5257857
Fax: +31-20-5257662
E-mail: hooghiemstra@science.uva.nl

HOUYUAN LU
Institute of Geology and Geophysics
Chinese Academy of Sciences
Beijing-100 029
China

JENS MINGRAM
GeoForschungsZentrum Potsdam (GFZ)
P.B. 3.3-Sedimente und Beckenbildung
Telegrafenberg, D-14473 Potsdam
Germany
Tel: +49-331-2881334
Fax: +49-331-2881302
E-mail: ojemi@gfz-potsdam.de

JIAQI LIU
Institute of Geology and Geophysics
Chinese Academy of Sciences
P.O. Box 9825
Beijing-100 029
China
Tel: +86-10-62040570
Fax: +86-10-62052184
E-mail: Liujq@mail.igcas.ac.cn

JOHN HEAD
State Key Laboratory of Loess and Quaternary
Geology
Institute of Earth Environment
Chinese Academy of Sciences
22-2 Xi Ying Road
Xian-710 054
China

JÖRG F.W. NEGENDANK
GeoForschungsZentrum Potsdam (GFZ)
P.B. 3.3-Sedimente und Beckenbildung
Telegrafenberg, D-14473 Potsdam
Germany
Tel: +49-331-2881301
Fax: +49-331-2881302
E-mail: neg@gfz-potsdam.de

JUDY R.M. ALLEN
Environmental Research Centre
University of Durham
United Kingdom

JUNKO KITAGAWA
International Research Center for Japanese
Studies
3-2 Oeyama-cho, Goryo
Nishikyo-ku, Kyoto-610 1192, Japan
Tel: +81-75-3352150
Fax: +81-75-3352090
E-mail: junkokit@nichibun.ac.jp

K. KRISHNAN
Department of Ancient History and Archaeology
The M.S. University of Baroda
Vadodara-390 002, Gujarat
India

KIRIT S. PARIKH
Indira Gandhi Institute of Development Research
Film City Road
Goregaon East
Mumbai-400 065, Maharashtra
India

KULDEEP KUMAR BHAN
Department of Ancient History and Archaeology
The M.S. University of Baroda
Vadodara-390 002, Gujarat, India
Tel: +91-265-2792436
E-mail: bhankuldeep@satyam.net.in

LAJWANTI SHAHANI
Department of Archaeology
Post-Graduate and Research Institute
Deemed to be a University
Deccan College
Pune-411 006, Maharashtra, India
Tel: +91-20-6120446
E-mail: lajush@hotmail.com

LALIT PANDEY
Institute of Rajasthan Studies
Rajasthan Vidyapeeth
Udaipur-311 001, Rajasthan
India
Tel: +91-294-418349, 424719
Fax: +91-294-492440

MADHUKAR K. DHAVALIKAR
33, Navaketan Society, 'Srivatsa'
Pune, Maharashtra, India

MANJIRI BHALERAO
Department of Archaeology
Post-Graduate and Research Institute
Deemed to be a University
Deccan College
Pune-411 006, Maharashtra
India
Tel: +91-20-6693794

MIROSLAW MAKOHONIENKO
Adam Mickiewicz University
Institute of Quaternary Research and Geoecology
Department of Biogeography and Palaeoecology
Fredry 10, 610701, Poznan
Poland
E-mail: makoho@main.amu.edu.pl

MUKUND D. KAJALE
Department of Archaeology
Post-Graduate and Research Institute
Deemed to be a University
Deccan College
Pune-411 006, Maharashtra, India
Tel: +91-20-6693794

NADEZHADA I. DOROFEYUK
Russian-Mongolian Biological Expedition
Institute of Ecology
Russian Academy of Science
Piatnitskaya 47, Stroenie 3
Moscow-109017, Russia

NORBERT NOWACZYK
GeoForschungsZentrum Potsdam (GFZ)
P.B. 3.3-Sedimente und Beckenbildung
Telegrafenberg, D-14473 Potsdam
Germany

P. AJITHPRASAD
Department of Ancient History and Archaeology
The M.S. University of Baroda
Vadodara-390002, Gujarat, India
Fax: +91-265-786627

PAVEL E. TARASOV
Alfred-Wegener-Institute for Polar and Marine
Research, Potsdam
Telegrafenberg A43
D-14473 Potsdam
Germany
Tel: +49-331-2882100
Fax: +49-331-2882137
E-mail: paveltarasov@hotmail.com

PRAMOD P. JOGLEKAR
Department of Archaeology
Post-Graduate and Research Institute
Deemed to be a University
Deccan College
Pune-411 006, India
Tel: +91-20-6693794

PURUSHOTTAM SINGH
Department of Ancient History and Archaeology
Banaras Hindu University
Varanasi-221 005, Uttar Pradesh, India
Tel: +91-542-318641
E-mail: psingh@banaras.ernet.in

ROBERT A. MARCHANT
Institute for Biodiversity and Ecosystem
Dynamics (IBED)
Faculty of Science
University of Amsterdam
Kruislaan 318, 1098 SM Amsterdam
The Netherlands
E-mail: marchant@science.uva.nl

SHARAD N. RAJAGURU
Department of Archaeology
Post-Graduate and Research Institute
Deemed to be a University
Deccan College
Pune-411 006, Maharashtra, India

SHOBHANA LAXMAN GOKHALE
Department of Archaeology
Post-Graduate and Research Institute
Deemed to be a University
Deccan College
Pune-411 006, Maharashtra, India

SHUBHANGANA ATRE
Department of Archaeology
Post-Graduate and Research Institute
Deemed to be a University
Deccan College
Pune-411 006, Maharashtra, India
Tel: +91-20-6693794

SHWETA SINHA DESHPANDE
Department of Archaeology
Post-Graduate and Research Institute
Deemed to be a University
Deccan College
Pune-411 006, Maharashtra, India

SUNIL GUPTA
Allahabad National Museum
Allahabad-211 002, India
Tel: +91-532-644373
E-mail: sunilcharu@hotmail.com

SWAYAM PANDA
Department of Archaeology
Post-Graduate and Research Institute
Deemed to be a University, Deccan College
Pune-411 006, Maharashtra, India
E-mail: negwest@vsnl.net

TAKESHI NAKAGAWA
Department of Geography
University of Newcastle
Newcastle upon Tyne, NE1 7RU, England
Tel: +44-191-2226436
Fax: +44-191-2225421
E-mail: takeshi.nagakawa@ncl.ac.uk

TAKESHI TAKEDA
International Research Center for Japanese Studies
3-2 Oeyama-cho, Goryo, Nishikyo-ku
Kyoto-610 1192, Japan
Tel: +81-75-3352150
Fax: +81-75-3352090
E-mail: yangtze@nichibun.ac.jp

VALENTINA T. SOKOLOVSKAYA
Russian-Mongolian Biological Expedition
Institute of Ecology
Russian Academy of Science
Piatnitskaya 47, Stroenie 3
Moscow 109017, Russia

VASANT SHINDE
Department of Archaeology
Post-Graduate and Research Institute
Deemed to be a University, Deccan College
Pune-411 006, Maharashtra, India
Tel: +91-20-6693794
E-mail: vshinde@pn3.vsnl.net.in,
shindevs@rediffmail.com

VIJAY SATHE
Department of Archaeology
Post-Graduate and Research Institute
Deemed to be a University
Deccan College
Pune-411 006
Maharashtra
India
E-mail: vijay_sathe@rediffmail.com

VISHWAS H. SONAWANE
Department of Ancient History and Archaeology
The M.S. University of Baroda
Vadodara-390 002
Gujarat
India

WEIJIAN ZHOU
State Key Laboratory of Loess and Quaternary
Geology
Institute of Earth Environment
Chinese Academy of Sciences
P.O. Box 17
22-2 Xi Ying Road
Xian-710 054
China
Fax: +81-29-5522566
E-mail: weijian@loess.llqg.ac.cn

XIANGJUN LUO
Institute of Geology and Geophysics
Chinese Academy of Sciences
Beijing-100 029
China

YOSHINORI YASUDA
International Research Center for Japanese
Studies
3-2 Oeyama-cho, Goryo
Nishikyo-ku
Kyoto-610 1192, Japan
Tel: +81-75-3352150
Fax: +81-75-3352090
E-mail: yasuda@nichibun.ac.jp

Index

Harappan and Chalcolithic, affiliation 121, 123, 124, 130–1
    subsistence strategies 256, 258–60
    village life 398–401
North America 326
Northeast Monsoon 134, 145
Northern Hemisphere 230, 237, 278
Northwest Monsoon 384
nuclear race 336
nucleated settlements 135
number of identified specimens and their state of preservation (NISP) 381
Nut, ancient Egyptin heavenly cow 293, 296

'oasis theory' 389
Obi river 13
obsidian 136, 148
obsidian flakes 143
occupational shift 170, 179–85
ocean-atmosphere phenomena 353–4
oceans 229
Ochre-colour pottery 250
oil-bearing palms 351
oilseeds 192, 193, 276
Old Kingdom (2700–2215 BC) 209
Old Testament 147, 320
Old World 234
Oman, Peninsula 152, 229, 232, 236
    coastal trade 135, 137, 138, 140–1, 145, 150
    shell middens 137
    millet 141
    pottery 143
    sorghum cultivation, 137–8
Oman-Persian Gulf 140
onagers 226
Oojeinensis and *Dialium* 86
Open area excavation, 270
Ophicephalus 130–1
Ophir 147–8
orbital eccentricity cycle 233
Orissa coast 147
Oriya Timbo, Saurashtra, Gujarat, India 225
    subsistence economy 255, 259, 260
Orkhon river, Central Mongolia 38
orthodoxy 319
Oryza-rice 70
Ostrya 70
otoliths and Late Holocene environment of North Gujarat 130–1
Ougeinia 86
overcrowding 254
oxen 201, 204
oxygen depletion 68

oxygen isotope values 58
ozone production 277–8

PWG 162
Pabumath, Gujarat, India
    domestic animals 226
Pachpadra lake, Barmer, Rajasthan 190
Pacific Ocean 25, 142
Pacific regions sea trade 148
paddy 384
Padri, Saurashtra, Gujarat 138, 252, 385
    Chalcolithic Phase 221, 222
    origin of village life 399, 400–1
padthar 220, 263
Painted Black Ware 192
Painted Grey Ware 192, 299
Painted Red Ware, 400
Pajjusana 301
Pakistan 83, 85, 116, 138, 217, 250, 384, 387, 389
palaeobotanical evidence 225, 260
palaeoclimate, palaeoclimatic
    of Andean Peru 372
    changes 57, 58, 275
    and changes in sea level 232–3
    of Rajasthan lakes 118
    of Western India 278, 384
    of Xian and Xianyang, China 78
palaeoclimatological explanation for dry-wet transition 275, 278
palaeoecoclimate
    of Western India 386–8
palaeoecological data, study
    of Huguang Maar Lake, China 51
    of North-western Latin America 352–66, 371–2
    records of vegetation change 351
    of fossil proboscidean assemblage 380
palaeoenvironment, palaeoenvironmental
    changes 57, 58
    in North Gujarat 117–9
    of North-western Latin America 366, 371–2, 377, 382
    of Thar desert, Rajasthan 85, 276
palaeoenvironmental reconstruction, 243
    of Central american civilizations 354–5
palaeoethnological studies 260
palaeogeographic distribution of man 110
palaeohydrological records 91
palaeolithic sites – Early Acheulian to Late Acheulian 86
palaeolithic tools and artefacts 88, 105
palaeomonsoonal oscillation 233
palaeontology 380–1
palaeosalinity, 58
palaeosol 88
Palermo Museum, Italy 209

plant macrofossil 41

plant milling stones 356

playas 84, 87, 89–91

Pleistocene environment 88–9, 92, 386, 389, 403
    Thar desert, Rajasthan 85–6, 92–3
    and evidence of prehistoric sites 86–7
    Late, and Palaeolithic Sites 87–9

Pliocene 85–6

Poa, 31

poaceae pollen, 26, 29, 31, 33, 36, 38, 190

points 397
    see also tools

Pola, 298

Polar front 25

polar regions 230

polar winds 230

polarization 167
    settlement rearrangement 182–5

political ambience 197

political competitioin 156n^{10}

political-religious system 357

political subjugation 186

political will 148

pollen analysis, 26, 57, 70, 72, 189–90, 243, 244, 323, 354–5, 386

polytheism 335–6
    and the forest, relation 321, 323–4
    protected by forests 326, 328, 330, 331

pools 87

Popatpura, North Gujarat, India 258, 270

population expansion, explosion 14, 15, 336–7, 330, 388

post-galcial 233

Post-glacial Climatic Optimum (6000–4000 BP) 232

post-Gupta era 210

Post-Urban Harappan culture 251

Potential of Evapo Transpiration (PET) 83, 89

Potentilla fruticosa 33

Potentilla 31

potsherds 106, 235, 253, 258

potshols 254

pottery 121, 123–4, 138, 147, 152, 192–3, 222–4, 235, 236, 250–1, 254, 260, 299, 342, 395–6, 398, 399–401

power and dominance 199, 341

Prabhas Patan, Gujarat, India 123, 145, 138, 252, 385
    Chalcolithic Phase 221
    Early Chalcolithic 126
    Late Harappan phase 253, 255
    origin of village life 400
    sea trade 143

Prajapati 287

Pratapgarh, Rajasthan, India 166

Pratapgarh, Uttar Pradesh, India 192

Pre Prabhas ceramics, 155n^3

precipitation, 23–24, 43, 57, 58, 115, 118, 122, 130, 168, 189, 196, 231, 248, 360, 362, 364, 371, 383

Pre-Harappan 162, 299

Pre-Hipsanic agriculture 359

prehistoric communities, culture 191–2, 219, 221, 270, 339–40
    adaptation of coastal environment 134–41
    in Pacific basin 371

Pre-Incan agricultural activity 359

Pre-Incan civilization 360

Pre-Narhan settlement 193

Pre-neolithic culture 196

Pre-Prabhas
    early Chalcolithic culture 123, 126, 130
    Chalcolithic Phase 221

primitive and barbaric civilization 16–17

proboscideans, 377

progeny 263, 287

Prosopis 90

Prosposis spicigera, 247

proterozoic age 85

Proto-Austronesian 148

Protohistoric culture 141, 193, 195–204, 276, 293, 298
    mode of production 198

Proto-urban cultures 404

Prsni 286

Prthvi 285–6

Pterocarpus santalinus 259

Ptolemaic Dynasty 143, 147, 150, 152, 332

Pullasakti 310, 314

pulses 192, 193, 202, 384

pulverization 381

Puna ecosystem 359

Punaka 315

Punjab, India 164, 225, 250

Punjab, West Pakistan 311

Punt, coastal exchange 136–7, 142–3
    see also Egypt-Punt

Puntola lake, Nagaur, Rajasthan, India 190

Puranic literature, symbolism 284, 285

Pusan 298

Pushkar lake, Rajasthan, India 90, 164, 165, 243
    Lacustrine sediments 189
    palynological and geochemical studies 117

Pusyavarman 311, 314

Pyrola, 33

Qana, 152, 154
    coastal exchange 150

Qataban 149

Qatar 140

Qilu Hu lake, southwest China 57

Qin Chuan, China 73